XVI International Scientific Conference and Young Scholars "Physical and Chemical Processes in Atomic Systems" (PCPAS 2017)

AA006188

Journal of Physics: Conference Series
Volume 1099

Moscow, Russia
20 - 22 November 2017

Editor:

Vladimir D. Borman

ISBN: 978-1-5108-7343-8
ISSN: 1742-6588

Printed from e-media with permission by:

Curran Associates, Inc.
57 Morehouse Lane
Red Hook, NY 12571

Some format issues inherent in the e-media version may also appear in this print version.

Printed by Curran Associates, Inc. (2018)

Institute of Physics
Dirac House, Temple Back
Bristol BS1 6BE UK

Phone: 44 1 17 929 7481
Fax: 44 1 17 920 0979

techtracking@iop.org

Additional copies of this publication are available from:

Curran Associates, Inc.
57 Morehouse Lane
Red Hook, NY 12571 USA
Phone: 845-758-0400
Fax: 845-758-2633
Email: curran@proceedings.com
Web: www.proceedings.com

Table of contents

Volume 1099

XVI International Scientific Conference and School of Young Scholars "Physical and Chemical Processes in Atomic Systems"

20–22 November 2017, Moscow, Russian Federation

Accepted papers received: 29 August 2018
Published online: 8 October 2018

Preface

Papers

I. ISOTOPES SEPARATION AND THEIR NEW APPLICATIONS

II. EXTREME GAS DYNAMICS

III. POROUS MEDIUM, SURFACE AND COLLOIDS

IV. THE MEMBRANE SEPARATION, WATER TREATMENT

Preface

This special volume contains the proceedings of XVI International Scientific Conference and School of Young Scholars "Physical and Chemical Processes in Atomic Systems". This forum was dedicated to the 75th anniversary of the nuclear Industry of USSR/Russia and Moscow Engineering Physics Institute (MEPhI) organized in 1942. The conference was held in Moscow, Russian Federation in 20–22 November, 2017.

The organizers of the conference are

- Institute of Nanotechnologies in Electronics, Spintronics, and Photonics at National Research Nuclear University "MEPhI".
- State Research Center of Russian Federation Troitsk Institute for Innovation and Fusion Research ("TRINITI").

This event has offered a platform in bringing together students, postdocs, innovative academics and industrial experts to exchange their ideas and contributing an integrative approach to research physical and chemical processes in various atomic systems.

The program of the conference consists of oral invited lectures of the leading experts on selected topics and numerous posters.

The wide range of fundamental and applied physical and chemical processes in various atomic systems has undergone consideration during the meeting. The main topics of the conference include the following tracks:

Track 1: physics of separation processes, including separation of isotopes in rapidly rotating flows.
Track 2: selective transport phenomena in membranes.
Track 3: water treatment.
Track 4: physics of surface and colloids.
Track 5: extremal gas dynamics.
Track 6. physical phenomena in porous medium.

As is well known, the study of physical and chemical processes in numerous atomic systems requires using the diverse physical and chemical methods. This will explain why a fair amount of reports are focused on the development of new instruments and techniques for the research of multiple phenomena in such systems.

Note that some reports presented in these proceedings have been considerably modified by their authors after discussions held during the conference. Also, the program committee of the conference came to the opinion of the possibility to include in the proceedings a few additional posters corresponding to the scope of the meeting as a result of the carried out round table discussion.

The interest shown by Russian and foreign scholars as well as students either to the forum format or the subjects of reports, gives us hope for a continuation of this meeting.

The special volume is supported by MEPhI Academic Excellence Project, contracts No 02.a03.21.0005, 27.08.2013.

Guest Editor
Professor Vladimir D. Borman
Head of the Department of Molecular Physics
National Research Nuclear University MEPhI, Russia

IOP Conf. Series: Journal of Physics: Conf. Series **1099** (2018) 011002 doi:10.1088/1742-6596/1099/1/011002

Peer review statement

All papers published in this volume of *Journal of Physics: Conference Series* have been peer reviewed through processes administered by the proceedings Editors. Reviews were conducted by expert referees to the professional and scientific standards expected of a proceedings journal published by IOP Publishing.

Applying enrichment capacities for multiple recycling of LWR uranium

A Yu Smirnov[1], G A Sulaberidze[1], V E Gusev[1], E A Andrianova[2], V Yu Blandinski[2], A V Grol[2], A A Dudnikov[2], V A Nevinitsa[2], P A Fomichenko[2]

[1] Department of molecular Physics, National Research Nuclear University "MEPhI", Kashirskoe shosse 31, Moscow, 115409, Russia
[2] National Research Center "Kurchatov Institute", Akademika Kurchatova pl. 1, Moscow, 123182, Russia

a.y.smirnoff@rambler.ru

Abstract. The use of fissionable materials accumulated during the open fuel cycle faces both challenges and opportunities for modern nuclear power engineering. A growing disproportion between the production and consumption of natural uranium draws attention to the problem of used nuclear fuel. The paper examines the current situation in the field of reuse (single or multiple) of reprocessed uranium (recovered from spent nuclear fuel), which is used for the production of low-enriched fuel (LEU) to fuel the fleet of light water reactors (LWR). The world experience in handling this material has been analysed. This study also gives an overview of the currently proposed methods for enriching the ^{235}U content in reprocessed uranium to the required level by means of gas centrifugation process, while simultaneously meeting the limitations on the presence of $^{232,234,236}U$ in commercial LEU. Savings of natural uranium was estimated for repeated recycling of VVER spent fuel. It was supposed that re-enrichment of reprocessed fuel would be done by arranging the most promising for such purposes cascade schemes. The obtained results can be used as a basis for further scientific, technical, and feasibility studies on the large-scale utilization of reusable materials in the different fuel cycles of LWR.

1. Introduction

1.1 Preamble

It is common knowledge that most of light-water-reactors (LWRs) use low enriched uranium fuel with enrichment of U up to 5%. And as nuclear industry poised to grow, we are waiting for a widening gap between uranium mining and its consumption [1]. Therefore, the problems with supply of this resource will appear and it will be difficult to provide enough fuel for LWRs. One of the possible solutions to this is using spent nuclear fuel (SNF) for producing fresh fuel assemblies.

Here we will focus on uranium recovered separately from used fuel. Worth noting, that vast bulk of SNF is represented by uranium (~66% of the total mass and ~95% of the residual fuel), which is effectively extracted in generally used conventional PUREX process [2]. The reuse of such reprocessed uranium (RepU) has a number of potential advantages: (1) natural uranium (NatU) saving; (2) avoidance of disposal costs for a vast volume of SNF that is reprocessed; (3) possible separative work reduction (depends on the ^{235}U assay of the RepU) [2].

IOP Conf. Series: Journal of Physics: Conf. Series **1099** (2018) 012001 doi:10.1088/1742-6596/1099/1/012001

1.2 Study objective

This paper provides an analysis of the factors involved in the multiple reuse of spent nuclear fuel. In our case, uranium is recovering from commercial light-water-reactor (LWR) – Russian VVER-type. It is assumed, that chemical reprocessing allows getting pure uranium from spent nuclear fuel. Such regenerated uranium undergoes re-enrichment and then returns to nuclear fuel cycle. Recycling of uranium in this way could be efficiently done several times [3].

In such a case, disadvantages of uranium recycling are connected mostly with presence of ^{232}U and ^{236}U minor isotopes [3-4] concentrations of which are strictly limited by the ASTM specifications for enrichment feed and product [5]. Complying with these standards, designed for uranium hexafluoride (a compound used in the uranium enrichment process [6]), ensures the fulfillment of the requirements of radiation safety during the manufacture of fuel elements (conditioned by ^{232}U presence), and preservation the neutron-physical characteristics of the fuel (owing to ^{236}U). Such limitations stem from ^{232}U and ^{236}U properties: ^{232}U is particularly dangerous contaminant because of intense gamma radiation emitted by Thallium (a daughter of ^{232}U, produced relatively quickly); ^{236}U acts as a parasitic neutron absorber that impedes the chain reaction (this means that a higher level of ^{235}U enrichment is required in the product) [4].

Once again, this study examines employing reprocessed uranium (RepU) – material, which is free of impurities and consists of only uranium isotopes. So, the scope of this paper was limited to describing the technical effects of the presence of the various isotopes of uranium in RepU. And this issue is considered mainly from an enrichment process point of view: how to deal with and overcome the harmful influence of $^{232,234,236}U$ [7]. In other words, the main purpose of the work was to illustrate the role of separation technologies, in particular, the applicability of centrifuge cascades to re-enrich reprocessed uranium efficiently. This needs to be done in order to sort out the general problem of closing the fuel cycle of light water reactors with uranium as the main component of nuclear fuel.

The impact of $^{232,234,236}U$ on re-enrichment process has been estimated using two key parameters: consumption of natural uranium and separative work requirements. It should be noted, that calculations were done while meeting the ASTM limitations on the presence of $^{232,234}U$ in enriched product [5]. In the same manner, additional enrichment in order to compensate the effect of ^{236}U should be done [8].

It is assumed that the spent fuel from a reactor pass is reprocessed into oxide, converted to UF_6, and used as enrichment feed along with sufficient natural-assay UF_6 feed to produce the required enrichment product to fuel the next reactor pass. In our case, this process is repeated over and over again (here, calculations were made for four consequent reactor passes). The whole process depicted in Fig.1, here, for the first reactor pass, the enriched uranium for the reactor fuel was produced in centrifuge enrichment plant using all-natural-uranium feed. Beginning with the first cycle on NatU, spent fuel from the reactor core goes to cooling and reprocessing. Then, starting with cycle 2, reprocessed fuel undergoes re-enrichment along with natural uranium in a cascade. Afterwards, LEU is used for fuel assembly manufacturing, these bundles are inserted into the reactor core. At the end of operating period, used fuel is placed in cooling ponds. After that, as the last step of each cycle, SNF is reprocessed and could be used over again.

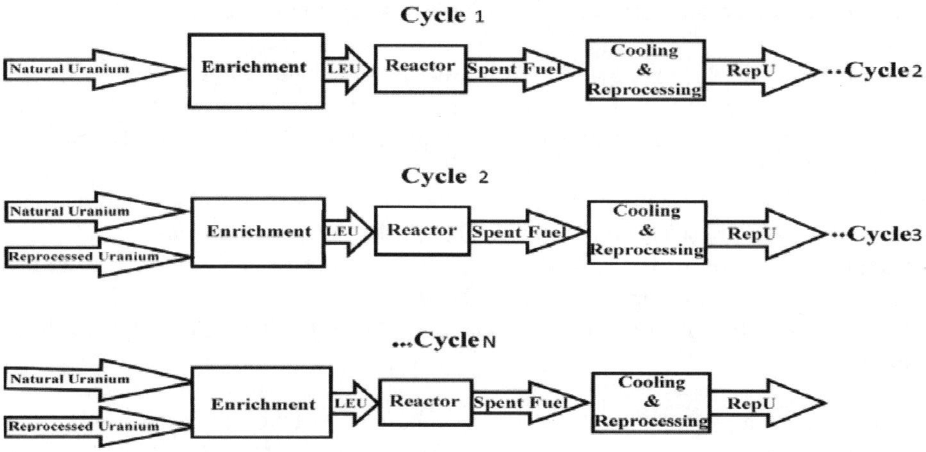

Figure 1. Multiple recycling of LWR uranium.

It should be pointed out, that in case of multiple recycling, we will face steady accumulation of undesirable isotopes $^{232,234,236}U$ [3]. Unfortunately, such consistent RepU re-enrichment inevitably leads to deterioration of fuel isotopic composition: build-up of minor isotopes significantly worsens the quality of reused material, because more NatU is needed to dilute these pollutants.

2. Main part

2.1 Brief review of cascade schemes for RepU re-enrichment

The industrial cascade for uranium enrichment is an expensive large-scale hi-tech structure. In this connection, it is particularly important to design such installations properly. That is why for more than seventy years along with the improvement of separation technologies, the theory of binary and multicomponent isotopic mixtures separation in cascades is developing. Mathematical modeling of enrichment cascades allows to make an adequate analysis of mass transfer processes in order to find their best parameters for solving the separation problems. Below we present an analysis of modern approaches to the solution of the RepU re-enrichment problem, based on the gas centrifuge separation technology and the achievements of cascade theory for the separation of multicomponent mixtures.

Lots of them are based on 'ordinary' (triple-flow) cascade, which could be applied, for instance, in following ways (Fig. 2) [8]:

- to enrich the mixture of natural *(F)* and reprocessed uranium *(E).*
- to produce enriched product from RepU *(E)* and then to dilute it with NatU *(F)* or with other uranium mixture.
- to over-enrich NatU (*F)* a bit, than it needed for reactor-grade fuel, and then to dilute with RepU *(E).*

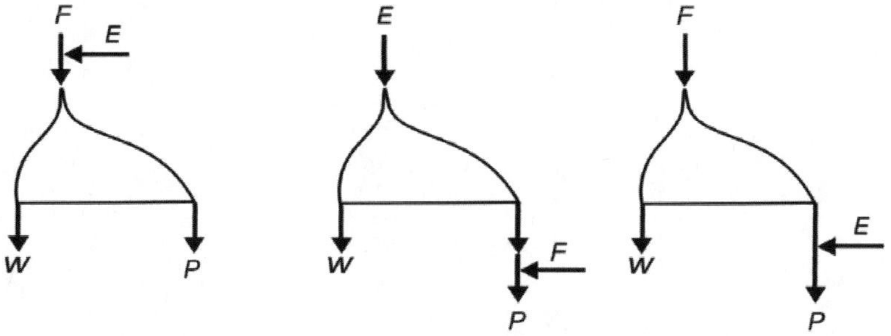

Figure 2. Schemes for RepU re-enrichment based on ordinary cascade.

IOP Conf. Series: Journal of Physics: Conf. Series **1099** (2018) 012001 doi:10.1088/1742-6596/1099/1/012001

The main drawback of all these schemes is losses of separative work, due to mixing flows with different concentration of ^{235}U.

Here, RepU needs to be somehow diluted by mixtures without harmful $^{232,236}U$ isotopes (usually NatU, but also it can be DepU). And as we can see on these figures, commercial LEU could be produced by watering down RepU at any outer point [8]. In this case, the ratio between the RepU flow and the diluent is determined by the limit of ^{232}U content in product *(P)*. At the same time, the concentration of ^{235}U should not be lower than required for LEU with specific properties.

The main advantage of such schemes is simplicity of implementation, because there is no need for a nonstandard modification of the cascade. Moreover, we already have necessary technologies for mixing gaseous uranium with different isotope content.

The main drawback here is losses of separative work due to mixing of flows with different isotopic concentration.

Apart from these simple methods there are other variants based on the use of multi-flow cascades with additional feeds. The advantage of using such cascades, comparing to the previous ordinary schemes, is that the RepU is directed into the cascade through a distinct feed point without losses of separative work [8-10]. However, in such schemes, a noticeable decrease in the content of $^{232,234,236}U$ isotopes in LEU is achieved, first of all, as in previous cases, due to dilution of the RepU with natural raw materials within the cascade.

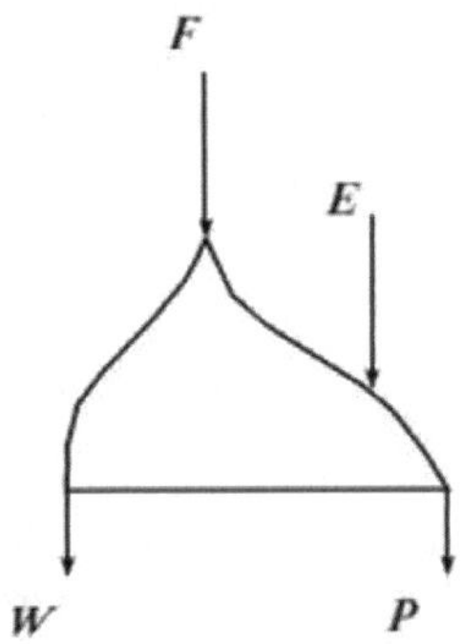

Figure 3. Cascade scheme with feeds of RepU *(E)* and NatU *(F)*.

In some cases, multi-flow cascades may have an additional external flow – additional product [8, 11]. Such a scheme of the cascade for purification of the RepU from minor isotopes is presented below (Fig. 4). Here, in (P_2) flow, a LEU is obtained, and in the intermediate (P_1) flow, a semi-product with a reduced content of minor isotopes is produced. The ^{235}U concentration in this "purified" product is about 0.8-1.1 of its value in the "purified" RepU and is regulated both by the ratio of the incoming fluxes (RepU to NatU) and by the relocation of intermediate outlet point (P_1) or additional feed point *(F)*. As a main disadvantage, this configuration is subject to radiation contamination, and the cleaning effect is caused, primarily, by the lesser stream of purified material (P_1). But in this scheme there are practically no loss of separative work and no extra cascade is required [11].

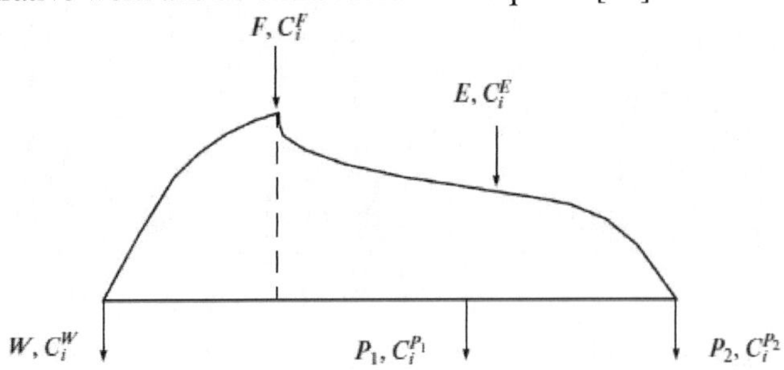

Figure 4. Cascade scheme with two feeds (NatU and RepU) and additional purified product (P_1).

Extensive NatU use for dilution of RepU is a common drawback of abovementioned cascade schemes (Fig. 2-4), because it is necessary, first of all, to decrease ^{232}U content in product. So, it is not helpful enough for NatU saving (no more than 15-18% economy is achievable, according to preliminary estimations).

To overcome this shortcoming, depleted uranium (DepU, byproduct of the enrichment process, so-called 'tails') could be employed, as alternative to NatU diluent, although this material often treated as 'waste'. The main advantage of its use is possibility of significant natural uranium saving. But such approach could also be helpful in reducing its own stored amounts.

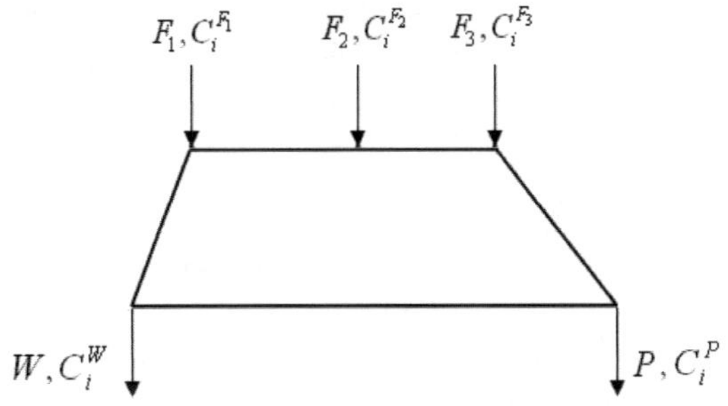

Figure 5. The schematic drawing of the cascade with three feeds.

In this paper, we consider a scheme, where DepU and NatU are used together as the cascade feeds to water down harmful $^{232,234,236}U$ isotopes in RepU (Fig. 5) [12]. F_1, F_2 and F_3 are flows with concentrations $C_i^{F_1}$, $C_i^{F_2}$, $C_i^{F_3}$ into feed points of DepU, NatU and RepU respectively. This cascade has product (P) and waste (W) outlets with a concentrations C_i^P, C_i^W (from now on, the concentration is expressed as a percent by mass; $i=\overline{1,m}$, i - index number of each isotope; m - the total number of components in mixture). Detailed description of mathematical model is given in [11].

This feature makes this scheme promising, as it combines the ratio of two diluents (DepU and NatU) to achieve proper natural uranium or separative work saving, which allows to "tune" the cascade for specific isotopic compositions of feed mixtures and current prices of NatU and separative work unit (SWU).

Yet there are other options, such as double cascade, which is formed by two 'ordinary' cascades through serial connection. It helps to re-enrich reprocessed uranium even without diluting it with raw materials, but such scheme usually considered to be 'dirty' because of polluting enrichment devices, as its main drawback [8]. Another shortcoming of this option is mounting concentration of fissile ^{235}U in a by-product, which could even reach the level of highly enriched uranium [8], and this is inappropriate for meeting the terms of non-proliferation treaty designed to control spreading of fissile materials. Finishing the brief review of the proposed methods for RepU re-enrichment, it should be noted that there is still no answer to the question which of the schemes is inherently better and more effective. And the final answer could not be given, because the advantages of every cascade is inseparable from the industry aims and fuel cycle short-term goals (connected with current market prices) or long-term strategies (which are basically based on specific national policy). So, for instance, in the first case, we could pick the right cascade, focusing on the immediate task after carrying out detailed technological and economic evaluations of the effectiveness of the schemes considered. And, in an ideal world, the outcome of the choice would always be closely connected with path that industry pursue, and the choice itself would be obvious due to comprehensive scenario analysis.

2.2 Multiple recycling

Let's look closely at the method employed in this research. The simulation of multiple recycling of LWR uranium was carried out in this study. Here, we used complex approach based on coordinated simulations of neutron-physical processes (during reactor operation and spent fuel storing in cooling pools at the reactor site) and molecular-selective transfer [13]. In this way, neutronic calculations were made at NRC KI for VVER-1000 fuel burnup of 46.9 GWd/tU and sequential 5 years cooling period for each cycle (Fig. 1). NRNU MEPhI, in its turn, performed computer simulation of molecular-selective transfer in studied cascades for each cycle.

Two of the above-mentioned cascade schemes (with two feeds (Fig. 3) and three feeds (Fig. 5)) were taken as reference objects (for convenience, denote them as scheme #1 and scheme #2, respectively). The basic ordinary cascade will serve as a benchmark for making comparisons using fundamental indicators: consumption of natural uranium, separative work requirements. To represent proposed methodology when considering the issue of RepU multiple recycling in VVER, these two schemes were picked for the following reasons: (1) the first configuration being the best option for RepU re-enrichment, comparing with cascades based on three-flows schemes [8, 9], could facilitate comparison between RepU employment and basic open fuel cycle; (2) the second scheme could show the ability of isotope separation technologies to attain significant saving of NatU. So, primarily, the schemes serve to illustrate the potential of gas centrifuge enrichment plants for increase of NatU saving.

The external parameters of the cascade were specified: the required concentrations of the target component ^{235}U in the waste and product flows, the ^{236}U penalty, and the maximum acceptable ^{232}U concentration in the product.

R-cascade has been considered as a calculation model for both schemes [14]. It is a special case of so-called "quasi-ideal" cascade – cascade, where condition of no mixing is met for chosen pair of components (in our cases: ^{235}U and ^{238}U isotopes). In the molecular-selective transfer calculations the following parameters were given: concentration of ^{235}U in product was 4.6% (not taking into account required additional enrichment in order to compensate the effect of ^{236}U); the ^{236}U compensation factor was 0.29; concentration of ^{235}U in waste end of the cascade was 0.1%. The concentration of ^{232}U in the product was limited by value $5 \cdot 10^{-7}$ %; and ratio of ^{234}U to ^{235}U must be lower than 0.02 [5]. DepU with 0.2% of ^{235}U (in scheme #2) and NatU (in both configurations) acted as diluents. Let us remind you that feeds of different compositions are directed into separated feed points of cascade. The locations of such inlets are placed where ^{235}U concentrations of external and internal flows coincide (external flows are feeds of the cascade, when internal flows are already circulating inside a set of gas centrifuges, in other words, cascade).

For instance, the calculations for cascade with three feeds were performed for the problem statement as follows: the concentrations and relations F_2/F_3, F_1/F_3 are given (n – the index number of the target component). Through calculations is necessary to compute remaining parameters of the cascade, including the concentrations of other components in outgoing flows and the number of stages in the cascade, etc. The problem is interesting from a practical point of view, because in this case clearly defined product quality allows comparing the effectiveness of different ways to produce commercial reactor-grade LEU.

As a benchmark for summarizing the results, the ordinary *R*-cascade is considered [14].

2.3 Results and discussion

Evolution of the isotopic composition during multiple recycling is shown in tables 1-2.

Table 1. Scheme #1 calculation results for a sequence of U recycles, taking into account ^{236}U compensation and ^{232}U limit

Cycle	Before enrichment				After enrichment			
Isotope / Recycle	^{232}U, %	^{234}U, %	^{235}U, %	^{236}U, %	^{232}U, %	^{234}U, %	^{235}U, %	^{236}U, %
1	2.44e-7	2.26e-2	1.16	6.38e-1	2.28e-7	5.18e-2	4.73	4.61e-1
2	5.25e-7	3.17e-2	1.26	1.07	4.92e-7	6.03e-2	4.83	7.84e-1
3	7.64e-7	3.75e-2	1.32	1.38	5.0e-07	5.75e-2	4.81	7.16e-1
4	7.40e-7	3.58e-2	1.31	1.31	5.0e-7	5.70e-2	4.80	7.03e-1

For cascade with two feeds starting from the third recycle, total employment of RepU becomes impossible, due to the deterioration of the isotopic composition of uranium and the presence of limitations on the isotopes $^{232,236}U$.

Table 2. Scheme #2 calculation results for a sequence of U recycles, taking into account ^{236}U compensation and ^{232}U limit

Cycle	Before enrichment				After enrichment			
Isotope / Recycle	^{232}U, %	^{234}U, %	^{235}U, %	^{236}U, %	^{232}U, %	^{234}U, %	^{235}U, %	^{236}U, %
1	2.44e-7	2.26e-2	1.16	6.38e-1	2.28e-7	4.54e-2	4.67	4.65e-1
2	5.27e-7	2.89e-2	1.26	1.07	4.94e-7	5.25e-2	4.83	7.84e-1
3	7.54e-7	3.31e-2	1.32	1.38	5.0e-7	4.88e-2	4.82	7.27e-1
4	7.45e-7	3.08e-2	1.31	1.32	5.0e-7	4.75e-2	4.81	7.03e-1

As we can see from the following tables, enrichment technologies could equip the nuclear industry with effective instrument for 'refining' isotopic composition of reprocessed fuel, even in case of multiple recycling. An increase in the number of consecutive cycles is also possible. And we could achieve even greater results for both schemes if we consider an extension of permissible limit of even-numbered isotopes, first of all ^{232}U. For such purpose ASTM standards need to be re-examined, or maybe fuel production should comply with other international or local specifications. Anyway, the role of ^{232}U in enrichment process needs to be scrutinized over whether it is possible to establish renewed proven limitations for gaseous UF_6. Restrictions similar in nature were already extended for fuel assembly manufacturing due to robotization of factories, which enables remote fabrication [15].

A synopsis of the results for studied schemes is displayed in the table 3.

Table 3. Deviations for prior results in NatU consumption and separative work (SW) requirements from ordinary cascade parameters (positive values – a potential saving, negative values overspending)

Feeds	NatU+RepU (Scheme #1)		NatU+RepU+DepU (Scheme #2)	
Consumption, %	NatU	SW	NatU	SW
1	19,15	4,47	50,90	-47,09
2	19,02	3,70	51,07	-48,11
3	13,11	2,10	50,90	-58,50
4	13,58	2,28	50,72	-58,22

As we can see from the table 3, the cascade scheme #2, in which DepU is applied for dilution of RepU, is the most effective for all recycles in terms of natural uranium savings. A stable, more than 50% saving of natural uranium per recycle pass is achieved due to 50% separation work increase. In principle, schemes of this type can, theoretically, provide any degree of NatU savings (in this paper, 50% is chosen for illustration). But, when choosing the ratio between the costs of natural uranium and the work of separation, one should also start from economic criteria, and also take into account that this ratio will depend on the parameters of DepU, in particular the residual content of ^{235}U. Worth mentioning, that with this scheme, even greater results could be achieved by employing depleted uranium with higher concentration of fissile ^{235}U, than used in current calculations. Such materials are abundant (world stock is about 1.6 million tons [16]), so detailed inventory of ^{235}U assay in stored casks of depleted UF_6 is needed. Summarizing the advantages of this approach, it can be said, that, in total, these factors can allow reduction in fuel cycle costs for VVER due to significant decrease in NatU demand and in dynamics of DepU accumulation, as well as a reduction in the SNF that needs to be buried.

It is shown that the schemes considered in the paper may have different applications, depending on the objectives sought. So, the cascade #1 with two feeds allows to simultaneously achieve saving of separative work and natural uranium, in comparison with the regular ordinary cascade, however, the maximum saving of natural uranium could not exceed 19-20%. As for scheme #2 with three feeds, it could provide stable and considerable (>50%) natural uranium saving, by spending more separative work. Thus, the demonstrated possibility of a substantial natural uranium saving might be economically attractive in case of uranium demand growth.

It was noted that with the cycle number there is a tendency to rise in concentrations of $^{232,234,236}U$ isotopes, in other words, to deterioration of isotope composition. This effect leads to an increase in separative work requirements and, in the absence of an additional diluent (DepU) to higher NatU input [3].

To sum up, implementation of considered 'advanced' schemes could lead to substantial reduction of VVER fuel cycle costs due to significant decrease in NatU demand (and in the accumulation of DepU for scheme with three feeds), as well as SNF that need to be disposed.

Carrying out comprehensive scenario analysis would also be important for more efficient fissile material allocation. It could be helpful in forecasting of RepU final destination.

3. Conclusion

Isotope separation technologies are crucial in efficient uranium multiple recycling. Enrichment advancements are primarily important at pushing the limits of RepU massive usage and limits of natural uranium saving.

If proposed advanced configurations are assembled (which is not that difficult), such benefits, as natural uranium economy and, in some cases, separative work economy, will be received right away.

Finally, technical and economic analysis is necessary for the right decision over cascade scheme. The right choice between proposed methods will play the main role in establishing the most optimal way of solving uranium re-enrichment task, which is vitally important for nuclear fuel cycle closing. Feasibility study also could help with a choice of best strategy and with establishing optimal share of different ways to close nuclear fuel cycles. Holistic approach would certainly be useful, if every entity in nuclear supply chain was willing to work toward shared goals.

References

[1] A Joint Report by the Nuclear Energy Agency and the International Atomic Energy Agency. Uranium 2016: Resources, Production and Demand (Paris: OECD, 2016) p 12, 75

[2] Del Cul G, Trowbridge L, Renier J, Ellis, R, Williams K, Spencer B, and Collins E 2009 *Analysis of the Reuse of Uranium Recovered from the Reprocessing of Commercial LWR Spent Fuel* (ORNL/TM-2007/207, Oak Ridge National Laboratory)

[3] Smirnov A Yu, Sulaberidze G A, Dudnikov A A et al. 2012 *Physics of Atomic Nuclei* vol 75 **13** 1616-25

[4] Proselkov V N, Aleshin S S, Popov S G, Sidorenko V D, Slavyagin P D, Tataurov A L, Milovanov O V, Mikheev E N, Anan'ev Y A, Pytkin Y N, Pimenov Y V 2003 *Atomic Energy* vol 95 **6** 829-34

[5] ASTM C787-15, Standard Specification for Uranium Hexafluoride for Enrichment, ASTM International, West Conshohocken, PA, 2015, www.astm.org

ASTM C996-96, Standard Specification for Uranium Hexafluoride Enriched to Less Than 5% 235U1, ASTM International, West Conshohocken, PA, 1996, www.astm.org

[6] Orlov A A, Tsimbalyuk A F, Malyugin R V 2017 *Separation and Purification Reviews* **46** vol 1 81-89.

Orlov A A, Tsimbalyuk A F, Malyugin R V, Glazunov A A 2016 *MATEC Web of Conferences* 72:01079

[7] Orlov A A, Kravchenco A V, Titov E S, Lebedev A Ya 2015 News of Higher Schools. Physics vol 58 **2-2** 35-40 (in Russian)
[8] Smirnov A Yu, Sulaberidze G A, Nevinitsa V A et al. 2012 *Nuclear physics and engineering* vol 3 **5** 396–03 (in Russian)
[9] Sulaberidze G A, Borisevich V D and Xie Q X 2006 *Theoretical Foundations of Chemical Engineering* vol 40 **1** 5-14
[10] G A Sulaberidze, V D Borisevich, Xie Quanxin 2004 *Proc. of IX All-Russia (International) Scientific Conference "Physical and Chemical Processes on Selection of Atoms and Molecules" (*Russia, October 4-8) 70
[11] Palkin V A 2010 *Prosp. Mater* **8** 11-14
[12] Smirnov A Yu, Sulaberidze G A 2014 *Atomic Energy* vol 117 **1** 44-51
[13] Dudnikov A A, Nevinitsa V A, Chibinyaev A V, Proselkov V N, Smirnov A Yu and Sulaberidze G A 2010 *Int. Conf. on Management of Spent Fuel from Nuclear Power Reactors, IAEA, Vienna, Austria*
[14] De la Garza A, Garrett G A and Murphy J E 1961 *Chem. Eng. Sci.* **15** 188–209
[15] Nyman, D H, Benson, E M, Yatabe, J M, Nagamoto, T T 1981 *Remote fabrication of nuclear fuel: a secure automated fabrication overview* (HEDL-SA--2410-FP). United States
Robert William Kupp 2005 *A Nuclear Engineer in the Twentieth Century*, p 233
[16] http://www.world-nuclear.org/information-library/nuclear-fuel-cycle/uranium-resources/uranium-and-depleted-uranium.aspx

IOP Conf. Series: Journal of Physics: Conf. Series **1099** (2018) 012002 doi:10.1088/1742-6596/1099/1/012002

Production of stable isotopes of light elements: past, present and future

A V Khoroshilov

Isotope Technology and Hydrogen Energy Department, D. Mendeleev University of Chemical Technology of Russia, 9, Miusskaya Sq, Moscow 125047, Russian Federation

avkhor@inbox.ru

Abstract. Taking into account the world trends in the production and use of stable isotopes of light chemical elements, the main directions of their application and the scale of modern production are considered. Particular attention is paid to stable isotopes, mainly used in clinical medicine and various biological studies. At examples for isotopes of ^{13}C, ^{18}O and ^{15}N are analyzed the change in the production volume of isotope products by years, the technologies used for isotope enrichment are considered and manufacturers in different countries of the world are indicated. Particular attention is paid to the appearance of relatively new industrial processes for the enrichment of ^{13}C and ^{18}O. The dominant role of physicochemical separation methods in the production of light isotopes is noted. In addition, from the technology development point of view, technical proposals for the improvement of physicochemical separation methods aimed at enriching ^{15}N and separation of boron isotopes are considered.

1. Introduction

What nomenclature of stable isotopes of light elements is most in demand now and will be in demand in the future? Which method and what isotope separation processes will dominate in the coming years, and, accordingly, on what processes of isotope enrichment are to be bet on?

Answers to these questions can be obtained on the basis of an analysis of the history of development and the scale of isotopes production and, of course, the technologies used.

2. Main applications of stable isotopes of light elements

2.1. Stable isotopes of light elements

As a rule, stable isotopes of light elements include isotopes of hydrogen, helium, lithium, boron, carbon, nitrogen and oxygen. Sometimes this list includes magnesium and sulfur. Without considering the special role of such isotopes as D, ^{6}Li and ^{10}B in the defense industry and nuclear power engineering, attention should be paid to the use of stable isotopes of light elements, mainly in the medical and biological fields.

At the same time, for some isotopes that have become instruments of large-scale clinical diagnostics, it is necessary to consider the change in the volume of production for a long time, and in particular for the last 10-20 years, as well as the technologies which are used for isotope enrichment.

2.2. *Main fields of uses*

In general, it is possible to single out, as noted in [1], six main directions for the use of stable isotopes. The first is earth sciences, that is, geochemistry, geology, geophysics, hydrology, and marine sciences. The second: biochemistry, biology, biotechnology, chemistry, physics, physiology, that is, fundamental and applied research. The third is food and agriculture. The fourth is nutrition. The fifth is toxicology. And, finally, health care. However, it is not difficult to see that the last five regions are combined and in one way or another connected with the last direction - health care.

2.3. *Healthcare application*

The application of stable isotopes of light elements in health care should be divided into a number of special areas. First of all, it is medicine, biology and pharmacology. The most common specific applications are listed in Table 1 [1-3], where is reflected the special role of ^{13}C (the breath tests for diagnosis of *H. Pilory* infection) and ^{18}O (the starting material in the production of ^{18}F for positron emission tomography) as diagnostic isotopes in clinical medicine [3-6].

Table 1. The main directions of the application of stable isotopes of light elements in the field of medicine, pharmacology and biology.

Isotope	Application
Deuterium	Vitamin research Noninvasive, accurate, and prolonged measurement of energy expenditures during everyday human activity
Lithium-7	Starting material for berylliosis studies (^{7}Be)^{a}
Boron-10	Boron neutron capture therapy (BNCT)^{a}
Boron-11	Starting material for positron emission tomography (PET diagnosis , ^{11}C)
Carbon-13	Noninvasive breath tests for metabolic research and diagnosis (*H. Pilory* infection) NMR research and diagnosis Starting material for positron emission tomography (PET diagnosis, ^{13}N)
Nitrogen-15	Large-scale uptake studies in plants Excrement test for *H. Pilory* infection Starting material for positron emission tomography (PET diagnosis, ^{15}O)
Oxygen-17	Measurement of the blood stream velocity
Oxygen-18	Noninvasive, accurate, and prolonged measurement of energy expenditures during everyday human activity Starting material for positron emission tomography (PET diagnosis, ^{18}F)

^{a} Research, not clinical practice

In addition to the applications indicated in the table, both proteomics and clinical pharmacology should be indicated, such as the assessment of drug products or drug delivery systems, assessment of patients and assessment of drug pharmacology, as well as numerous studies aimed at the prospect of use in clinical practice, including new opportunities of ^{13}C-breath tests [6-9].

3. Scale of production and its changes

The technology of stable isotopes of light elements and their production have a relatively short time interval - less than 100 years. The very discovery of many stable isotopes dates from 1913-1920 [10]. Naturally, experimental-industrial installations for the enrichment of isotopes began to be created in the 1930s-1940s [11]), and the first grams of highly enriched ^{13}C (99.8% at.), ^{15}N (99.8% at.) and ^{18}O (99.75% at.) were obtained in the 50s of the XX century in K. Clusius laboratory by the thermal diffusion method [12]. Therefore, considering the change in the production scale of stable isotopes in time, the starting point can be taken in 1940, when the output was estimated at hundreds of grams [11]. Twenty years later, by 1960, isotope production could provide tens of kilograms, and approximately in 1980 the total scale of production was estimated at about 100 kg annually and by the beginning of the third millennium overcame the 1,000 kilogram mark (Fig. 1). In this case, the dependence of the production volume of stable isotopes of light elements, more precisely ^{13}C, ^{15}N and ^{18}O, in semilogarithmic coordinates can be represented by a straight line with a high correlation coefficient ≈0.98.

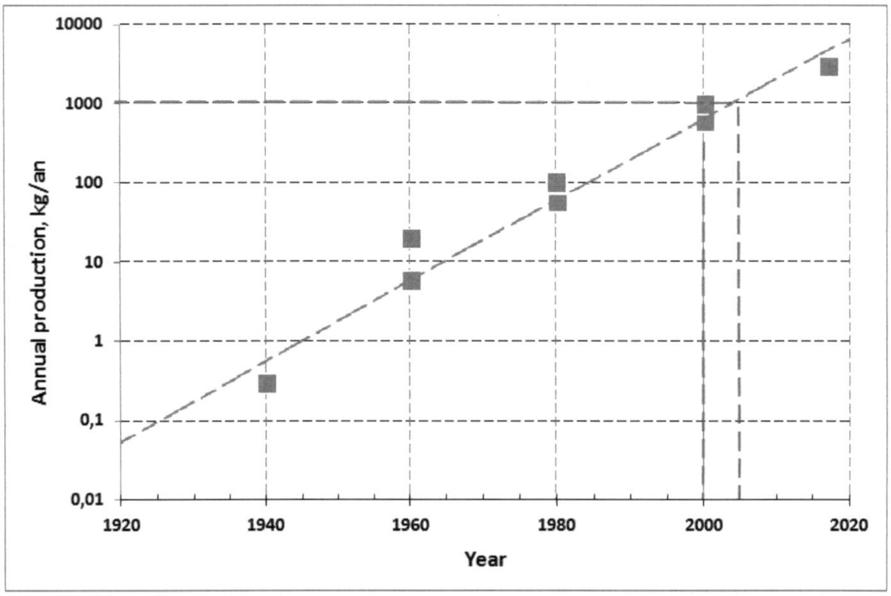

Figure 1. The production scale of ^{13}C, ^{15}N and ^{18}O highly enriched stable isotopes

As follows from the figure, every 20 years there is an increase in the global production volume by approximately one order of magnitude, that is, 10 times. It can be calculated that annually the production volume (ΔV_{an}) increases on average by

$$\Delta V_{an} = \sqrt[\tau-1]{\Delta V \times 100} = \sqrt[10-1]{\frac{10-1}{2} \times 100} \approx 2\,\% \qquad (1)$$

and this increase is characterized by an enviable constancy. The forecast for the change in the volume of the global market for stable isotopes of light elements and labeled compounds for the next 5 years to 2022 points to 2.94% of the annual increase and the reaching of $ 294.2 million at the end of this period as noted on several websites. It is noted that the main factors that stimulate the growth of this market are an increase in research activities in the pharmaceutical and biotechnological sectors, the growth of research in proteomics and the expansion of the incidence of cancer. However, as noted repeatedly, the high cost of labeled compounds will to some extent impede market growth during the forecast period. Nevertheless, the history of production development and the forecast for the next five years show an exponential growth in the production and consumption of isotope production, which is clearly shown in Fig. 2 for isotopes of ^{13}C and ^{18}O.

It should be noted that for the last 10-15 years a new quantitative and qualitative transition has occurred – the volume of production has overcome the tonnage boundary (Fig. 1, 2a). It can assume that the ^{13}C and ^{18}O isotopes from the group of "research isotopes" passed into a new group of "diagnostic". It is interesting to note the decline in production of ^{18}O in the period 1980-90, which is clearly seen in Fig. 2b. The most likely explanation is the stop of ICON production in the Los Alamos laboratory in the US, as well as the accident of the installation of low-temperature rectification of NO in the company Isotec, Inc., which led to a deficit in the market for heavy-oxygen water in this period of time.

IOP Conf. Series: Journal of Physics: Conf. Series **1099** (2018) 012002 doi:10.1088/1742-6596/1099/1/012002

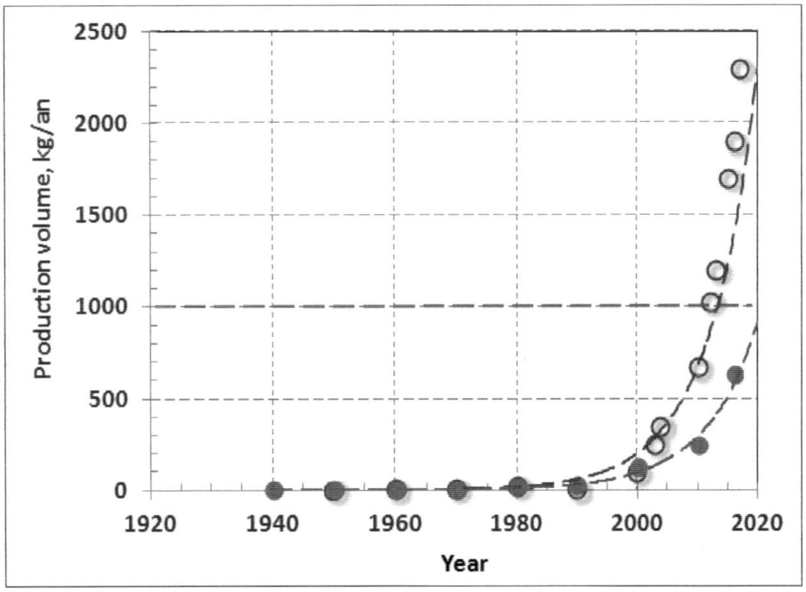

a

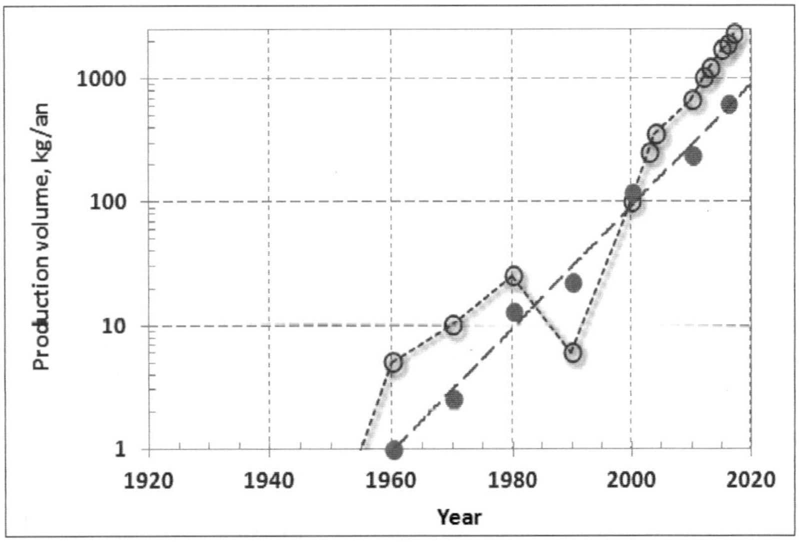

b

Figure 2. Change in the annual production volume of ^{13}C (●) and ^{18}O (○) isotopes by years: *a*- on an ordinary scale; *b*- in semilogarithmic coordinates

4. Manufacturers and technologies

For the two "diagnostic" isotopes ^{13}C and ^{18}O (in the form of heavy-oxygen water), their manufacturers and the technologies used are shown in Tables 2 and 3 respectively, and Figures 3 and 4 reflect the contribution of this or that technology to the total production volume of isotope products. Analysis of the tables and figures shows that the dominant amount of light isotopes is produced by physicochemical methods.

Producers and technologies for the ^{15}N isotope are considered separately in Table 4. Production of this isotope for many years remains at the level of several tens of kilograms annually and the main production volumes based on the chemical exchange between nitrogen oxides and nitric acid (Nitrox process).

Table 2. Production of ^{13}C in the world.

Country	Manufacturer	Method of production	Volume, kg ^{13}C/an
USA	Isotec, Inc.	CO Low-temperature rectification	100
	CIL, Inc.	CO Low-temperature rectification	420
China	Jiangsu Zhengneng Isotope Co., Ltd.	CO Low-temperature rectification	*Project 600*
Russia	RRC “KI”[a]	CO Low-temperature rectification	Stand by
	Gas-Oil, Ltd.	Laser / Gas centrifugal separation	Does not exist
	SC “PA Electrochemical Plant”	Gas centrifugal separation	≈ 5
Japan	Tokyo Gas Company	CH_4 (LNG) Low-temperature rectification	100

[a] Russian Research Center "Kurchatov Institute"

Table 3. Production of heavy-oxygen water $H_2{}^{18}O$ in the world.

Country	Manufacturer	Method of production	Volume, kg $H_2{}^{18}O$/an
Romania	INCDTIM[a]	Chemical Exchange NO-HNO_3	?
Georgia	NHTCG[b]	NO Low-temperature rectification	10
USA	Isotec, Inc.	NO Low-temperature rectification	120
	CIL, Inc.	Vacuum rectification of water	430
	ICON, LANL	NO Low-temperature rectification	Stand by
Israel	Marshall Isotopes, Ltd.	Vacuum rectification of water	150
	Rotem Industries Ltd.	Vacuum rectification of water	220 or 250
China	SRICI[c] Co., Ltd.	Vacuum rectification of water	100
	Jiangsu Zhengneng Isotope Co., Ltd.	Vacuum rectification of water	400
Russia	CJSC “GST”[d]	Vacuum rectification of water	60-80
South African Republic	Klydon	Vacuum rectification of water	200-230
Japan	Taiyo Nippon Sanso Corp.	O_2 Low-temperature rectification	600

[a] National Institute for Research and Development of Isotopic and Molecular Technologies
[b] National high technology centre of Georgia;
[c] Shanghai Research Institute of Chemical Industry;
[d] CJSC “Global Scientific Technologies”

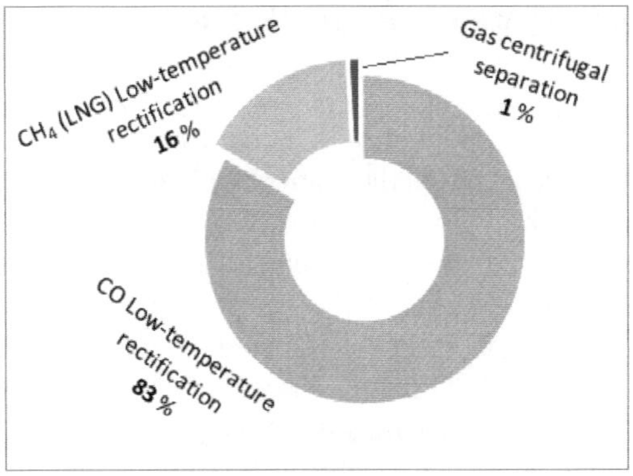

Figure 3. ^{13}C isotope technologies and their share in the total production scale

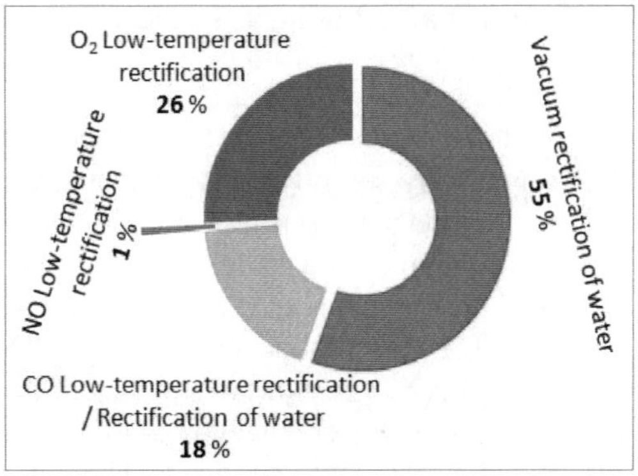

Figure 4. The isotope ^{18}O technologies and their share in the total production scale (in the form of $H_2{}^{18}O$)

Table 4. Production of ^{15}N in the world.

Country	Manufacturer	Method of production	Volume, kg ^{15}N/year
Georgia	NHTCG[a]	Chemical Exchange NO-HNO_3	
USA	Isotec, Inc.	NO Low-temperature rectification	
	CIL, Inc.	NO Low-temperature rectification	16
	ICON, LANL	NO Low-temperature rectification	Stand by
Romania	INCDTIM[b]	Chemical Exchange NO-HNO_3	?
China	SRICI[c] Co., Ltd.	Chemical Exchange NO-HNO_3	30
Japan	SHOKO Science Co., Ltd (SI Science Co., Ltd.)	Chemical Exchange NO-HNO_3	5

[a] National High Technology Centre of Georgia;
[b] National Institute for Research and Development of Isotopic and Molecular Technologies
[c] Shanghai Research Institute of Chemical Industry

Attention is drawn to the production of ^{13}C in Russia (Table 3) where, in contrast to other countries of the world, physical separation methods are developing - laser and gas centrifugal. In this case, the first one does not allow economical production of highly enriched ^{13}C [13], and the second one is characterized by a relatively low productivity [14]. In addition, the combination of these two methods, when used consecutively for the production of highly enriched heavy carbon isotope [15], is likely to lead to expensive ^{13}C, which can not contribute to the widespread introduction of diagnostic procedures with its use in the practice of clinical medicine. The separation of light-mass isotopes is most economical when using physicochemical separation methods (rectification, chemical isotope exchange). Existing publications [16, 17] demonstrate this clearly, and gas centrifuging, for example, is theoretically characterized by two orders of magnitude of higher energy costs compared to the process of rectification [18].

5. Features of technological development of isotope separation of light elements

5.1. Analysis of technological changes

Analysis of the technologies used for separating carbon isotopes on a large scale shows that, along with the relatively recently used CO low-temperature rectification, the enrichment coefficient $\varepsilon \approx 7\cdot10^{-3}$), a new methane rectification technology appeared. The ^{13}C methane distillation plant has been in operation since 2000 as a result of lengthy studies initiated in the first half of the 1990s by the Japanese company Tokyo Gas, which produces liquefied natural gas (LNG) [19]. But the main thing is an example of the production realization of the physicochemical isotope separation process, characterized by the value of the enrichment factor $\varepsilon \approx (3\text{-}4)\cdot10^{-3}$. Other similar technologies were not implemented either for rectification or for chemical exchange to the company Tokyo Gas.

In the case of ^{18}O enrichment, the dominant technology is rectification of water under vacuum. At the same time, the transition from an environmentally dangerous technology is seen (NO low-temperature rectification) to a more pure process, which is the rectification of molecular oxygen.

5.2. New opportunities

Technological successes achieved in Japan allow us to optimistically consider the possibility of producing a heavier nitrogen isotope ^{15}N, which can be claimed as a component of nitride fuel in nuclear power engineering in fast neutron reactors [20]. In addition, an increase in the volume of its production will undoubtedly lead to a spreading of the application of ^{15}N in medicine and biology (Table 1). Features of large-scale production of ^{15}N were considered repeatedly [16, 21], but the technologies used are the chemical exchange between NO, NO_2 and HNO_3; NO low-temperature rectification and the methods developed [17, 22, 23] are either environmentally hazardous or economically unsuitable, which is probably explained by the low production of ^{15}N. The most environmentally friendly and therefore scalable technology can be a low-temperature rectification of molecular nitrogen, realized by a transit scheme in combination with air separation plants producing liquid nitrogen. The value of the enrichment factor in such a process is at the same level as in the rectification of methane, $\varepsilon = 4\cdot10^{-3}$ at T = 77 K [11]. To obtain a highly enriched ^{15}N, such a process must provide for a homomolecular isotopic exchange (HMIEX)

$$2\ {}^{14}N^{15}N \rightleftarrows {}^{14}N^{14}N + {}^{15}N^{15}N \quad (2)$$

and the stage of additional enrichment of ^{15}N at the extraction of lighter molecules of $^{14}N^{14}N$ from nitrogen, as is the case with the production of ^{13}C by CO low-temperature rectification [11, 24]. One of the main problems in the use of physicochemical processes with a small value of ε – is a large steady-state time, possibly measured in years.

In order to overcome this disadvantage, it is proposed, at the initial stage of ^{15}N enrichment, to fill the distillation columns by the previously ^{15}N enriched and produced using the isotope separation method, which is most effective from the single isotopic effect point of view, that is, using the Nitrox process. For this purpose a starting small-scale installation is created, producing ^{15}N with a concentration (3-30)% at. in the form of a mixture of NO and NO_2, which is then easily reduced by hydrogen to molecular nitrogen on the catalyst

$$NO, NO_2 + H_2 \rightarrow N_2 + H_2O \quad (3)$$

The formed nitrogen passes through a stage of deep drying, is liquefied and poured into the columns, forming in them a fluid hold-up enriched by ^{15}N (Fig. 5). Calculations show that these techniques should ensure that the steady-state time decreases by approximately an order of magnitude. The proposed scheme can be considered as a separation process, based on the joint use of two physicochemical methods for isotope separation.

The development of physicochemical methods is based not only on their combination, but also on the application of new working systems in the chemical isotopic exchange, and on the use of a new type of equipment. Let us consider these two directions, for example, at the separation of boron isotopes. It is known that the production of boron isotopes is based on chemical isotope exchange between gaseous BF_3 and its liquid molecular complex, for example, with $CH_3OC_6H_5$ (methylphenyl ether or MPE) in mass-exchanging columns with a height of ≈ 60 m [25].

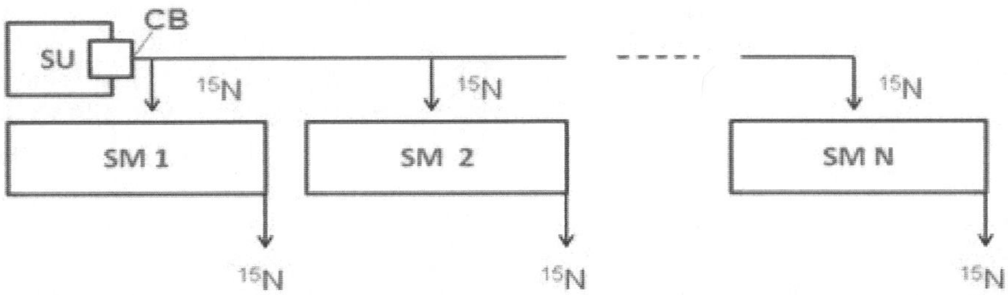

Figure 5. The large-scale production of ^{15}N with starting unit (production of low enrichment ^{15}N by Nitrox process) for the formation of ^{15}N hold-up in columns for low-temperature rectification of N_2: SU – starting unit; CB – conversion block (восстановление смеси NO и NO_2 до N_2 и глубокая осушка N_2); SM 1 … SM N – separation modules for ^{15}N enrichment by N_2 rectification

This requires the use of high production facilities, which leads to large capital and operating costs. The solution of the problem is the use of horizontal rotor mass-exchange columns, which have the same efficiency in comparison with classical vertical equipment. In Fig. 6 shows the results of a horizontal rotor column tests both in rectification mode and in the regime of chemical isotope exchange at the separation of boron isotopes [26]. In the latter case, based on the achieved *S* value, the number of theoretical separation plates (NTP) turned out to be equal 22±2, and the height of the theoretical plate was (3.6±0.4) cm.

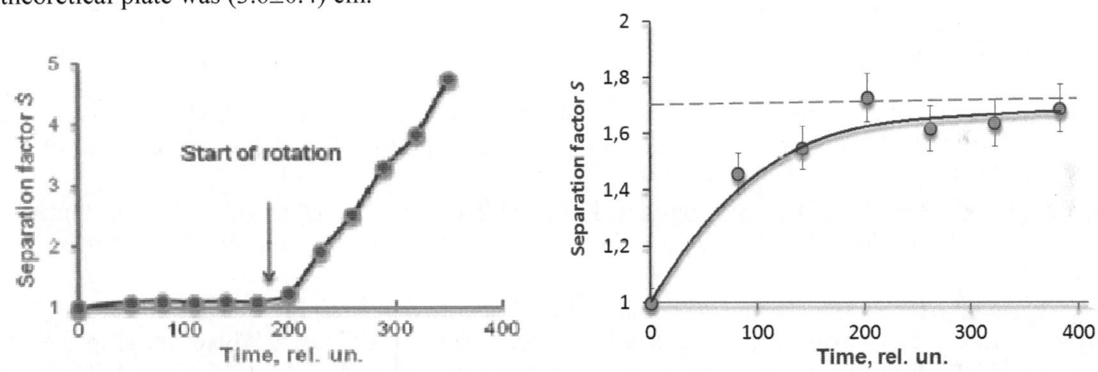

a *b*

Figure 6. The results of a horizontal rotary column tests with 80 cm long and spiral-prismatic packing 2,5×2,5×0,2 mm: *a*- change the separation degree of the components of the model mixture in the rectification mode without rotation and with rotation of the mass transfer part; *b*- kinetic curve for separation of boron isotopes in the regime of chemical exchange between BF_3 and its complex with MFE

More impressive results were obtained with the first experimental separation of boron isotopes between BF_3 and its complex with nitromethane CH_3NO_2 [27-29], that is, in the two phase system, the singly stage separation factor in which is measured in [29]. In a column with a height equal to only 1.4 m, the value of the separation degree was achieved S = 3.8 (Fig. 7) with a difference in ^{10}B concentration at the ends of the column equal to ≈ 30% at. [28]! It is interesting to note that the same separation degree was achieved by rectification of BCl_3 in an experimental plant with columns of a total height of 22 m [30].

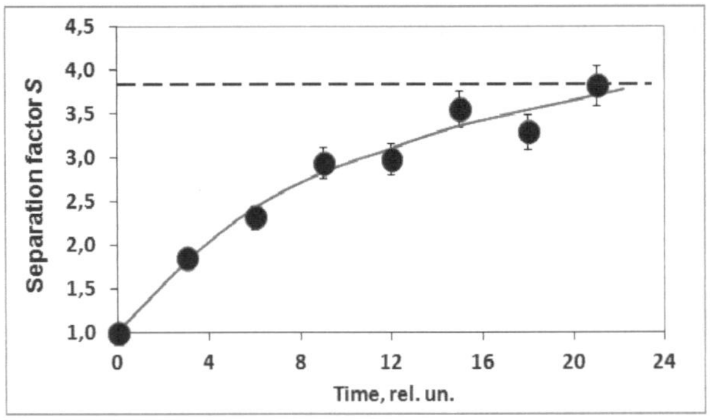

Figure 7. The change of the separation degree of boron isotopes during the chemical exchange between BF_3 and its complex with CH_3NO_2 in a mass-transfer column with 140 cm height and a spiral-prismatic packing of 1.2×1.2×0.2 mm

This more than clearly demonstrates the advantage of the chemical exchange method at the separation of boron isotopes in comparison with the rectification method and, moreover, in comparison with the physical methods of isotope separation.

6. Conclusion

During the last 15-20 years there have been significant changes in the production of stable isotopes of light elements. The production scale of ^{13}C and ^{18}O isotopes is measured in thousands of kilograms that was achieved by the development and use of physicochemical methods of isotope separation. These methods - rectification and chemical isotope exchange - have been developing in the world and Russia. Both new working systems and a new type of equipment were studied, new technical solutions were proposed to create more cost-effective technologies for production of stable isotopes of light elements.

7. References

[1] Adelstein S J and Manning F J 1995 Enriched Stable Isotopes *Isotopes for Medicine and the Life Sciences* ed S J Adelstein and F J Manning (Washington, DC: The National Academies Press) pp 19-34

[2] Meer K, Roef M J, Kulik W and Jakobs C 1999 In vivo Research with Stable Isotopes in Biochemistry, Nutrition and Clinical Medicine: An Overview *Isotopes in Environmental and Health Studies* **35** (1-2) pp 19-37

[3] *Beneficial Uses and Production of Isotopes* 2005 (Paris: OECD NEA) No. 5293 pp 20-22, 57

[4] Reinout C A Schellekens R C A, Stellaard F, Woerdenbag H J, Frijlink H W and Kosterink J G W 2011 Applications of stable isotopes in clinical pharmacology *Br. J. Clin. Pharmacol.* **72** (6) pp 879-97

[5] Sharp P F and Welch A 2005 Positron Emission Tomography *Practical Nuclear Medicine* ed P F Sharp (London: Springer) pp 35-48

[6] Fan T W-M, Lorkiewicz P K, Sellers K, Moseley H N et al 2012 Stable isotope-resolved metabolomics and applications for drug development *Pharmacology & Therapeutics* **133** (3) pp 366-91

[7] Butler R N, Kosek M, Krebs N F, Loechl C U et al 2017 Stable Isotope Techniques for the Assessment of Host and Microbiota Response During Gastrointestinal Dysfunction *Journal of Pediatric Gastroenterology and Nutrition* **64** (1) pp 8-14

[8] Tando Y, Matsumoto A, Matsuhashi Y, Tanaka H et al 2011 Carbon-13 and Its Clinical Application *Gas Biology Research in Clinical Practice* ed T Yoshikawa and Y Naito (Basel: Karger International) pp 112-18

[9] Nicholson B D, Abel L M, Turner P J, Price C P et al 2017 Point-of-care *Helicobacter pylori* testing: primary care technology update *Br. J. Gen. Pract.* **67** (665) pp 576-77

[10] Thoennessen M 2016 The Discovery of Isotopes: A Complete Compilation (Springer

International Publishing Switzerland) Chapter 3 Isotopes of Stable Elements pp 23-37
[11] *Separation of Isotopes of Biogenic Elements in Two-phase Systems* 2006 ed B M Andreev (Amsterdam: Elsevier Science & Technology) pp 217-97
[12] Bigeleisen J 1969 Isotope separation practice *Advances in Chemistry* **89** pp 1-24
[13] Apatin V M, Laptev V B, Petin A N and Ryabov E A 2003 Study on the Capabilities of One- and Two-Stage Separation Schemes for Producing Highly Enriched Carbon-13 by the Laser Method *High Energy Chemistry* **37** (2) pp 101-07
[14] Borisevich V D, Vlasov A L, Kozlova E I et all 2005 Isotope enrichment of uranium *Isotopes: properties, production, application* ed V Yu Baranov (Moscow: Fizmatlit) vol 2 pp 128-145 (*in Russian*)
[15] Baranov V Yu, Astakhov A V, Zinchenko A K et al 2001 Technological complex for laser separation of carbon isotopes *Russian Chem J* **45** (5-6) p 89-95 (*in Russian*)
[16] Borisevich V D, Khoroshilov A V, Gorshkov V I and Ivanov V A Comparative estimation of separation methods for production of highly enriched isotope of nitrogen-15 *Separation Phenomena in Liquid and Gases. 7th Workshop Proc.* (Moscow: MEPHY) pp 218-28
[17] Borisevich V D, Khoroshilov A V and Cherednichenko S A 2005 Thermal conductivity of silicon-28 and separation technologies for its production *Proc. of the 5th Intern. Conf. on Isotopes (5ICI)* 25-29 April 2005 Brussels, Belgium pp 193-201
[18] Letokhov V S 2005 Principles of laser isotope separation *Isotopes: properties, production, application* ed. V Yu Baranov (Moscow: Fizmatlit) vol 2 pp 357-74 (*in Russian*)
[19] Haga K and Soh H 1994 CH_4 Distillation Plant *5th Intern. Symp. The Synthesis and Applications of Isotopes and Isotopically Labelled Compounds Symposium Handbook and Collection of Abstracts* (France: Strasburg) p 144
[20] Boner A A 1972 Nitride Fuels: Properties and Potentials *Reactor Technology* **15** (2) pp 87-104
[21] Khoroshilov A V 1999 Technology of nitrogen isotopes: problems and solutions *Khimicheskaiy promishlennost* 4 (241) pp 37-46 (*in Russian*)
[22] Aoki E, Kai T and Fujii Y 1997 Theoretical Analysis of Separating Nitrogen Isotopes by Ion-Exchange *J. Nucl. Sci. Technol.* **34** (3) pp 277-82
[23] Borisevich V D, Morozov O E, Zaozerskiy Yu P, Shmelev G M and Shipilov Yu D 2000 On the enrichment of low-abundant isotopes of light chemical elements by gas centrifuges *Nuclear Instruments and Methods in Physics Research A* **450** pp 515-21
[24] Zelvensky Ya D and Khoroshilov A V 1999 Production of stable carbon isotopes by cryogenic rectification of carbon monoxide *Khimicheskaiy promishlennost* 4 (241) pp 25-31 (*in Russian*)
[25] Katalnikov S G and Ivanov V A 2001 Physico-chemical and engineering principles of boron isotopes separation by using BF_3 - anisole·BF_3 system *Separation Science and Technology* **36** pp 1737-68
[26] Khoroshilov A V and Sizov A P 2011 Isotope Separation in the Horizontal Rotating Columns 7th International Conference on Isotopes *The collection of abstracts* (Moscow: Atominform) p 46
[27] Khoroshilov A V, Lizunov A V, Stepanov A V and Cherednichenko S A 2009 Thermal dissociation of the complex $BF_3 \cdot D$ and boron isotope separation in the system BF_3–$BF_3 \cdot CH_3NO_2$ *Radiochemistry* **51** (4) pp 400-02
[28] Khoroshilov A V, Stepanov A V, Lizunov A V and Zernova E V 2010 The first separation of boron isotopes by chemical exchange at a lower temperature in system the boron trifluoride - its complex compound with nitromethane *Perspectivnie materiali* Special issue (8) pp 258-62 (*in Russian*)
[29] Herbts R S and McCandless F P 1994 Improved Donors for the Separation of the Boron Isotopes by Gas-Liquid Exchange Reactions *Separation Science and Technology* **29** pp 1293 – 310
[30] Andrietz S P, Guschin A A, Kalashnikov A L, Kozirev A S et al Creation and testing of a pilot plant for separation of boron isotopes by BCl_3 rectification *Perspectivnie materiali* Special issue (8) pp 193-198 (*in Russian*)

Physical and chemical processes on the ^{212}Pb radionuclide production for nuclear medicine

P P Boldyrev[1], B V Egorova[2], K V Kokov[1], Yu A Perminov[1], M A Proshin[1] and D Yu Chuvilin[1]

[1] Kurchatov Complex of Physical and Chemical Technologies, National Research Center "Kurchatov Institute", 1, Akademika Kurchatova pl., Moscow, 123182, Russia
[2] Faculty of Chemistry, Lomonosov Moscow State University, 1-3, Leninskiye Gory Street, Moscow, 119991, Russia

kvkokov@yandex.ru

Abstract. An automated ^{212}Pb radionuclide generator with ^{228}Th as a parent isotope has been designed. The generator operation principle is based on transport by an airflow of gaseous ^{220}Rn radionuclide emanating from DOWEX ion exchange resin containing ^{228}Th into a helix-shaped collector vessel where post-decay ^{212}Pb is collected and deposited on the collector walls. After a 72-hour operation cycle of the generator, sampling of ^{212}Pb in the form of solution in 0.1M HCl is executed with approximately 30% yield of ^{212}Pb. The generator is remotely controlled by a PC-based automatic system. The generator is reloaded once in a few years. The generator is intended for medicobiological and radiochemical investigations in the field of developing antineoplastic drugs of targeted delivery.

1. Introduction

The development of Kohler and Milstein, the future Nobel laureates (1984), of methods for creating monoclonal antibodies opened a new era in biology and medicine [1]. A promising method for diagnosis and treatment of cancer began to be developed on its basis – immunotherapy (or targeted delivery). Artificially created antibodies for certain markers of cancer cells have a specific ability of directed delivery to malignant cells the attached drugs. Such drugs, or therapeutic agents, can be realized by radionuclides, which radiation during its decay can lead to the cell death. In addition to the radioimmunotherapy, it is possible to develop radiopharmaceuticals based on synthetic peptides, such as, for example, octreotide and its modifications [2,3].

Radionuclide therapy with α-particle emitters is a promising tool in cancer therapy with ^{211}At, ^{212}Bi, ^{213}Bi and ^{212}Pb (*in vivo* generatior of ^{212}Bi) as the most promising radionuclides [4]. α-particles are highly cytotoxic, namely, it has been shown *in vitro* that a single α-particle traversal of a cell nucleus has a 20–40% probability of killing the cell [5-7].

^{212}Pb is of particular concern, since it is a β-emitter, but its daughter radionuclides ^{212}Bi and ^{212}Po undergo α-decay, whereby ^{212}Pb is to be considered as an *in vivo* α-particle generator [8]. ^{212}Pb is a member of decay chain of the parent ^{228}Th and can be produced by generators.

^{233}U is produced by the neutron irradiation of ^{232}Th, but its production invariably produces small amounts of ^{232}U as an impurity, because of parasitic (n,2n) reactions on ^{233}U itself, or on protactinium-

233, or on ^{232}Th. The amount of ^{232}U in the accumulated ^{233}U increases with increasing neutron flux density and time of thorium irradiation and depends on the neutron spectrum in the reactor. After recirculation of the irradiated mixed oxide fuel (Th, U)O_2, the amount of ^{232}U in ^{233}U can reach 0.08-0.1%. Further, the α-decay of ^{232}U occurs with a probability of almost 100% with ^{228}Th formation. ^{228}Th decay chain is presented on figure 1 (scheme from [9]), one of its daughter nuclide is ^{212}Pb.

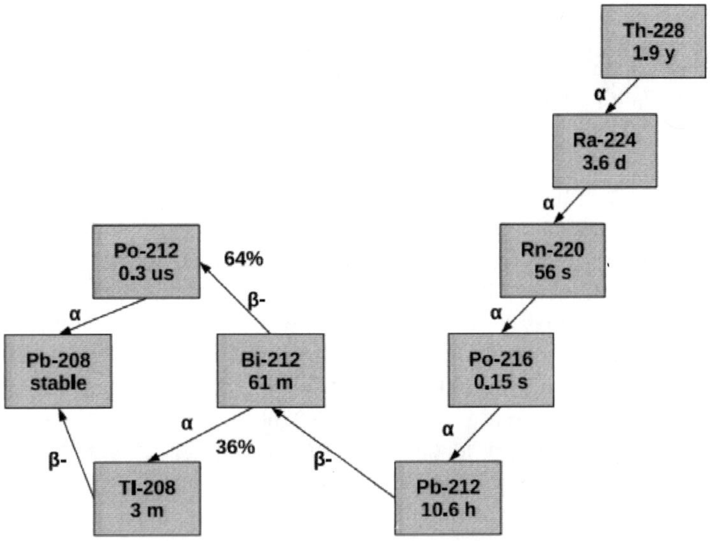

Figure 1. ^{228}Th decay chain.

For efficient therapeutic use of ^{212}Pb/^{212}Bi 10 GBq should be readily available at the treatment site [10]. The current generator principle for producing ^{212}Pb/^{212}Bi involves elution of a ^{224}Ra-loaded cation-exchanger column with HCl or HI [11]. If the ^{224}Ra–based generator could accommodate radioactivity levels of 10–20 GBq, generator replacement would be needed every 3–6 d. Such frequent generator replacement would be expensive and inconvenient, due to the high-energy γ-rays emitted from the decay of ^{208}Tl. If the longer-lived ^{228}Th was used as the primary source instead of ^{224}Ra, replacements would only be necessary every 2–4 yr. Thus, a significant reduction in production and transportation costs, and reduction in the radiation dose to the workers involved, could be achieved.

Previous efforts to construct ^{228}Th-based generators have failed to achieve therapeutic levels of radioactivity due to radiolytic damage of the parent nuclide materials [12,13]. Elimination of this obstacle was described in paper [10], where destructive radiolytic effect was reduced due to procuring low radioactivity-to-mass ratio by special source compartment construction. Namely, dry air was passed through a number of thin source frames with [^{228}Th][Ba($C_{18}H_5O_2$)], carrying emanated ^{220}Rn out into a radon collector, where radon gas was rapidly extracted into the organic solvent at temperatures below -71^{0}C and solidified. At the end of a collection period, the required decay product ^{212}Pb could be recovered.

Other scheme for producing ^{212}Pb and ^{212}Bi [14] was proposed. The technology is based on two generators operating in tandem: ^{228}Th/^{212}Pb and ^{212}Pb/^{212}Bi. The first one is based on separation from an initial solution containing thorium isotopes the gaseous ^{220}Rn. After conditioning, the ^{212}Pb solution from the ^{228}Th/^{212}Pb generator was used to charge a column, which functioned as a ^{212}Pb/^{212}Bi generator, with a cation exchanger.

In this paper a novel ^{228}Th-based generator for the ^{212}Pb production for nuclear medicine investigations is proposed. It is based on the principle of collecting ^{212}Pb, daughter nuclide of ^{220}Rn emanating from ^{228}Th-containing ion-exchange resin.

2. Materials and methods

2.1. Production of ion exchange resin with ^{228}Th

The starting material in the generator was powdered ^{228}Th-containing ion-exchange resin. The resin was produced by adding DOWEX ion-exchange resin to the solution containing a mixture of ^{228}Th, ^{229}Th, ^{232}Th radionuclides and their daughter products. The isotope composition of such solution extracted from ^{233}U of long-term holding (about 50 years) was as follows: ^{229}Th – 6.81%; ^{230}Th ≅ 0.08%; ^{228}Th – traces; ^{232}Th – 93.11%.

20 ml of the solution of the mixture of ^{228}Th, ^{229}Th radionuclides and their daughter products in 8M HNO_3 were mixed with 5-6 ml of DOWEX-1 anion-exchange resin making use of the thorium ability to bind tightly with the functional group of the anion-exchange resin. After 1-hour holding, practically all thorium is sorbed on the resin. Then the solution was decanted, and wet resin was dried up.

According to the values of the constants of formation of thorium anion nitrate complexes [15] in the environment of 8M HNO_3, all thorium is in the form of anion complexes $Th(NO_3)_5^-$ (2%), $Th(NO_3)_6^{2-}$ (98%) which replace chloride-ion (nitrate-ion) on the DOWEX-1 resin. When dried, thorium anion nitrate complexes, similarly to chloride ions, remain bound with ammonium groups on the resin which permits subsequent use of this dry resin as a "radon generator".

This ^{228}Th-containing resin was produced some time ago, so specific activity namely at the time of the generator testing (July of 2017) was approximately 1,2 MBq/g. The works were conducted with the total activity values of about 7 MBq (i.e. 5 g of resin).

2.2. Generator principle of operation

The layout diagram of the ^{212}Pb generator based on the distinction feature of the ^{228}Th decay chain consists of two interconnected module loops (see figure 2), namely:

– an "air loop" consisting of a ^{220}Rn source (thorium reactor – TR), an air pump – MP1 and a ^{212}Pb collector vessel – HC;

– a "fluid loop" containing vessels for ^{212}Pb rinsing and after-rinse collection of the solution, a fluid pump – MP2 and the collector.

Such modular layout makes it possible to use different TRs and collectors independently of one another. Ion-exchange resin with the ^{228}Th isotope was located in a special vessel referred to as a thorium reactor (figure 3). A membrane pump continuously pumped air through the system during the accumulation cycle. The air passed through the reactor and carried away gaseous ^{220}Rn emanating from ^{224}Ra into the helix-shaped collector vessel. The ^{220}Rn decay results in the formation of ^{212}Pb which deposited on the collector surface and when accumulation ceased it was flushed with 0.1M solution of hydrochloric acid. Owing to such separation of the solid phase with the parent nuclide and the gaseous phase with ^{220}Rn, a probability of long-lived parent isotopes getting into the final preparation is eliminated, thereby ensuring high radionuclide purity of the preparation and permitting to avoid operations of its purification. Elimination of long-lived parent radionuclides was confirmed by γ-spectrometric measurements.

IOP Conf. Series: Journal of Physics: Conf. Series **1099** (2018) 012003 doi:10.1088/1742-6596/1099/1/012003

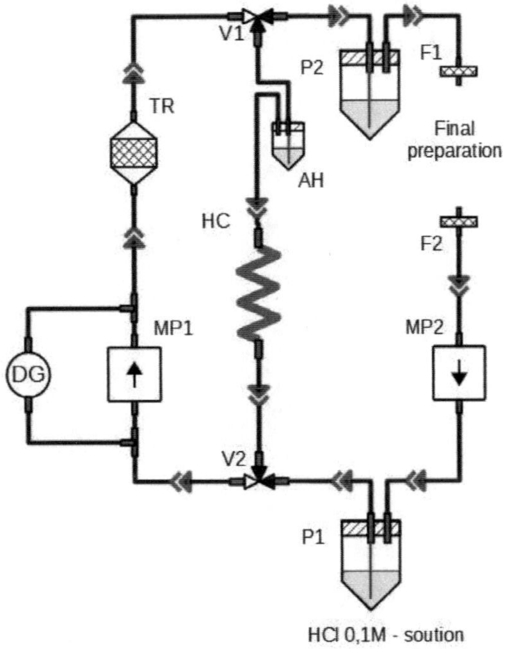

Figure 2. Process flowsheet of the generator. TR — thorium reactor, MP1-MP2 — membrane pumps, HC — helical collector, AH – air humidifier, DG — differential pressure gauge, F1-F2 — gas filters, V1-V2 — solenoid valves, P1-P2 – vessels for chemical agents.

To ensure effective deposition of lead on the collector walls, firstly, a saturated steam condition was maintained in the system by providing the system with an air humidifier (AH) containing 0.1M solution of hydrochloric acid, and secondly, a condition of essential excess of the collector volume over parasitic volumes (those of the thorium reactor and communications) was satisfied.

All the elements of the installation were connected among themselves with silicone tubes and all the connections were made with leak-tight joints.

2.3. Thorium reactor

The thorium reactor (see figure 3, all sizes are given in mm) was formed by welding metal elements to form a cylinder-shaped container, with two openings for air inlet and outlet. All elements, except a rubber sealing disc, were made of circonium because of its high chemical resistance.

On the one hollow cylinder (1) face an opening was made, the other side cut off; inside this structure a glassful (3) was placed for immersion in it ^{228}Th-containing ion-exchange resin. Further, from the side of the cut face, the construction was covered with a cover (2); the tightness was ensured by sealing with radiation-resistant rubber (6). In order to prevent the entrainment of solids from the reactor, Petryanov filters (7) were placed at the inlet and outlet, which were attached to the sides of the glass with special support disks (4).

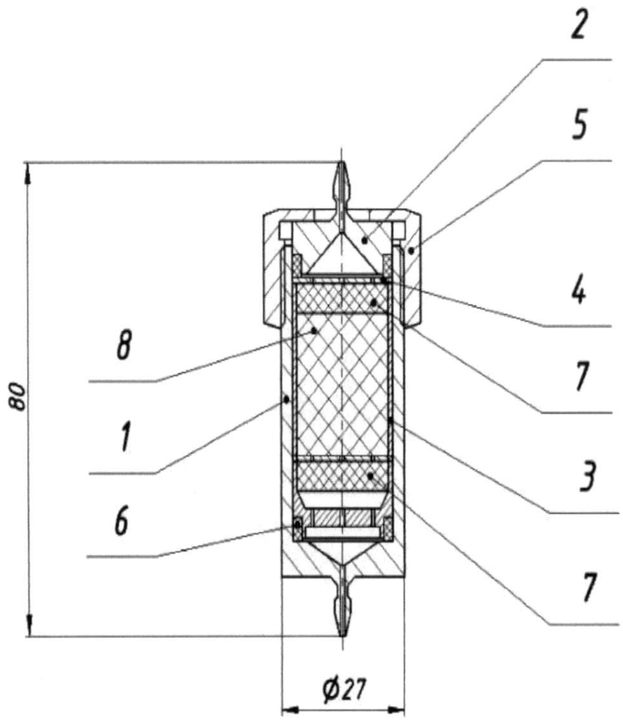

Figure 3. Thorium reactor. 1 — vessel, 2 — cover, 3 — glassful, 4 — support, 5 — nut, 6 — sealing disk, 7 — Petryanov filters, 8 – resin.

2.4. ^{212}Pb collector vessel

A helix-shaped collector vessel is intended for collecting the ^{212}Pb target radionuclide. It is made as a coiled thin fluorinated ethylene propylene (FEP) tube, wrapping around a hollow metal drum. The tube is 30 m long with the inner diameter of 2 mm, providing the 94 ml total volume of the collector. Radionuclide ^{212}Pb formed on a decay of ^{220}Rn passing through the collector with an air flow is deposited on its walls. FEP has been chosen because of its convenient chemical and physical properties, mainly, high chemical resistance and low wettability. The latter circumstance has favored washing-down of the deposited ^{212}Pb.

In the 0.1M HCl medium, according to constants [16], Pb^{2+} is present in the solution as $PbCl^+$ (60%), Pb^{2+} (20%) and $PbCl_2$ (20%), which ensures its rinsing from the collector walls. In view of a high value of the solubility product of $PbCl_2$, the formation of insoluble compound in the presence of 0.1M Cl^- may take place only with the Pb^{2+} content $(Pb^{2+}) > 1 \cdot 10^{-3}$M which is by 7 orders of magnitude larger than that being formed in the generator under discussion.

Washing-down of ^{212}Pb from the collector surface is made by pumping the 0.1M solution of hydrochloric acid, HCl, of the total volume of 10 ml through the helix using a membrane pump, connected separately and in parallel with the first pump pumping air under working conditions.

To ensure minimum fluid losses, helix mounting conditions (helix slope angle, winding density) were very carefully calculated in order to avoid the formation of areas where liquid might get accumulated. It ensured a uniform ascend of the helix along the cylinder surface without sudden discontinuities and artificial waterlocks.

An air humidifier containing about 5 ml of 0.1M diluted hydrochloric acid, HCl, was mounted in front of the helix-shaped collector. In the course of a complete generator working cycle, the humidifier evaporated liquids in the volume of up to 2 ml which allowed the saturated vapor condition to be sustained and a thin liquid film to be formed on the inner surface of the collector. The latter circumstance has favored an increase of the deposited ^{212}Pb.

Radionuclides ^{212}Pb and ^{212}Bi which are produced resulting from ^{220}Rn decay are deposited on the helix walls, but increased deposition of ^{212}Pb in view of a liquid film of 0.1M HCl on the walls of the FEP tube used as a collector may be attributed to an affinity of the Pb^{2+} formed to aqueous chlorine-containing solution due to its providing for the formation of chloride complexes [16] in addition to the non-volatile nature of Pb^{2+} under the normal conditions employed.

2.5. Generator performances and radioactivity measurements

The generator is remotely controlled by a PC-based automatic system. The air flow rate is measured with a flow rate meter (Microbridge Airflow Sensors). Experimental capabilities of the pump prevented gradual variation of the pumping rate, Q, in a wide range, therefore, two Q values, 270 ml/min and 420 ml/min, were specified. The generator service time is arbitrary, till a hardware fault.

After the accumulation cycle had been finished the pump switched off, and the valves were swapped so that the collector vessel got connected to the rinsing loop. Solution of 0.1M HCl was pumped through the collector tube followed by sampling of a 0.5 ml specimen and taking measurements of the activity for further determination of the activity in the final preparation.

The activity of the radionuclides were determined by the integrals of the photo-peaks of the 239, 583 and 727 keV γ-lines from decays of ^{212}Pb, ^{208}Tl and ^{212}Bi, respectively, measured with a Ge-semiconductor detector coupled to a multi-channel analyzer (Ortec).

3. Results

3.1. Experiments on generator operation

The 3-day operation of the generator produced the total value of ^{212}Pb activity of 2.01 MBq with the ^{228}Th activity of 7.05 MBq (as for July'3 of 2017) and the air flow rates of 414 ml/min totally in the humidifier and the helix-shaped collector. As was experimentally proven, a further increase of the accumulation time, i.e., to 5 days, did not result in an increase of the produced isotope amount.

On reducing the flow rate down to 273 ml/min the activity became 1.73 MBq. Thus, an increase in the flow rate from 270 ml/min to 420 ml/min ensured nearly 15% gain in yield of the isotope. Thus, it is possible to drive at a conclusion that the ^{212}Pb collection efficiency is at least about 40%.

3.2. Extraction of ^{212}Pb

After having been pumped through the helix-shaped collector and air humidifier, the solution was supplied into a vial for the product with nearly 1 ml of the rinsing solution remaining in the helix. Pumping during 12 minutes was carried out by the membrane pump; subsequent pump operation gave no decrease of the fluid loss in the helix. The vial with the product is dismounted for further ^{212}Pb use both for direct application and for the production of alpha-emitting ^{212}Bi. The specific activity of the final preparation is about 200 kBq/ml for the 10-ml volume of eluent and 1-ml volume of losses.

Radiometric measurements of the collector have proven that about 30% of the isotope remain on the walls of the tube. It can be related to the $^{212}Pb^{2+}$ sorption on carboxyl groups of the FEP surface. Furthermore, in the case under consideration, an increase or decrease of the acid concentration produced no noticeable effect on the amount of the lead washed-down from the tube which can testify to spatial inaccessibility of the surface-connected cation, $^{212}Pb^{2+}$, for the dynamically incoming acid. The total generator efficiency with considering this circumstance achieves, therefore, approximately 30%.

4. Discussion

4.1. Generator performance and prospectives

The collection efficiency increase following air flow rate increase can be explained by growth of resin particles total surface area, contacting with air flow, which provides to carry radon away from reactor volume more effective. The presence of a fluidized bed is an ideal case.

Thus, it is possible to increase the total efficiency by optimal configuration and modernization of the thorium reactor. Aside from the creation of the fluidized bed, cassette design of radon source could be realized, where ^{228}Th-containing resin could be distributed in a thin (1-2 mm) layer. Investigations with such a reactor type would allow to study humidity impact to radon emanation from the resin. It is also prospectively to investigate the use of different materials in collector vessel, in particular, for study lead deposit mechanism on its walls.

Despite the relative low ^{212}Pb production efficiency, the proposed design has several advantages. Closed circuit provides lead collection at room temperature, which fundamentally improves an operation process of the generator. Besides, ^{228}Th-containing resin production is expected to be more technologically and simpler than [^{228}Th][Ba($C_{18}H_5O_2$)]. Current levels of specific activity (up to 200 kBq/ml) can be used for *in vitro* experiments, as well as for preclinical trials within certain limits (for example, see [17,18]). For wide ^{212}Pb-generator use in preclinical trials thorium reactor is assumed to be filled with new ^{228}Th-containing ion-exchange resin followed by significant specific activity increasing. Simple and cheap generator design provides continuous ^{212}Pb production for the investigations with synthesis of radio conjugates and its *in vivo* and *in vitro* studies.

5. Conclusions

The generator was designed for experimental works aimed at the synthesis of radiopharmaceuticals and *in vitro* and *in vivo* trials of the synthesized radiopharmaceuticals; for this reason, the attained activity reaches the level of 2 MBq, and it could be easily significantly increased by resin renewal. The question of applicability of the suggested process for clinical practice which would require an increase of the product activity up to 1 GBq demands further investigations.

References

[1] Kohler G and Milstein 1975 *Nature* **256** 495–7
[2] Miao Y, Hylarides M, Fisher D, Shelton T, Moore H, Wester D, Fritzberg A, Winkelmann C, Hoffman T and Quinn T 2005 *Clin. Cancer Res.* **11** 5616–21
[3] Miao Y and Quinn T 2008 *Crit. Rev. Oncol. Hematol.* **67** 213–28
[4] Brechbiel M 2007 *Dalton Trans.* **21** 4918-28
[5] Soyland C and Hassfjell S 2000 *Int. J. Radiat Biol.* **76** 1315–22
[6] Pugliese M, Durante M, Grossi G, Montforti F, Orlando D, Ottolenghi A, Scampoli P and Gialanella G 1997 *Int. J. Radiat Biol.* **72** 397–407
[7] Hei T, Wu L, Liu S, Vannais D, Waldren C and Randers-Pehrson G 1997 *Proc. Natl. Acad. Sci.* **94** 3765–70
[8] Edem P, Fonslet J, Kjaer A, Herth M and Severin G 2016 *Bioinorg. Chem. Appl.* **2016** 6148357
[9] Ford R, Gerbier-Violleau M and Vazquez-Jauregui E 2010 *Phys.Lett.* A **374** 701-3
[10] Hassfjell S 2001 *Appl. Radiat Isot.* **55** 433-9
[11] Atcher R, Friedman A and Hines J 1988 *Appl. Radiat. Isot.* **39** 283–6
[12] Zucchini G and Friedman A 1982 *Int. J. Nucl. Med. Biol.* **9** 83–4
[13] Hassfjell S, Hoff P 1994 *Appl. Radiat. Isot.* **45** 1021–25
[14] Boldyrev P, Bortash A, Zagryadskii V, Zakharov A, Nikolaev V, Proshin M, Chuvilin D, Shatrov A and Vesnovskii S 2012 *Atomic Energy* **111** 422-7
[15] Danon J 1960 *J. Inorg. Nucl. Chem.* **13** 112-8
[16] Luo Y and Millero F 2007 *Geochim. Cosmochim. Acta.* **71** 326–34
[17] Kasten B, Oliver P, Kim H, Fan J, Ferrone S, Zinn K and Buchsbaum D 2018 *Int. J. Mol. Sci.* **19** 925
[18] Kasten B, Gangrade A, Kim H, Fan J, Ferrone S, Ferrone C, Zinn K and Buchsbaum D 2018 *Int. J. Nucl. Med. Biol.* **58** 67-73

IOP Conf. Series: Journal of Physics: Conf. Series **1099** (2018) 012004 doi:10.1088/1742-6596/1099/1/012004

Adaptive system model of stabilization of the power factor of actuators of centrifuges for uranium enrichment with structural dynamic model of electric drive with parallel reactive power compensation

V M Juromskiy

Department of molecular Physics, National Research Nuclear University "MEPhI", Kashirskoe shosse 31, Moscow, 115409, Russia

juromskiy@mail.ru

Abstract. Presents an engineering model of adaptive system of automatic stabilization of the power factor Cos φ = 1 as the condition of resonance at the frequency of power supply with parallel compensation of reactive power of electric drive of centrifuges battery of capacitors. The resonance in the power circuit corresponds to an extremum (minimum) of total current was taken as the goal of the algorithm automatic search. The operation of the adaptive control system in conjunction with a structural dynamic model of the object of optimization is presented modeling in terms of a Simulink (MATLAB). It is shown that in the process of automatic stabilization Cos φ = 1 excluded "overcompensation" in the local electric power network, including when disconnect blocs of the centrifuges. The adaptive control system is insensitive to additive errors of the sensor current I, can be realized general technical means and applied in general power networks.

1. Introduction

It is known that the energy efficiency of the electric drive is characterized by a power factor Cos φ = *P/S*, where $S = \sqrt{P^2 + Q^2}$ the total power, $P = U\, I$ Cosφ - active power, $Q = V\, I$ Sin φ is the reactive power creates the electromagnetic field in the windings of the electric drive. Reactive power performs no useful work, additionally load chains power transmission, compelling to increase the size and power of generating devices.

Reactive power compensation in electrical power circuits is a significant economic factor [1] separation of isotopes of uranium high-speed gas centrifuges with driven on the basis of the hysteresis motors /HM/ with power factor Cos φ = (0.2 – 0.4). HM is powered from static frequency Converter /SFC/ forming a local grid "HM – SFC" with non-standard frequency of the power network [2]. The increase in the power factor Cos φ in the energy system is enabling the parallel composition HMs of the compensating capacitors, which can provide a standard value Cos φ = 0.9 – 0.95. But in this case, the proportion I_Q of the reactive component of the total current I is (50– 30)%.

IOP Conf. Series: Journal of Physics: Conf. Series **1099** (2018) 012004 doi:10.1088/1742-6596/1099/1/012004

2. Statement of problem

In [3] introduces the concept of automatic stabilization of the power factor Cos φ = 1 motor centrifuge, based on automatic search [5] of minimum of extreme dependence of the total current I in functions capacity C (figure 1*a*) corresponding to the resonance in the power circuit on the supply frequency.

Extreme characteristics $I(C)$ in [3] are represented as inertia-free, built based on electric calculations. When changing the number of operating centrifuges and other factors affecting the inductance L and the resistance R multimotor electric drive, it is necessary analytically to re-present and tabulate new condition $I(C)$ in the model control system presented in [3].

3. Building a structural dynamic model of the actuator with compensation

In this paper, the electric drive with parallel compensation appears to be a dynamic model constructed based on methodology of structural modeling of dynamic processes in *L-R-C* circuits [6].Model can be easily represented by means of visual programming Simulink (MATLAB) with the ability to interactively change parameters of electric drive simulation. The transformation of electrical structure of figure 1*a* in the dynamic model figure1*b* suggests SFC as the "source voltage" U_{sup} = 380 V and the ratio of the currents I_1, I_2, educated, diverse reactive impedances in the form I_1, + $I_2 = I$, (see figure 1*a* A and figure 1*b*, the adder No. 1).

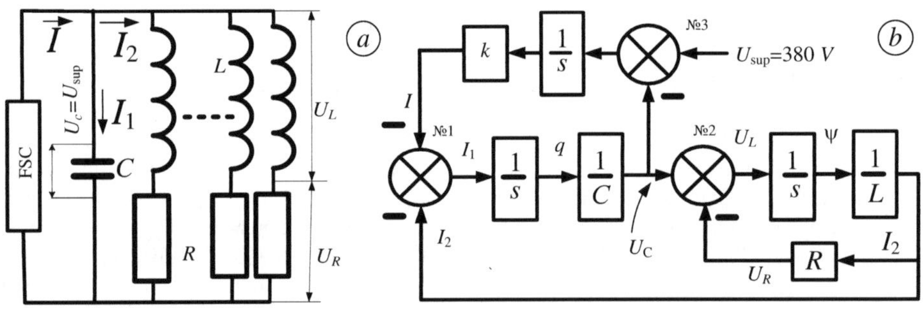

Figure 1. The electrical circuit (*a*) and dynamic model (*b*) power system "HM– FSC" of compensation of the reactive component of the total power parallel by connection of the capacitor C: *s* – Laplace operator, 1/*s* – block integration. *L-R* – inductance and ohmic resistance of the parallel-connected motors.

In a capacitive branch of the structure of figure 1*a* the charge of the condenser $q = \int I_1 dt$ (integrator 1/*p*, figure 1*b*) when the voltage on the capacitor $U_C = q / C = U_{sup}$. For *R-L* branch of figure. 1*a*, the voltage on the inductance $U_L(t)$ determines the linkage $-\psi(t) = \int_0^t U_L(t)dt$ (integrator 1/*p*) and the flux linkage – current $I_2(t) = \psi(t)\frac{1}{L}$ (unit is 1/*L*). The current $I_2(t)$ generates a voltage drop $U_R(t)$ on the resistance *R*. This voltage is the output for *R-L* circuit and is subtracted from the input $U_{sup} = U_C$ (adder No. 2, figure 1*b*) in determining the U_L. Thus is manifested the internal feedback determines the development of electrical processes in *R-L* – chain. [6]).

Negative feedback circuit formed by the element of comparison No. 3 voltage $U_C(t)$ c U_{sup} = 380 V in conjunction with block integration k/p, modify the output current I until the $U_c(t) = U_{sup}$. [4]. The structure of figure 1 assume that the electric drive operates in the regime close to the idle.

The family of characteristics according to the total current I from the compensating container for the composition HM, connected in parallel with the resulting parameters L = 50 μH, R = 0.15 Ω, Cos φ = 0.25, when the supply frequency f = 1800 Hz is shown in figure 2, where C = 147 F corresponds to the resonance *R-L-C* circuit in the normal condition of the cascade fits (see the curve 2).

IOP Conf. Series: Journal of Physics: Conf. Series **1099** (2018) 012004 doi:10.1088/1742-6596/1099/1/012004

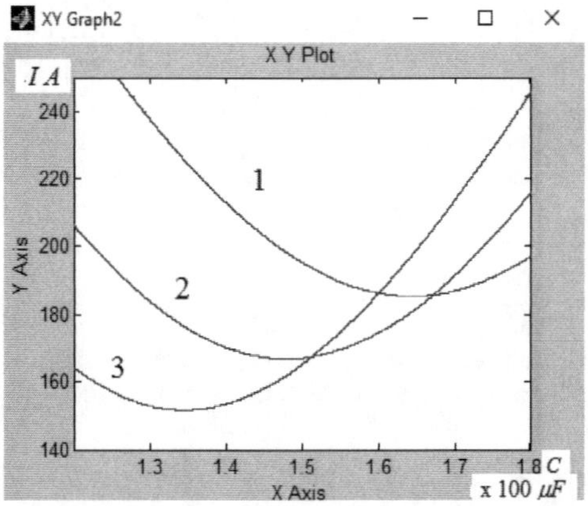

Figure 2. The family of static characteristics *I*(*C*) of the actuator with capacitive reactive power compensation: curve 2 – during normal condition of the cascade of centrifuges ($L = 50\ \mu H$, $R = 0.15\ \Omega$, $f = 1800\ Hz$); curves 1/ 3 – when the increase/decrease in the number of centrifuges by 10%.

A symptom of power factor Cos $\varphi = 1$ of actuator accepted the minimum of *I*(*C*) where $dI/dC = 0$ [3], which corresponds to the resonances of currents *R-L-C* circuits in a specific technological conditions

The management task is solved by the use of the stepper algorithms of automatic search $dI/dC = 0$ extreme *I*(*C*) [5] (figure 3).

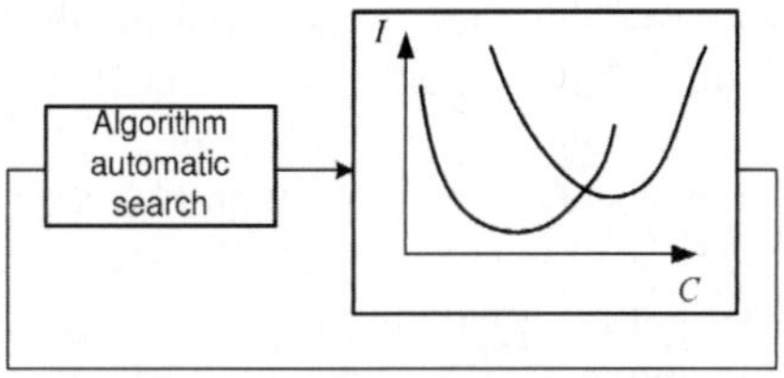

Figure 3. The system of automatic search of the of the power factor Cos $\varphi = 1$.

4. Control law

The algorithm is based [3, 5] on the determination of the direction of movement of the search coordinates $C(t)$ to $dI/dC = 0$ the extreme dependence $I\ (C)$ in finite differences: $\Delta I_n/\Delta C_n = 0$ where $n = 1, 2\ \ \infty$; $\Delta C_n = C_n - C_{n\text{-}1}$, where ΔC_n – amplitude search step in the interval T the duration of the n – th search step; C_n, $C_{n\text{-}1}$ – capacity values of C at time T_n, T_{n-1}; $\Delta I_n = I_n - I_{n-1}$ – the increment of output of the total current I in T_n, $T_{n\text{-}1}$ in the logic:

$$\text{Sign}\ [\Delta C_n \sqcup \Delta I_n] < 0 \rightarrow (\Delta C_{n+1} > 0);$$
$$\text{Sign}\ [\Delta C_n \sqcup \Delta I_n] > 0 \rightarrow (\Delta C_{n+1} < 0);$$
$$C_{n+1} = \sum_{1}^{\infty} C_n\ .$$

The algorithm is characterized by self-oscillation process is relatively $\Delta I_n/\Delta C_n = 0$.

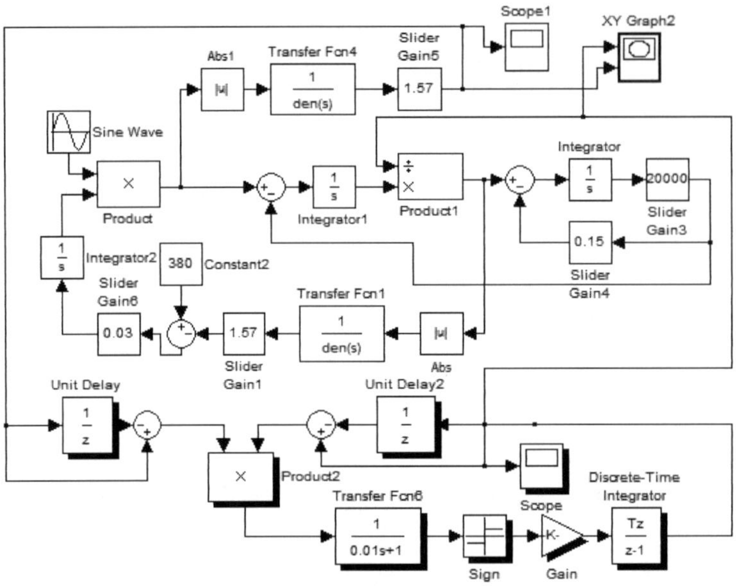

Figure 4. Engineering model of system of automatic stabilization of the power factor Cos φ = 1 of electric drive of centrifuges with capacitive parallel compensation: the upper part – the dynamic model of the drive: “Slider Gain3” correspond to $1/L$, “Slider Gain4” – R ($L = 50\ \mu H$, $R = 0.15\ \Omega$, $f = 1800\ Hz$); “Prodact1” unit performs the operation of dividing the input variable of the unit on the capacitance C; the lower part – the control algorithm.

Measurement of the total current I at the output of the model is managed FSC (“Sine Wave” together with “Prodact”) carried out by rectification (block “Abs1”) and filter the variable component “Tranfer Fcn4” with $W(s)= 1.57/(0.01^2s^2+0.2s+1)$ with correction matching $I(t)$ of the amplitude value of the high frequency variable. Value of I is automatically configured by a negative feedback loop consisting of measuring the amplitude $|U_C|$ of the high-frequency variable UC(t), of element comparisons $U_{sup} = 380$ V (bloc “Constant2”) and $|U_C|$ of bloc “Integrator2” with transfer ratio $k = 0.03$, of the model managed FSC with setting up “Sine Wave” ($\omega = 11309\ s^{-1}$) and amplitude "250", more of current I in the structure of figure 1 at resonance at the frequency of the supply. Unit “Product1” implements the operation $1/C_{n+1}$. The settling time of transients in the model of the actuator figure 4 is $t = 0.2$ c.

Algorithm of automatic search is implemented as digital in terms of Digital Simulink with the period of quantization of discrete blocks is equal to T. To work stepping algorithm for static [5] increment $I(C)$ made of $T = 0.2\ s$. Magnitudes ΔC_n and ΔI_n are computed using elements of “Unit Delay” with transfer function $W(z) = 1/z$, where “z” is the argument of the mathematical apparatus of analysis of digital automatic systems (“Z – transform” [7]). Control function C_{n+1} is generated by a discrete integrator c $W(z) = Tz/(z-1)$ [7]. The amplitude of search step accepted $\Delta C_n = 3.6\ \mu F$ (setting "Gain" is "-1.75 10^{-5}"). The terms of the reverse, according to the logic of the search algorithm are implemented by blocks “Produkt2” and the operator “Sign”. The initial conditions of the system operation $C = 165\ \mu F$ installed in integrator.

For local power "HM-FSC" the most dangerous is the factor of "overcompensation" that is associated with the disconnecting of centrifuges ((including centrifuge with enhanced performance [8]) in the restructuring of the separation of cascades and risk of transition through the state Cos φ = 1, causes a change in the nature of the power supply from inductive to capacitive. Figure 5 illustrates the operation of the system, included at $C = 165\ \mu F$ when you increased the number of centrifuges in the cascades (the extreme characteristics curve 1). The search algorithm finds the condition Cos φ = 1 of the energy system corresponds to the resonance at $C = 161\ \mu F$. At the moment $t = 2$ s restored staffing number of centrifuges (curve 2), with the exit power grid in mode "overcompensation". Targeted search procedure finds a new value of the resonance restoring the Cos φ = 1 at C = 147 μF. Similar situation with further reduction in the number (at time t 4.5 s) centrifuges, providing Cos φ = 1 at C = 132 μF.

Drift of the electrical parameters of the electric drive of centrifuge is mostly due to temperature changes of the production floor and slightly changes the magnitude R of the windings of the electric drive, which almost corresponds to the small offset $I(C)$ on axis I.

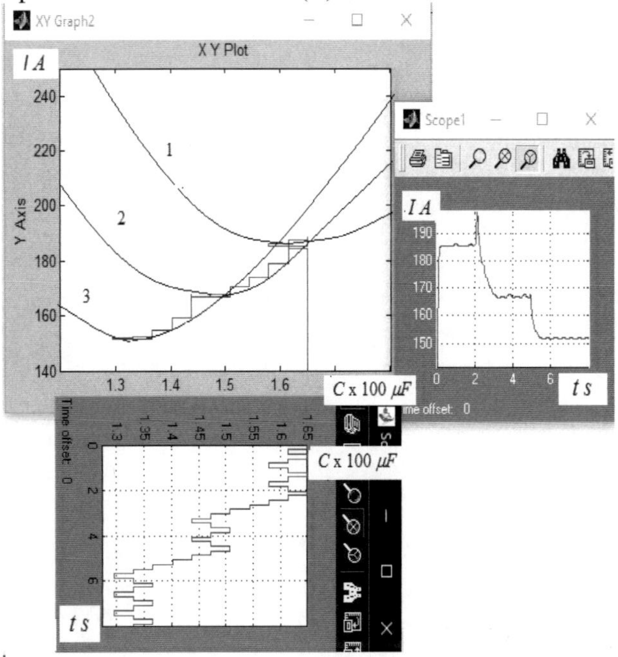

Figure 5. Automatic stabilization Cos φ = 1 if you change the number of centrifuges in the cascade: curve. 2 – regular, curves 1/3 – more/less 10% of the centrifuges in the cascade. Symbols correspond to the text.

In the process the self-oscillating mode of search in the neighborhood of the values dI/dC = 0 the deviation from the extremum does not exceed (1..2) A and can be reduced by choosing ΔC. In of this collection of articles presents an algorithm of automatic search $dI/dC \rightarrow 0$ for the inductive nature of the power supply with the improved characteristics of the work.

5. Other applications

Adaptive search system of automatic stabilization with a structural dynamic actuator model in the vicinity of the resonant condition at the frequency of the supply current is an alternative to increase the efficiency of power factor in industrial technological systems, as well as in the autonomous /board/ with power-limited (renewable) resources.

Acknowledgement

The author would like to express his gratitude to Professor V.D. Borisevich whose recommendations have improved the paper.

Conclusion

1. Presents a model of adaptive system of stabilization of power factor Cos φ = 1 of the electric drive of high-speed gas centrifuges for separation of uranium isotopes with a structural model of dynamic processes in the drive with capacitive parallel compensation of the reactive component of the total power in the vicinity of the resonance frequency of the supply.
2. The system operation is based on the automatic search for extremum of the minimum of the total current I on the output terminals FSC in functions of capacity C, when dI/dC = 0. Excluded state of the power system in the mode of "overcompensation"».

IOP Conf. Series: Journal of Physics: Conf. Series **1099** (2018) 012004 doi:10.1088/1742-6596/1099/1/012004

3. To operate the system enough measure of the total current I at the terminals FSC.
4. The control algorithm is insensitive to additive errors in measurement of the total current I.
5. The control system can be implemented on the basis of General industrial equipment

References

[1] Grossland I (Ed) *Nuclear Fuel Cycle and Engineering*, Oxford: Woodhead 2012.
[2] www.uralpribor.com
[3] Juromskiy V M 2015 *Physics Procedia* **72** 146
[4] Juromskiy V M 2016 *Nucl. Phis. Eng.* **7** 156 (in Russian).
[5] Juromskiy V M, System static optimization. MEPhI, Moscow, 1978 (in Russian).
[6] Aleksakov G N, Gavrilin V V, Fedorov V A, Structural models of dynamic processes, Engineering center of MEPhI, Moscow, 1991 (in Russian).
[7] Dabney D B, Harman T L, *Mastering Simulink 4*, Prentice Hall, Upper Saddle River, New Jersey, 2001.
[8] Borisevich V D, Juromskiy V M and Trushkin N S 2013 *Nucl. Phys. Eng.* **4** 434 (in Russian).

IOP Conf. Series: Journal of Physics: Conf. Series **1099** (2018) 012005 doi:10.1088/1742-6596/1099/1/012005

Algorithm of automatic search $\cos\varphi = 1$ for electric drive of high speed centrifuges with minimum limit cycle in region of the inductive nature of the power supply

V M Juromskiy

Department of molecular Physics, National Research Nuclear University "MEPhI", Kashirskoe shosse 31, Moscow, 115409, Russia

juromskiy@mail.ru

Abstract. System stabilization of cos φ is an adaptive system of automatic search in area of the resonance of power circuit, which is form by including a compensating capacitance *C* in parallel with the electrical drives (inductive - active load) of centrifuges. The resonance corresponds to the extremum - minimum of total current *I* as a function of the capacitance *C*. The aim of working an adaptive system is ensured of automatic search $dI / dC < 0$ in a small neighborhood of the extremum *I* (*C*), which provides of the inductive nature of electric power. The operation of the system in a neighborhood of a given $dI / dC < 0$ is improved by modifying the switching conditions search impact separately for positive / negative steps. The operation of the adaptive control system was represent by simulation results in terms of Simulink (MATLAB). In the process of automatic stabilization Cos $\varphi \sim 1$ is excluded state of "overcompensation", the search period is reduced 2-fold relative to the classic of search procedures.

1. Introduction

It is known that the energy efficiency of the electric drive is characterized by a power factor Cos $\varphi = P/S$, where $S = \sqrt{P^2 + Q^2}$ is the total power, $P = U\,I\,\mathrm{Cos}\,\varphi$ is active power, $Q = U\,I\,\mathrm{Sin}\,\varphi$ is the reactive power. Reactive power performs no useful work, additional load circuit of the power transmission and forces to increase the size and power generating devices.

Reactive power compensation of the electric drive of centrifuge separation systems is a substantial part of the to reduce the cost of electricity supplied by nuclear power plants [1].

Traditionally, the compensation of reactive component of the hysteresis motors (HM) in the local electric power network (HM – static frequency converter (SFC)) cascade high-speed centrifuge is the inclusion of capacitor bank.

2. Statement of the problem

In [2 - 4] considered the automatic stabilization system of Cos $\varphi = 1$ for the local electic power grid "HM – SFC" as providing resonance at a frequency of power *R-L-C* circuit (*L-R* inductance / resistance of the electric drive) based on the adaptive search algorithm of minimum of the total current *I*(*C*).

In the present work is considered as the continuation of a theme [2] (see in this volume), the algorithm of high precision automatic search in a small neighborhood of the minimum of *I*(*C*) in the field of the inductive nature of the power supply.

3. Structural dynamic model of the actuator

For further discussion on figure1 presents with brief explanations of the result of converting *R-L-C* electrical circuit [5] in the structural dynamic model as an object of extreme control [6] in according to changes of the capacitance *C*.

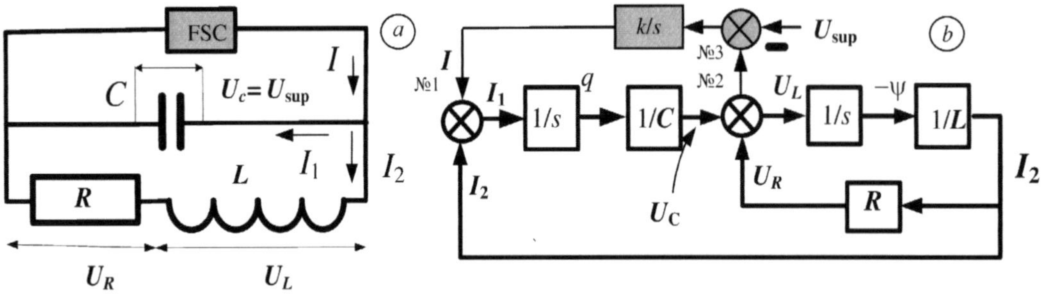

Figure 1. Schematic diagram (a) and a structural dynamic model (b) power system "GD– SFC" with the compensation of the reactive component as part the parallel connection of the condenser *C*: *s* –

Laplace operator, $1/s$ – blocks of integration, *L-R* is inductance and active resistance of motors. The voltage on the capacitor $U_C = q/C$, where the charge $q = \int I_1 dt$ in figure 1*a* (or $q(s) = I_1(s)/s$ in figure 1*b*). For *R-L* circuit figure 1*a*, the voltage on the inductance $U_L(t)$ determines the flux linkage $-\psi(t) = \int_0^t U_L dt$ (or $-\psi(s) = U_L(s)/s$) and the flux linkage $\psi(t)$ is current value $I_2(t) = -\psi(t)/L$ ($I_2(s) = -\psi(s)/L$).

The current I_2 creates a voltage drop U_R on the resistance *R*, which is subtracted from the input U_C (adder No 2, figure 1*b*) when determining the U_L [5]. Circuit negative feedback with integral control action (highlighted in dark color in Figure 1*b*) simulates the SFC as the "generator voltage", providing automatic stabilization $U_C = U_{пит}$ due to the change of the total current *I* [6].

4. Modified algorithm of automatic search of the minimum total current

Because of the physical discrete (ΔC is units-tens μF) of compensating capacitance change *C* in the adaptive algorithms [2...4] applied to the stepping principle of the automatic search. When the duration *T* of the search step ΔC more end-time transients $\Delta I(t)$ in the neighborhood of $\Delta I/\Delta C = 0$ occurs classic [6] the self-oscillating process with a limit cycle in 4*T* and every 3*T* the unwanted transition through an extremum in the region of the capacitive nature of the electric power on value ΔC. To stabilization $\cos\varphi$ → in the region of the inductive nature of electric power self-oscillating steady-state of search must be carry out for a values *C* less than its resonant value in a specific technological condition of the electric drive. This is done by setting the characteristic of the search $\Delta I/\Delta C = -\gamma$. At $-\gamma \to 0$ power factor Cos $\varphi \to 1$. Improving the quality of automatic search is achieved by modification of the conditions of the reverse impact of *C* for steps search ΔC with different sign. [3, 4].

The stepping algorithm automatic search $\Delta I/\Delta C = -\gamma$ [7] based on the determination of the direction of change search coordinates C_n in finite differences $\Delta C_n = C_n - C_{n-1}$, $\Delta I_n = I_n - I_{n-1}$ in the interval of durations $T = T_n - T_{n-1}$ ($n = 1, 2 \ldots \infty$). The algorithm logic in the area $\Delta I/\Delta C = -\gamma$ is:

$$(\Delta C_n < 0, \Delta I_n \leq \alpha) \to (\Delta C_{n+1} < 0)$$
$$(\Delta C_n < 0, \Delta I_n > \alpha) \to (\Delta C_{n+1} > 0) \quad (1)$$

works in conjunction with the algorithm

$$\text{Sign}\ (\Delta C_n \sqcup \Delta I_n) > 0 \to (\Delta C_{n+1} < 0),$$
$$\text{Sign}\ (\Delta C_n \sqcup \Delta I_n) < 0 \to (\Delta C_{n+1} > 0), \quad (2)$$

$$C_{n+1} = \sum_{n=1}^{\infty} \Delta C_n \text{ where } \alpha = \Delta C^2 \gamma ,$$

ensuring the convergence of the search procedure to $\Delta I/\Delta C = -\gamma$ with initial conditions on the right or left of the extremum.

Limit cycle of the oscillatory process in the area to the left of the minimum of $I(C)$ has duration $2T$. For the algorithm to work enough fact of existence of the extremum, regardless of its shape and lateral offsets $\{I; C\}$. Additive error of sensor I does not affect the performance of the search algorithm, based on the determination of the differential signals ΔI. Possible use of the sensor I with a nonzero scale.

5. Modeling

Model figure 2 adaptive system is performed in terms of modelling dynamic systems Simulink (MATLAB) [8].

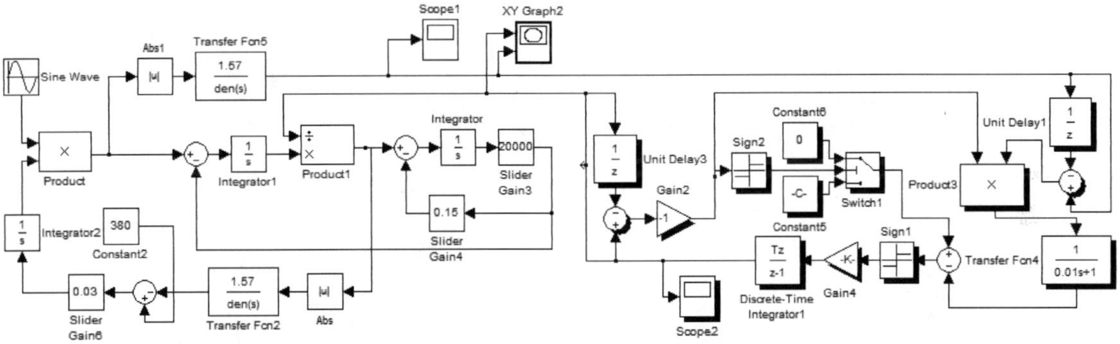

Figure 2. The model of adaptive system of automatic stabilization of the power factor Cos $\varphi \rightarrow 1$ in the field of the inductive nature of the electric powers. The left part is a structural dynamic model of the electric drive. Settings "Slider Gain3" correspond to $1/L$ ($L = 50\ \mu H$) "Slider Gain4" $\div R$ ($R = 0.15\ \Omega$). Configure a "Sine Wave" is $\omega = 11309\ s^{-1}$ [3]. "Prodact1" unit performs the operation of dividing the input variable for C_n. The right part is the control algorithm.

Measurement of the total current I at the output of the model is managed SFC ("Sine Wave" together with "Prodact") are used to detect high-frequency variable $I(t)$ ("Abs1", filter "Tranfer Fcn5" with $W(s) = 1.57/(0.01^2 s^2+0.2 s+1)$. The value of $|I(t)|$ to be configured by a circuit of negative feedback, consisting of the same detector (with filter) $U_C(t)$, of element comparisons $U_{\text{sup}} = 380$ $[V]$ ("Constant2") with amplitude $|U_C(t)|$, "Integrator2" and "Sine Wave". Unit "Product1" implements the operation $1/C_n$. The settling time of transients in the model of the actuator is $t = 0.2\ s$. Algorithm of automatic search is implemented in terms of Digital Simulink [8] with the period of quantization $T = 0.2\ s$. The magnitude ΔC_n and ΔI_n are computed using elements of the delay $W(z) = 1/z$, where "z" is the argument of the mathematical apparatus of analysis of digital automatic systems ("Z-transform" [8]). The control action C_n is formed by of the discrete integrator $W(z) = Tz/(z-1)$ [7] with the configuration $\Delta C_n = 4\ \mu F$ ("Gain4" $\div$ 2 10^{-5}) and the initial conditions "171 10^{-6} "($C = 171\ \mu F$). Block "Produkt3" and "Sign1" implement the logic of the algorithm (2).The operators "Sign2", logic ("Switch1") provide separate conditions of reverse C_n for $\pm\ \Delta C_n$ according to algorithm (1)

While modeling accepted $|-\gamma| = 6\ 10^{+5}$ $[A / F]$ the corresponding Cos $\varphi \sim 0.99 - 0.98$. When $\Delta C = 4\ 10^{-6}\ F$ according to for $\Delta C > 0$ value $\alpha = 9.6\ 10^{-6}$ $[F A]$ ("Constant5").

The purpose of the simulation is a proof of work algorithm (1) in the most adverse technological situation – a sharp decrease in the inductive – resistive load. This corresponds to the off part of the multi-engine electric drive when rebuilding or emergency in the cascade of centrifuges, especially significant in the case fits centrifuges high performance [9]. In such a situation, the possible transition of the energy system "HM – SFC" crossing the condition Cos $\varphi = 1$ in the region of the capacitive nature of power supply.

Figure 5 illustrates the operation of the adaptive system. Extreme characteristic curve 1 corresponds to the increased number of centrifuges in the cascade relative to the nominal condition (curve 2). The control system is enabled when the value $C = 171\ \mu F$ that corresponds to the "overcompensation" in energy systems "HM – SFC".

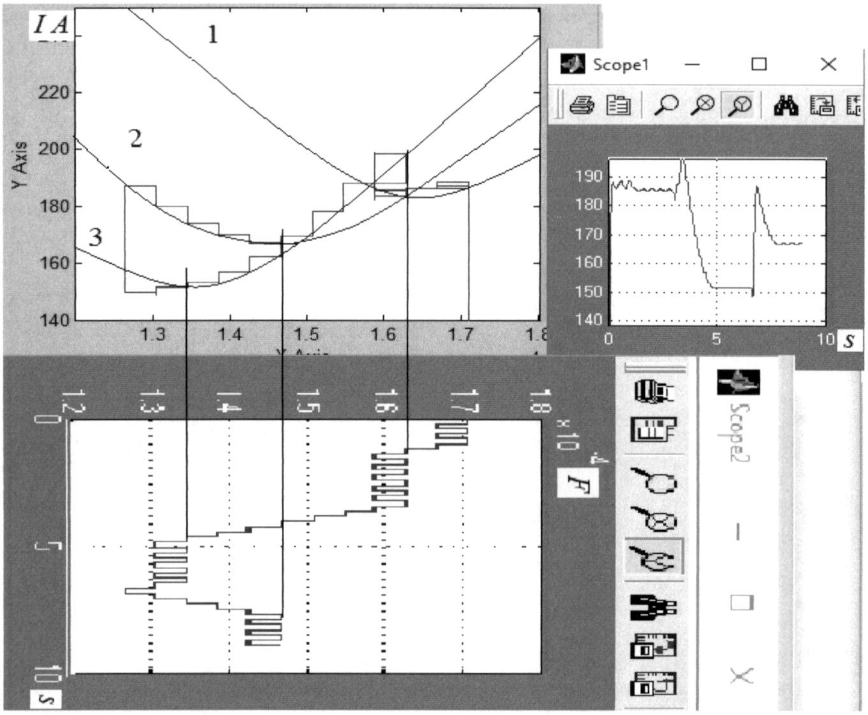

Figure 3. Working of the adaptive algorithm of automatic search Cosφ → 1 in the region of the inductive nature of the electrical power supply, to the left of the extrema curve 1...3.

The search algorithm finds the condition Cosφ → 1 to the left of the minimum of $I(C)$ (resonance at $C = 165\ \mu F$) when the oscillatory process in the range (158…162) μF in the field of the inductive nature of the load. At time $t = 3\ s$ switches off parts centrifuges and load SFC sharply decreased (Figure 3) (curve 3). When the capacitance value $C = 158...162\ \mu F$ characteristic and the resonance characteristic in a state of "curve 3" the nature of power meets "overcompensation". Targeted search procedure finds the value of Cos φ → 1 to the left of the new condition of the minimum of $I(C)$ in the range (130…134) μF (the resonance at $C = 134\ \mu F$). At time $t = 7\ s$ restored the nominal number of centrifuges (curve 2). The search algorithm finds $\Delta I/\Delta C = -\gamma$ to the left of the new extreme values $C = 145\ \mu F$, providing Cos φ ~ 0.98 – 0.99 in the field of the inductive nature supply of the load. In all cases the duration of a limit cycle is $2T$. Change of the total current I in the process of stabilization of a set of values Cos φ (figure 3, the graphics "Scoup1") are shares, units of amperes.

Acknowledgement
The author would like to express his gratitude to Professors V. D. Borisevich whose recommendations have improved the paper.

Conclusion

1. The algorithm of the system of automatic stabilization Cos φ ~ 0.99 - 0.98 in the field of the inductive nature of the power of multi-engine electric drive of centrifuges for uranium separation in a parallel capacitive compensation of reactive component of the total power is presented. Removed ""overcompensation".

IOP Conf. Series: Journal of Physics: Conf. Series **1099** (2018) 012005 doi:10.1088/1742-6596/1099/1/012005

2. The work of the algorithm is based on the automatic search procedure in small neighborhood minimum full electric current, corresponding to the phenomenon of resonance at the frequency of the power supply.
3. For the adaptive operation of the system is only measuring total load current. The control algorithm is insensitive to additive errors of current sensor, does not require the measurement of active / reactive, apparent power, calculation of capacitance compensating capacitors, assessments of the nature of the supply (inductive / capacitive) of the power system, insensitive to additive errors of the sensor total current.
5. Limit cycle in a neighborhood of Cos $\varphi \rightarrow 1$ in 2 times less in comparison with the classical procedure of automatic search of extremum.
6 The system can be realized based on of general technical devices of automatics. In a production environment, the algorithm can be activated periodically in depending on the specific technological situation or to be performed by the operator.
7. The system can be used for automatic stabilization of the Cos $\varphi \rightarrow 1$ in general electrical networks with variable inductive – active load.

References

[1] Grossland I (Ed) *Nuclear Fuel Cycle and Engineering*, Oxford: Woodhead, 2012
[2] Juromskiy V M, Adaptive system model of stabilization of the power factor of actuators of centrifuges for uranium enrichment with structural dynamic model of electric drive with parallel reactive power compensation
[3] Borisevich V D, Juromskiy V M 2016 *Journal of Physics: Conference Series* **751** 1
[4] Juromskiy V M 2015 *Physics Procedia* **72** 146
[5] Aleksakov G N, Gavrilin V V, Fedorov V A, *Structural models of dynamic processes*, Engineering center of MEPhI, 1991 Moscow (in Russian).
[6] Juromskiy V M 2016 *Nucl. Phis. Eng.* **7** 156 (in Russian)
[7] Juromskiy V M, System static optimization, MEPhI, Moscow, 1978 (in Russian).
[8] Dabney D B, Harman T L, *Mastering Simulink 4*, Prentice Hall, Upper Saddle River, New Jersey, 2001
[9] Borisevich V D, Juromskiy V M and Trushkin N S 2013 *Nucl. Phys. Eng.* **4** 434 (in Russian).

IOP Conf. Series: Journal of Physics: Conf. Series **1099** (2018) 012006 doi:10.1088/1742-6596/1099/1/012006

Separation of boron isotopes by chemical exchange in liquid-liquid systems

A V Khoroshilov, P I Ivanov

Isotope Technology and Hydrogen Energy Department, Mendeleev University of Chemical Technology of Russian, 9 Miusskaya Square, Moscow 125047, Russian Federation

avkhor@inbox.ru

Abstract. The currently used technologies for the production of boron isotopes for nuclear power and nuclear medicine are based on the application of toxic boron halides. The report is devoted to alternative processes for enrichment of ^{10}B and ^{11}B by chemical exchange in extraction systems using boric acid, that is, without of boron halides. The results of studying phase and isotope equilibrium in two liquid-liquid systems are considered: an aqueous solution of boric acid – an organic phase in the form of tributyl phosphate or a solution of trioctylamine in o-xylene. The obtained values of the separation factor for boron isotopes characterize the predominantly phase isotope effect in the first system, and in the second system the effect of chemical isotope exchange. Accordingly, the values of the single stage separation factor are 1.0067 ± 0.0035 and 1.025 ± 0.005 and are close to similar values for rectification of boron trifluoride and chemical exchange in the anisole process of boron isotope separation. The practical use of the studied extraction systems may allow the separation of boron isotopes in the horizontal cascades of centrifugal extractors in the absence of an environmental hazard.

1. Introduction

Industrial separation of stable boron isotopes (^{10}B и ^{11}B) is carried out by methods of low-temperature rectification of boron trifluoride BF_3 [1, 2], chemical exchange rectification in the system gaseous BF_3 – complex compound BF_3, for example, with dimethyl ether [3] and chemical isotope exchange in gas – liquid systems [4-6] (for example, BF_3 - $BF_3 \cdot C_6H_5OCH_3$). The described methods are based on the use of relatively expensive and dangerous compound – boron trifluoride BF_3 (relatively recently proposed new gas-liquid systems for boron isotope separation, unfortunately, are also based on the use of BF_3 [7-9].

As an alternative method of boron isotope separation can be considered the process of boric acid (H_3BO_3) extraction by organic liquid from aqueous solution. The first study of boron isotope separation by extraction was performed in the early 1960s in the system "aqueous solution H_3BO_3 – isoamyl alcohol" [10], for which the single stage separation factor of boron isotopes was $\alpha = 1.0027$. Higher values of α were obtained by extraction of boric acid by 1,1,7-trihydrododecafluoroheptanol and trioctilamine [11]: as the authors indicate, α was found to be not larger than 1.005.

It is possible that the use of other extractants can lead to the formation in the organic phase of boric acid complex, with the participation of which the isotope enrichment coefficient $\alpha = \varepsilon - 1$ will be greater.

The results of experimental studies of phase equilibrium and measurement of a single stage separation factor for boron isotopes in two extraction systems are considered in the article: system 1 - aqueous solution of boric acid - tributyl phosphate (TBP); system 2 - aqueous solution of boric acid – trioctylamine (TOA) in xylene.

2. Method of extraction and analysis

2.1. Experimental method for the extraction of boron

Boric acid extraction was carried out from the H_3BO_3 aqueous solution by an organic liquid (in the presence or absence of a solvent). If the solvent was required in the organic phase, the required volume of extractant was diluted and brought to a certain volume by the solvent (o-xylene) to obtain the required concentration of extractant in the organic phase. Extraction was carried out at room temperature in separatory funnels at different volume ratio of the organic and aqueous phases (org:aq or aq:org). Also in the experiments, the initial concentration of boric acid C_{aq} was varied over a wide range of values. After the separation funnel was filled with aqueous and organic phases, its contents were vigorously stirred and held until complete phase separation, after which the phases were successively poured into separate containers. The aqueous phase was investigated for the determining of boron content and boron isotope composition, and the organic phase was re-extracted by distilled water.

Based on the results of quantitatively analysis and the equation of material balance, the extraction coefficient K_D as the ratio of boron concentration C^*_{org} in the equilibrium organic phase to the boron concentration C^*_{aq} in the equilibrium aqueous phase was determined by equation (1):

$$K_D = \frac{C^*_{org}}{C^*_{aq}} \quad (1)$$

2.2. Isotope analysis

Isotope analysis of boron in boric acid both in initial and in equilibrated aqueous phase was carried out by ICP-MS method on the mass-spectrometer XSeries^{II} (Thermo Scientific) relative to the isotope boron standard SRM 951 according to the previously developed method [12].

The influence of the total measurement time I_t at a different values of the sweep frequency S (from 100 to 3000) and the time of counting pulses D_t (from 10 to 200 ms) was determined within the used method of isotope analysis. The choice of measurement optimal total time I_t is illustrated in figure 1.

The sequence of measurements of the isotope ratio included the following steps: determination of the isotope ratio for the standard solution, flushing out by deionized water in the measurement mode, measurement of the boron isotope ratio in the sample under study, again flushing out by deionized water, re-determination of the isotope ratio to account the effect of mass discrimination

2.3. Quantitative analysis of boron in aqueous phase

Quantitative determination of boron in the aqueous phase (C_{aq} and C^*_{aq}) was carried out using a scanning spectrometer Cintra 303 (GBC) and the test set "MET-B-PC" (MedEcoTest) similarly to method in [13].

It was determined the value of the maximum absorption A observed at a wavelenght 410 nm (figure 2). The calculation of boron concentration in the aqueous phase was made on a pre-established calibration dependence of absorption on the boron concentration (figure 3).

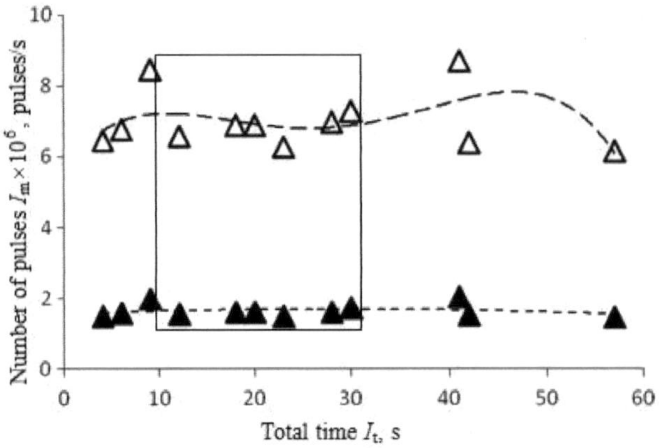

Figure 1. To the definition of the optimal total time of measurement I_t: Δ- **^{11}B;** ▲- **^{10}B.**

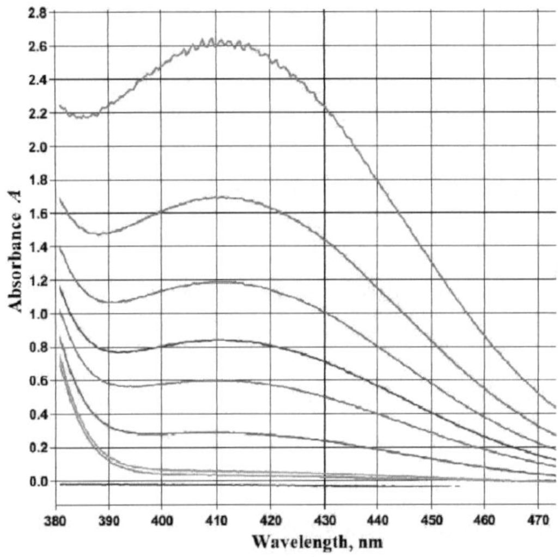

Figure 2. Maximum absorbance A observed at 410 nm.

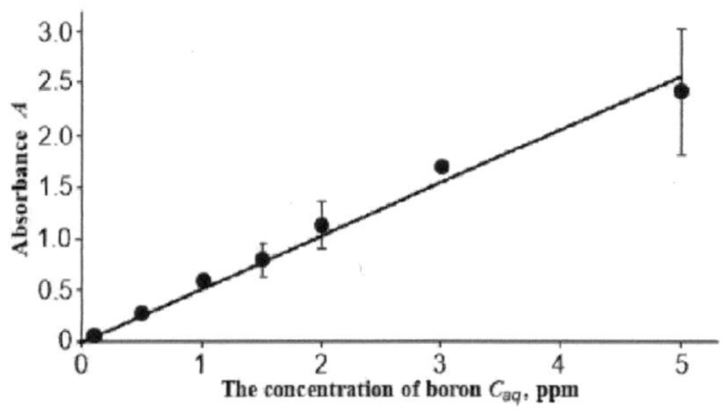

Figure 3. The calibration line for the determining of boron concentration C (values of absorbance at 410 nm).

3. Results and discussion

3.1. Phase equilibrium

As a result of the study of phase equilibrium in liquid – liquid systems, where TBP (system 1) and TOA (system 2) were used as an organic phase, dependences of extraction coefficients on the initial concentration of boric acid (boron) in the aqueous phase (figure 4 and 5) were obtained.

For system 1, there is a weak increase in the values of extraction coefficients from the initial concentration of boric acid in the aqueous phase, and this trend is observed regardless of the ratio of the volume of phases throughout the studied boric acid concentration range. And yet the value of the extraction coefficient is very small and does not exceed 0.2 and the dependence of the extraction coefficient on the initial concentration of boric acid in the aqueous phase for system 1 described by equation (2):

$$K_D = 3.18 \times 10^{-2} \times \ln\left(C^{\circ}_{aq}\right) + 0.214 \qquad (2)$$

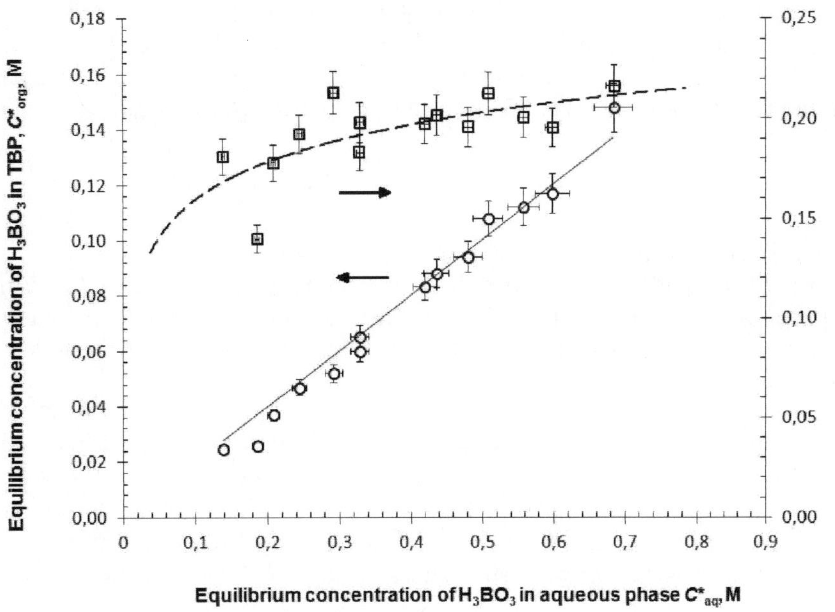

Figure 4. The concentration of boric acid at equilibrium in organic and aqueous phases and extraction ratio for system 1 (single dependencies for different volumetric phase ratios org:aq: 1:1; 2:1; 3:1)

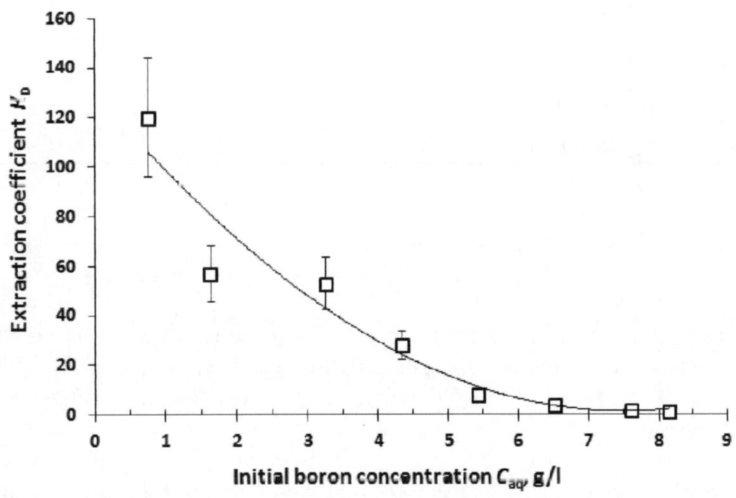

Figure 5. The change in the extraction coefficient as a function of initial concentration of boric acid in aqueous phase for system 2.

The small values of the extraction coefficient for system 1 can be explained by the lack of formation of a sufficiently stable boric acid – TBP complex under these experimental conditions, which may imply a boric acid extraction mechanism based on partial mutual dissolution of phases in each other.

For system 2, there is a characteristic change in the values of the extraction coefficient for extraction processes, which manifest itself in a decrease in the extraction coefficient with an increase in the initial boron concentration in the aqueous phase. It should be noted that the presence of tartaric acid in the system 2 is due to the fact that in the absence of tartaric acid, there was no observed transition of boron into the organic phase, which can be explained by the presence of a steric effects between TOA and boric acid. The dependence of the boric acid extraction coefficient on the initial boron concentration in the aqueous phase for system 2 is described by equation (3):

$$K_D = 2.322 \times \left(C^{\circ}_{aq}\right)^2 - 34.743 \times C^{\circ}_{aq} + 131.36 \quad (3)$$

3.2. Flow reflux: re-extraction of boric acid

For the considered liquid – liquid systems, re-extraction of boric acid with distilled water was considered as a method of flow reflux of boron from the organic phase to the aqueous phase. As a parameter characterizing the re-extraction efficiency, the re-extraction degree was used, which was calculated according to equation (4):

$$\Gamma_{\text{Re}-ex}, \% = \frac{n^{*}_{aq}}{n^{\circ}_{org} + n^{\circ}_{org}} \times 100 = \left(1 + \frac{K_D \times V_{org}}{V_{aq}}\right)^{-1} \times 100 \quad (4)$$

The figures 6 and 7 show that the increase in the volume of re-extractant, that is distilled water, during re-extraction leads to a characteristic increase in the re-extraction degree.

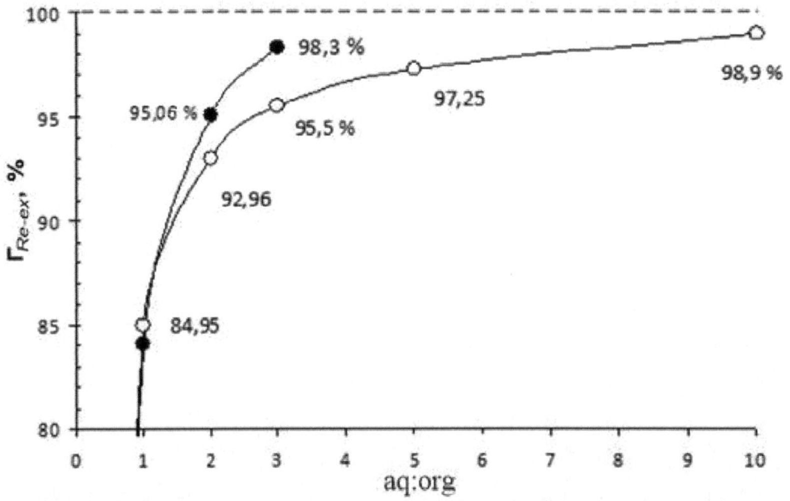

Figure 6. Boric acid re-extraction degree $\Gamma_{\text{Re-ex}}$, depending on the volume ratio of re-extractant (aq) to the organic phase (org) for system 1 at room temperature: • – multiplicity of re-extraction; ○ – single re-extraction by increasing the volume of re-extractant

For example, the system 1 is shown (figure 6) that the increase of multiplicity of re-extraction with equal volumes of the phases leads to the achievement of higher values of the degree of re-extraction with a lower total use of distilled water than the volume increase re-extractant in a single-step balances. At the same time, relatively high values of the degree of re-extraction are observed for system 1 already at room temperature.

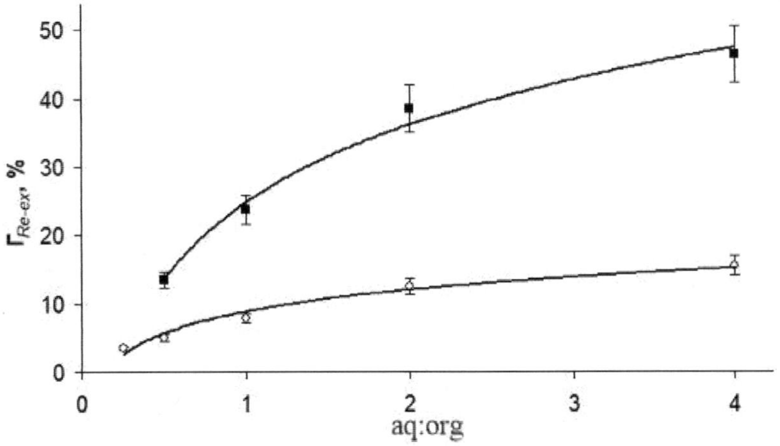

Figure 7. Boric acid re-extraction degree $\Gamma_{Re\text{-}ex}$, depending on the volume ratio of re-extractant (aq) to the organic phase (org) for system 2: ○ – room temperature; ■ – 75.5 °C.

For system 2, the increase on the re-extractant volume, as for system 1, leads to an increase in the degree of re-ectraction. However, a relatively significant transfer of boron from the organic phase to the aqueous phase is observed only with an increase in temperature, which, like the data in the figure 7, may indicate the formation in the organic phase of sufficiently strong complexes of boric acid – tartaric acid – TOA.

3.3. The values of the single isotope effect and its comparison with the separation factors of the industrial methods of boron isotopes separation

To determine the value of a single isotope effect in systems 1 and 2, the values of the isotope ratio R of the initial aqueous phase and the aqueous phase after equilibration with the organic phase were determined. Isotope ratio R is the ratio of isotope concentrations of boron-11 to boron-10, which looks like equation (5):

$$R = \frac{\left[{}^{11}B\right]}{\left[{}^{10}B\right]} \tag{5}$$

In this case, taking into account the isotope ratio for SRM 951 (R^{st}), the true value of the isotope ratio, for example for the equilibrium aqueous phase, is determined by equation (6):

$$R^*_{true} = R^*_{measured} \times \frac{R^{st}_{true}}{\left\langle R^{st}_{measured} \right\rangle} \tag{6}$$

According to the results of the isotope ratio measurements, it was found that during the extraction of boric acid the ^{10}B is concentrated in the organic phase, and the isotope ^{11}B, respectively, in the aqueous phase. In accordance with the equation of material balance, the value of a single stage isotope effect was determined by equation (7):

$$\alpha = \frac{R^*_{true}}{R^{\circ}_{true}} \tag{7}$$

Table 1 presents a comparison of the single stage separation factors of boron isotopes for the methods of boron trifluoride rectification, chemical exchange rectification and chemical isotope exchange in the system with anisol, and α values in the investigated extraction systems: system 1 and system 2.

Table 1. The single stage separation factors for gas – liquid (industrial technology) and liquid – liquid (R&D technology) systems.

Industrial technology		
Name of technology	*T*, K	Single stage isotope effect α
Low-temperature rectification BF_3	170	1.0085
Chemical exchange rectification $BF_3 - CH_3OCH_3{\cdot}BF_3$	≈400	1.015
Chemical isotope exchange $BF_3 - C_6H_5OCH_3{\cdot}BF_3$	293	1.0300
	338	1.0260
R&D technology		
Extraction in the liquid – liquid system	*T*, K	Single stage isotope effect α
System 1	293-295	1.0067 ± 0.0035
System 2		1.0250 ± 0.0050

The measured value of the single stage separation factor of boron isotopes in system 1 is equal to 1.0067 ± 0.0035 and is comparable with the same value for low-temperature rectification of boron trifluoride, and the value of α = 1.0250 ± 0.0050 in system 2 corresponds to the isotope effect of chemical exchange between BF_3 and its complex with methyl phenyl ether or anisole. On the basis of the measured alpha values, it can be assumed that in the first system the isotope effect corresponds to phase isotope exchange, but in the second, to predominantly chemical isotope exchange.

4. Conclusions

As a result of phase and isotope equilibrium studies in two extraction systems, an aqueous solution of boric acid – boric acid in the organic phase, where the organic phase is tributyl phosphate (system 1) or trioctylamine in o-xylene (system 2), is established:.

- at room temperature, the extraction factor in system 1 is much less than unity, equal to 0.2 and practically independent of the concentration of boric acid in water;
- in system 2, the extraction factor is significantly higher (as a minimum by an order of magnitude) and sharply changes with a change of boric acid concentration in the aqueous phase, increasing as it decreases;
- the lighter boron isotope (boron-10) in the studied systems is concentrated in the organic phase;
- the measured value of the single stage separation factor of boron isotopes in system 1 is equal to 1.0067 ± 0.0035 and is comparable with the same value for low-temperature rectification of boron trifluoride;
- the value of a single boron isotope separation factor in system 2 was 1.0250 ± 0.0050, which corresponds to the isotope effect of chemical exchange between boron trifluoride and its complex with methyl phenyl ether or anisole;
- the isotope effect in system 1 corresponds to the phase isotope effect, and in system 2 – to the effect of chemical isotope exchange;
- from the flow reflux point of view, in order to achieve the required concentration of boron-10 isotope, either multistage or cross re-extraction of boric acid from the organic phase is necessary.

Acknowledgments
Authors express their gratitude to the D.I. Mendeleev Center for the collective use of scientific equipment for assistance in carrying out research.

IOP Conf. Series: Journal of Physics: Conf. Series **1099** (2018) 012006 doi:10.1088/1742-6596/1099/1/012006

References

[1] Nettley P T, Cartwright D K and Kronberger H 1958 *Proc. of the Intern. Symp. on Isotope Separation* (Amsterdam: North Holl. Pub. Co.) p 385
[2] Amirkhanova I B, Asatiani P Ya, Borisov A B, Gverdtciteli I G et al. 1967 *Atomnaya Energiya*, **23** No. 4 336 (*in Russian*)
[3] Kaminskiy V A, Karamian A T, Giorgadze I A et al. 1973 *Production of isotopes* (Moscow: Atomizdat) pp 466-468
[4] Panchenkov G M, Makarov A V, Pechalin L I 1960 *Vestnic moskovskogo universiteta* № 2 p 3-12 (*in Russian*)
[5] Voloshuk A M, Dzhidzheishvili Sh I, Katalnikov S G 1973 *Production of isotopes* (Moscow: Atomizdat) pp 443-450 (*in Russian*)
[6] Katalnikov S G and Ivanov V A 2001 *Separation Science and Technology* **36** 1737
[7] Herbst R S and McCandless F P 1994 *Separation Science and Technology* **29** 1293
[8] Khoroshilov A V, Lizunov A V, Stepanov A V and Cherednichenko S A 2009 *Radiochemistry* **51** 400
[9] He N, Wu X, Guo X and Bai P 2015 *Asian Journal of Chemistry* **27** 444
[10] Kuznetsova E M, Panchenkov G M, Filippova R S and Malakhov V F 1960 *Russian Journal of Physikal Chemistry* **34** 2370 (*in Russian*)
[11] Demin S V, Zhilov V I and Tsivadze A Yu 2015 *Russian Journal of Inorganic Chemistry* **60** 663
[12] Khoroshilov A V, Silakova E L and Ivanov P I 2015 *Uspekhi v himii i himicheskoy tehnologii*. **29** No. 6 pp 56-58 (*in Russian*)
[13] Ivanov P I, Khoroshilov A V, Garipova L Kh and Makoveeva K A 2016 *Uspekhi v himii i himicheskoy tehnologii* **30**, No. 6 pp 86-88 (*in Russian*)

On the physical validity of the proposed separation potentials for multicomponent mixtures

A Yu Smirnov[1], G A Sulaberidze[1], A R Mustafin[1] and E Yu Goloulin[2]
[1] National Research Nuclear University "MEPhI", Department of molecular Physics, 115409, Russia, Moscow, Kashirskoe shosse, 31
[2] RUDN University, 117198, Russia, Moscow, Miklukho-Maklaya str.6

E-mail: a.y.smirnoff@rambler.ru

Abstract. The object of the study is the "generalized" separation potential for multicomponent mixtures. A critical assessment of its validity from the point of view of thermodynamics is given and the assumptions made in its derivation are analyzed. The work is of a debatable nature and aims to develop a unified point of view on such a problematic topic of the theory of separation of multicomponent molecular and isotopic mixtures as the separation potential and separating capacity.

1. Introduction

One of the open questions of the modern theory of separation of multicomponent isotope and molecular mixtures is the search for a universal separation potential (value function), which would become a complete analogue of the corresponding potential widely used in the case of separation of binary mixtures. The basic idea of using the separation potential in the binary case is that this function allows, based only on the external parameters of the separation facility (concentration at the product and waste), to characterize the "separation forces" associated with obtaining the desired product (for example, the mixture of enriched uranium). In practice, this makes it possible to introduce a single scale of separation efficiency, not taking into account the features of the specific method used to solve the problem of obtaining a product with specified characteristics, which turned out to be extremely convenient for carrying out practical, including economic, calculations [1].

However, despite the active development in recent decades in the theory of separation of multicomponent mixtures, it has not yet been possible to introduce a separation potential that would allow us, without calculating the parameters of separation devices, to estimate the complexity and cost of the separation process only at specified concentrations of the target isotope in the ingoing and outgoing flows. Nevertheless, until the moment several variants of separation potentials have been proposed [2-15]. At the same time, in some works, an attempt was made not only to introduce the separation potential but also to relate it to the thermodynamics of the process [2, 3].

The object to study within the framework of the present paper is the potential for separating a multicomponent mixture of a kind $\Phi(\vec{x}) = \Phi(x_1,...,x_m) = \sum_{j=1}^{m} x_j \ln(\frac{x_j}{1-x_j})$, (adhering to the author's notation x_j - the concentration of the jth component, m is the number of components of the mixture) considered in [2, 3] and originally proposed by P.G. Kuznetsov and others in 1963 [4] without

derivation and physical justification. The choice of this type of separation potential is largely determined by the fact that it is most often used in the works of other authors [4-15]
The authors of [2, 3] attempted to derive and interpret this relation, based on the following assumptions:

- the separation work that characterizes the "efforts" made by the cascade in the separation of a mixture W_k can be represented as a sum
$$W_k = \sum W_{k,ij}, \quad (1)$$
where - refers to each, in the terminology of the authors, the "selected" pair of components that make up the multicomponent mixture;
- the efficiency of the cascade is determined by the "symmetry" of the operation of its stages, the authors suggest the possibility of non-mixing at the entrance to each stage by several relative concentrations from the selected pairs of components.

In view of the fact that the assumptions and statements made by the authors seem, in our opinion, to be insufficiently correct, and also in order to develop a concept coordinated with the authors [2, 3], within the framework of this article, we give our remarks and point of view on the principal problems touched upon in [2, 3].

2. An analysis of the problem

Let us consider in more details the results of [2, 3] for the purpose of carrying out a critical analysis. Below are some comments on the essence of these works.

1. The formula $\sum W_{k,ij}$ implies, obviously, that the separation of a multicomponent mixture occurs as an independent binary separation of a mixture of vapor-containing components. Since each component of the mixture "interacts" with each of the other (m-1) components, the work of separating the 4-component mixture (A, B, C, D-components of the mixture) [2, 3]
$$W_k = W_{k,AB} + W_{k,AC} + W_{k,AD} + W_{k,BC} + W_{k,BD} + W_{k,CD}, \quad (2)$$
$$W_k = \sum_{\substack{j=1 \\ j=1, i<j}}^{m} W_{k,ij}, \quad (3)$$

However, in [3], $\sum W_{k,ij}$ they speak of the sum of the "selected" pairs of components of the mixture, without giving the principle that the "choice" of these pairs is made. In this case, the work of separating the 4-component mixture is represented as:
$$W_k = W_{k,AB} + W_{k,CD}, \quad (4)$$
those "selected" are two pairs (AB, CD) of 6 possible. As a result, it remains unclear the criterion for choosing such pairs and the validity of this assumption in principle.

2. It is logical to assume that in [2, 3] the authors arrange the components of the 4-component mixture in the series: A (1) -B (2) -C (3) -D (4) according to the increase (or decrease) of "activity" components. This means, for example, that $\alpha_{ij} > 1, \alpha_{ji} = \alpha_{ij}^{-1}$, when $i < j$ and $\alpha_{ii} = 1$. Taking into account the obvious relation $\alpha_{ij} = \alpha_{il} \cdot \alpha_{lj}$, it can be stated that the equations given in [3] $\alpha_{BD} = \alpha_{CA}$ $(\alpha_{24} = \alpha_{31})$ and $\alpha_{AD} = \alpha_{CB}$ $(\alpha_{14} = \alpha_{32})$ can not be fulfilled.
3. During the past decades has been introduced many different variants of separation potential for multicomponent mixtures based on axiomatic assumptions (in particular, a new treatment Peierls-Dirac axioms different interpretation representation of the separation process in the form of an independent separation binary mixtures consisting of pairs of components or aggregates and etc. [4-11]). Detailed information can be found in [15]. The results of numerical experiments on the optimization stages show

that all the known potential for multicomponent mixtures corresponds to particular cases depending on the type of mixture to be separated, which does not allow to determine which one is best for a particular purpose. The latter circumstance leaves open the question of the expediency of using them in practical calculations.

What is the reason for the absence of a single consistent theory of the potential for separation of multicomponent mixtures? Let us consider the analytical form of the separation potential for the case of binary mixtures:

$$\Phi(x) = (2x-1)\ln(\frac{x}{1-x}), \tag{5}$$

where x is the concentration of the "target" component.

This type of potential was obtained in the classical Peierls-Dirac theory on the basis of the following axioms:

- the separation potential Φ(x) is only a function of the concentration of the mixture;
- separating capacity of an individual element $\delta U = \sum_{j=1}^{n} q^j \Phi(x^j) - q\Phi(x^0)$, (n - number of species of mixture obtained as a result of the process of separation of the initial mixture of components, indices 0 and j determine, respectively, the parameters of the initial and enriched mixtures: q^0 and q^j - the quantitative fluxes of the initial and enriched mixtures per unit time) is independent of the mixture concentration [16,17] .

Expression (5) strictly justifies the theory of the "non-mixing" (ideal) cascade which, under certain conditions (small or large symmetrical separation on the steps), corresponds to the optimal separation conditions - the minimum number of separating elements (minimum of the total flow) to fulfill the given program of separation [16-18] The function $\Phi(x) = (2x-1)\ln(\frac{x}{1-x})$, can be considered as a universal and generally accepted characteristic of the process of separation of a binary mixture. In particular, it is one of the basic concepts of the theory and practice of separating uranium isotopes [19].

In the case of separation of multicomponent mixtures, the mathematical model of mass transfer of components in the cascade is a system of differential equations (for "weak" enrichment) [20], or a system of difference equations (for the case values of separation coefficients α_{ij} which are not close to unity) [9, 21,22] the impossibility of constructing an ordinary multicomponent cascade, which, like the ideal cascade for a binary mixture, would have "non-mixing" all concentrations at the entrances to the separation stages simultaneously. In other words, the impossibility of introducing a universal separation potential for a multicomponent mixture is a consequence of the absence in this case of a "reference" cascade, whose role in the separation of binary mixtures is played by an ideal cascade. At the same time, the multiplicity of possible boundary conditions in the optimization problem (the minimum of the total flow), whose solution is possible in the general case only by numerical methods [23]) also indicates the absence of a universal characteristic for an arbitrary separation process. Therefore, the use of physically unreasonable relationships derived from formal considerations (for the potential of P.G. Kuznetsov and others) in calculating and interpreting such characteristics as the thermodynamic work of separation, in our opinion, the total flow of the "best" cascade is inappropriate.

The procedure for calculating the total cascade flow with constant relative separation coefficients α_{ij} is presented in [3], with the subsequent use of the result obtained to justify the relation determining the separation forces performed during processing of the multicomponent mixture. During the calculation, the condition was used that the "symmetry" of the separation at the stages of the cascade is possible for more than one pair of components in the separated mixture.

In this connection, it should be noted that in the case of constancy of the separation coefficients α_{ij}, the system of transport equations has an analytical solution connecting "external" and "internal" parameters (since the "quasi-ideal" cascade) [9, 22, 24]. In other words, there is no need to calculate such a cascade by using the appropriate recurrence relations, as was done in [3]. In a particular case, this solution allows the construction of a cascade with no mixing of one (and only one) pair of selected "key" components $R_{nk} = \frac{x_n}{x_k}$, (n, k are the key component numbers) at the entrance to the steps [8, 9, 22, 25, 26].

The impossibility of simultaneously mixing at several relative concentrations (for example R_{AB} and R_{CD}, in [3]) follows both from a rigorous analysis of the system of transport equations, so from the next consideration: when the m-component mixture is divided (the number of independent equations describing the transfer of components (m-1)) the only parameter that ensures the "symmetry" of the separation on the stage is the function of the distribution of the power flow of the stages L(s) (s - stage number), which for each pair of "key" components will have its own individual character.

3. Conclusion

The paper gives a critical assessment of the physical validity of the separation potential described in [2, 3]. Inaccuracies in the determination of the separation work by the authors [2, 3] were found using the analyzed potential. It is shown that the considered potential, as well as the other variants of separation potentials proposed so far, can not act as a full-fledged analog of the corresponding function widely used in the case of binary mixtures separation.

The work is of a debatable nature and aims to develop a unified point of view on such a problematic topic of the theory of separation of multicomponent molecular and isotopic mixtures as a separation potential and separating ability.

4. References

[1] Isotopes. Properties. Receiving. Application. Ed. V.Yu. Baranov, vol. 1, Moscow: *Fizmatlit*, 2005 (in Russian).
[2] Chizhkov V P, Boytsov V N 2013 *Zh. fiz. Chemistry* **87** 52
[3] Chizhkov V P, Boytsov V N 2015 *Zh. fiz. Chemistry* **89** 316
[4] Kuznetsov P G, Sokolov V A, Sedin I P Separation and analysis of hydrocarbon gases M.: *Izd-vo ANSSSR*, 1963, p. 114 (in Russian).
[5] Lehrer – Ieamed Y 1969 *Nucl. Technol.* **23** 559
[6] Kolokoltsov N A, Minenko V P, Nikolaev B I, Sulaberidze G A, Tretyak S A 1970 *Atomic Energy* **29** 426
[7] Yamamoto I and Kanagawe A 1979 *J. J. Nucl.Sci.Technol.* **16**(1) 43
[8] Kai T 1983 *Ibid.* 491
[9] Zhigalovsky B.V., Lecture materials on multicomponent mixtures. Novouralsk, *UEIP*, 1999 (in Russian).
[10] Sazykin AA, In: isotopes: properties, production, application. Under. Ed. V.Yu. Baranova M. *Izd. At.*, 2000, p. 72 (in Russian).
[11] Palkin V A 2003 *Atomic Energy* **95** 373
[12] Palkin V A, Gadelshin V M, Aleksandrov O E, Seleznev V D 2014 *IFH* **87** 501
[13] Yamamoto I and Kanagawa A 1977 *J. J. Nucl.Sci.Technol.* **14** 565
[14] Palkin V A 2013 *Atomic Energy* **115** 230
[15] Palkin V A 2013 *Atomic Energy* **114** 211
[16] Rosen AM The theory of isotope separation in columns. Moscow: *Atom Izdat*, 1960, p.440 (in Russian).
[17] Benedikt M., Pigford T. Chemical technology of nuclear materials. Moscow: *Atomizdat*, 1960, p. 560
[18] Sulaberidze G A, Borisevich V D 2013 *Atomic energy* **114** 331
[19] Sinev N M Economics of Nuclear Energy. M .: *Atomizdat*, 1987, 480 p. (in Russian).

[20] Kucherov R Ya, Minenko V P 1965 *Atomic Energy* **19** 360
[21] Yamamoto I and Kanagawa A 1978 *J. J. Nucl. Sci. Tech.* **5** 28
[22] Sulaberidze G A and Borisevich V D 2001 *Sep. Sci and Technol.* **36** 1769
[23] Palkin V A, Sbitnev N A, Frolov Ye S 2002 *Atomic Energy* **92** 130
[24] Yamamoto I, Kanagawa A J 1987 *J. Nucl. Sci. and Technol.* **15** (8) 28
[25] De la Garza A, Garrett G A and Murphy J E 1961 *Chem. Eng. Sci.* **15** 189
[26] De la Garza A 1963 *Ibid* **18** 73

IOP Conf. Series: Journal of Physics: Conf. Series **1099** (2018) 012008 doi:10.1088/1742-6596/1099/1/012008

A ^{208}Pb coolant for a multipurpose nuclear reactor

G L Khorasanov, D S Samokhin and A S Zevyakin

Obninsk Institute for Nuclear Power Engineering of the National Research Nuclear University "MEPhI" Department of Nuclear Physics and Technologies 1 Studgorodok, Obninsk, Kaluga region, 249040 Russia

E-mail: khorasanow@yandex.ru

Abstract. Natural lead containing 52.3% of stable isotope ^{208}Pb can be applied in the nuclear small power engineering as such a coolant for fast reactors of various purposes. In this case unique consumer properties of ^{208}Pb might be realized which are the following: the low inelastic scattering and small absorption of neutrons by ^{208}Pb nuclei.

In the paper is shown that a high-enough neutron spectrum can be achieved.in a small dimensions core fueled with a plutonium-zirconium alloy (Pu-Zr) and cooled with natural lead. In fulfilling these conditions, an average neutron energy in the core center is equal to 0.869 MeV .and a share of hard neutrons (E_n>0.8 MeV) reaches the value of 35%.

The replacement of the lead coolant with a ^{208}Pb coolant leads to further increasing the average core neutron energy up to the value of 0.991 MeV and increasing the share of hard neutrons in the neutron spectrum up to 42%.

An interest for creating hard neutron spectrum reactors is high-enough mainly thanks to their ability to transmutate minor actinides (Np, Am, Cm) into fission products.

Production of radio isotopes for medicine and investigation of radiation damages in materials of nuclear engineering may be other applications of hard neutron spectrum reactors.

1. Introduction

Last time at the Obninsk Institute for Nuclear Power Engineering a concept of a small power fast reactor (FR) with high-enough core neutron energy has been developed [1-5]. The goal desired was achieved by selecting a core of small dimensions, innovative metallic fuel and low neutron moderating lead coolants containing natural lead (Pb) or enriched lead (^{208}Pb).

An interest for creating hard spectrum reactors is high-enough mainly thanks to their potential ability to transmutate (to inverse into fission products) the most dangerous from the radiotoxicity point of view minor actinides (MA – Np, Am, Cm). As is generally known, a hard spectrum of neutrons in FR core is preferable for incineration of MA.

Some other possibilities of the hard spectrum reactor usage exist, for instance, production of radio isotopes for medicine and investigation of radiation damages in materials of nuclear engineering.

2. Design parameters of hard neutron spectrum reactors

At the Obninsk Institute for Nuclear Power Engineering a series of reactors named BRUTS was designed under conceptual study. The first reactor BRUTS [1] was of 0.5 MW thermal capacity and of core dimensions D x H = 0.62 x 0.62 m^2. Uranium dioxide as such a fuel and natural lead as such a coolant were used. A low-enough (26%) share of coolant in the core was selected. The calculated average neutron energy in the core center was equal to 0.554 MeV with a share of hard (E_n>0.8 MeV) neutrons equal to 18,11%. The replacement of the natural lead coolant with the ^{208}Pb coolant did not change dramatically the core neutron spectrum.

In the BRUTS-M (0,5 MW of thermal capacity, D x H = 0.7 x 0.7 m^2) [2] fueled with U-Zr10wt% and cooled with natural lead (70% of coolant shares in the core) the average core neutron energy was equal to 0.592 MeV with the share of hard neutrons in the neutron spectrum equal to 20.17%. The replacement of lead with ^{208}Pb led to increasing the average core neutron energy up to 0,643 MeV and increasing the share of hard neutrons up to 23.06%.

As the next step, design of BRUTS-M2 [3-5] fueled with U53wt%-Pu30wt%-Zr17wt% was performed. At thermal capacity of 15 MW, coolants of Pb/^{208}Pb the average core neutron energy grew up to 0.724/0.874 MeV and the share of hard neutrons increased up to 28,45/34,14%.

At last, the hardest core neutron spectrum was obtained in the project named as BRUTS-M5 – the reactor fueled with an uranium free plutonium-zirconium alloy (Pu52wt%-Zr48wt.%). At thermal capacity of 25 MW, neutron flux density of $3 \cdot 10^{15}$ cm$^{-2} \cdot$s^{-1}, coolants of Pb/^{208}Pb the average core neutron energy was equal to 0.869/0.991 MeV and the share of hard neutrons was equal to 34.61/42.34%.

Thus, using the ^{208}Pb coolant instead of the natural lead coolant allows increasing the average core neutron energy at 14% and increasing the share of hard neutrons at 22%.

A harder spectrum, than spectra in known fast sodium and molten salt reactors, has been obtained thanks to the selection of relatively small core dimensions, metallic fuel and lead (Pb or ^{208}Pb) as such a coolant.

In the calculations of these compositions, increased average neutron energy and a high share of hard neutrons in the spectrum (with energies greater than 0.8 MeV) are achieved that is caused by the low inelastic neutron-fuel and neutron-coolant interactions thanks to the usage of a fuel without light chemical elements and a coolant containing ^{208}Pb, a low moderating isotope.

An interest for creating hard neutron spectrum reactors is explained by the fact that such reactors can be practically used as burners of minor actinides as well as isotopic and research reactors with new consumer properties.

With the uranium oxide fuel (UO_2) replaced by the plutonium fuel (Pu-Zr) in lead fast reactors, the one-group (averaged upon the neutron spectrum) fission cross section of ^{241}Am increases from 0.3 to 0.6 barn and the probability of ^{241}Am fission increases from 22 to 45%.

It is suggested that fuel resulting from regeneration of irradiated fuel of sodium fast reactor to be used in future burner reactors. It contains unburnt plutonium isotopes and some 1% of minor actinides which transmutate into fission products in the process of being reburnt in a harder spectrum. This will make it possible to reduce the MA content in the burner spent fuel and facilitate so the long-term storage conditions for high-level nuclear waste in dedicated devices.

Additionally, the hard neutron spectrum reactor can be used for production of radio isotopes for medicine via nuclear reactions of (n, p), (n, 2n) and (n, α) types which do not run in modern isotopic reactors with thermal or intermediate neutron spectra.

At last, the hard neutron spectrum reactor may be also used as such a research reactor thanks to the high share (35-42%) of hard neutrons which are able to cause stronger damages in the materials of nuclear engineering than modern research reactors can do it.

3. Conclusions

- In fast reactors of small core dimensions (D x H ≈ 0.5 x 0.5 m^2) fueled with a plutonium-zirconium alloy and cooled with natural lead, the average neutron energy at the core center reaches the value of 0.8 MeV with a high share (35%) of hard neurons (E_n>0.8 MeV) in the neutron spectrum.
- The replacement of the natural lead coolant with the ^{208}Pb coolant leads to increasing the average core neutron energy and the share of hard neutrons in the neutron spectrum at 14% and 22%

respectively. Thus, for the hard neutron spectrum reactor, about 42% of neutrons in the core center have energies above 0,8 MeV, while for known sodium fast reactors, the share of such hard neutrons in the core center does not exceed 15%.

- The hard neutron spectrum reactors have some consumer properties while no such properties exist in the modern power and research reactors:
  - In using the hard neutron spectrum reactor for transmutation minor actinides, is possible to reach the ^{241}Am fission probability as high as 45% while in sodium fast reactors it does not exceed 15%.
  - In using the hard neutron spectrum reactor for production of radio isotopes for medicine, is possible to obtain some rare radio isotopes via nuclear reactions (n, p), (n, 2n) and (n, α) which run only in hard neutron spectra.
  - In irradiating materials of nuclear engineering in the hard neutron spectrum reactor, is possible to obtain in them greater radiation damages as compared with these materials irradiating in modern research reactors.

Acknowledgments
The reported research was funded by Russian Foundation for Basic Research and the Government of the Kaluga region, grant № 18-48-400004.

References

[1] Samokhin D S, Khorasanov G L, Tormyshev I V, Zemskov E A, Gostev A L, Terehova A M and Kuz'michyov S A 2015 *Izvestiya Visshikh Uchebnikh Zavedenij. Yadernaya Energetika* issue 3 pp 135–141 (in Russian)
[2] Khorasanov G L, Samokhin D S and Zevyakin A S 2017 *Proc. Int. Conf. on Ecological, Industrial and Energetical Safety* (Sevastopol': Sev. State University) pp 1467–71
[3] Khorasanov G L and Samokhin D S 2017 *Theses of the 2nd Int. Conf. of Young Scientists, Specialists, Postgraduates and Students on Innovative Nuclear Reactors of Small and Ultra-Small Power* (Obninsk: INPE NRNU MEPhI) pp 19–21 (in Russian)
[4] Khorasanov G L, Samokhin D S and Zevyakin A S 2017 *Theses of the All-Russian Scientific and Practical Conf. on Materials of Nuclear Engineering* (Moscow: AS VNIINM) pp 43–5 (in Russian)
[5] Khorasanov Georgy, Samokhin Dmitry and Zevyakin Alexander 2017 *Transactions of the American Nuclear Society* **117** pp 1083–85

Optimization of cascades with variable overall separation factors by various efficiency criteria

G A Sulaberidze[1], A P Mustafin[1], A Yu Smirnov[1], V D Borisevich[1], S Zeng[2] and D Jiang[2]

[1] National Research Nuclear University MEPhI, 31 Kashirskoe Shosse, Moscow 115409, Russian Federation
[2]Tsinghua University, Department of Engineering Physics Beijing 100084, P.R.China

Abstract. Construction and operation of separation facilities require considerable investments. That is why the special attention in the theory of isotope separation in cascades is paid to their optimization. The installation allowing obtaining the required quantity of a commercial isotope product by a minimum unit cost it is customary to call as the optimum cascade. However, the use of such the criterion in theoretical calculations is inconvenient, in view of the fact that the cost coefficients are generally inaccessible and can vary depending on external factors. Therefore, theoretically the minimum possible number of separation unit or elements, for instance, gas centrifuges, is often used as the optimization criterion, since its value directly indicates the unit costs. In addition, such criteria as the minimum of the total cascade separation capacity or the minimum of the deviation sums of the maximum possible separation capabilities are also applied in the optimization calculation. In this connection for the case of the cascade with a variable from stage to stage overall separation factor of a centrifuge, the parameters of the cascades optimized by means of various optimization criteria may differ largely. In this paper, we compare the parameters of cascades with the same external conditions, which are optimized by the above criteria. As the criterion to compare all cascades under investigation is chosen the minimum number of separation elements in the installation. The chosen approach allows us to estimate the deviation in the number of gas centrifuge in a case of optimization of a separation plant by different efficiency criteria.

1. Introduction

As is well known, for separation of isotopes in cascades of gas centrifuges, one of the key criteria for determining the efficiency of such an installation is the minimum financial cost for obtaining the required product. However, in view of the specifics of the nuclear industry, it is not always possible and expedient to use the minimum of financial costs as a criterion in theoretical studies devoted to optimization of separation processes in cascades of gas centrifuges (GC) [1]. Therefore, other criteria that indirectly determine financial costs are often applied. One of these quantities is the total number of separation elements (centrifuges) in the cascade [2], since it basically determines the unit costs for obtaining the isotope product.

In the theory of cascades it is proved that if the separation factor of a single separation element depends only on a feed flow (capacity), then the number of separation elements in the cascade is directly proportional to its total flow, and for the same performance of all elements at cascade stages, the criterion of the minimum separation elements in the cascade and the minimum of the total flow in a cascade coincide. Assuming that all cascade elements are identical and operate in identical mode, the total number of separation elements is directly proportional to the value of the total flow in a cascade. Therefore, often in calculation proceed from the minimum of the total flow as an efficiency criterion for cascade [3]. However, the use of the total flow as a criterion is justified only in the case of identical operating modes of the elements. In the case of variable overall separation factors and cuts at cascade stages of the use of this criterion may lead to errors in the calculation of the optimal parameters [4]. That is why in the works of some researchers it was suggested to use other efficiency criteria. For example, in [5], when optimizing the cascade of centrifuges, the criterion of the minimum of the deviations for the separative powers of cascade stage stages from their maximum possible values was used. In this regard, it is necessary to analyze this criterion in order to determine its place among many others, as well as to indicate the limits of its applicability. The results of such studies are called up, in many ways, to dot the "i" in the question of choosing the correct criteria for optimization of cascades with variable overall separation factors that depend on the parameters of a single centrifuge machine.

The purpose of this research is to develop a technique to optimize a cascade with variable separation factors at its stages and as well as comparison of the cascade parameters optimized by means of various efficiency criteria reviewed in this study. The efficiency of various criteria is defined by comparison with the results obtained by optimization procedure using the total number of separation elements as the efficiency criterion.

2. Description of the cascade model

As the object of research a symmetrical countercurrent cascade for uranium enrichment is considered (Figure 1). The working substance is uranium hexafluoride (UF_6) [6-7].

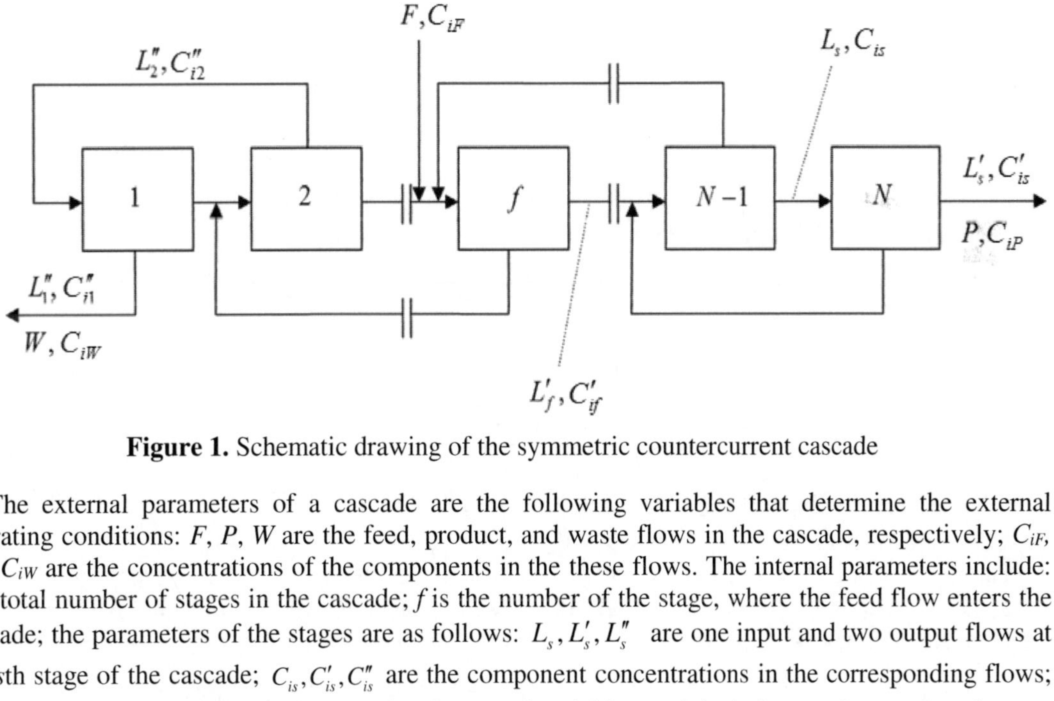

Figure 1. Schematic drawing of the symmetric countercurrent cascade

The external parameters of a cascade are the following variables that determine the external operating conditions: *F*, *P*, *W* are the feed, product, and waste flows in the cascade, respectively; C_{iF}, C_{iP}, C_{iW} are the concentrations of the components in the these flows. The internal parameters include: *N*th total number of stages in the cascade; *f* is the number of the stage, where the feed flow enters the cascade; the parameters of the stages are as follows: L_s, L'_s, L''_s are one input and two output flows at the *s*th stage of the cascade; $C_{is}, C'_{is}, C''_{is}$ are the component concentrations in the corresponding flows; the overall separation factor and separation factors of enriching and depletion sections at the *s*th stage of the cascade $q_{ik}, \alpha_{ik}, \beta_{ik}$ and the corresponding cuts at cascade stages θ_s (*s*=*1*,*2*...*N*).

In the absence of losses of the working substance at cascade stages and in the stationary mode of cascade operation, the external parameters must satisfy the equations of the material balance equation:

$$F = P + W, \tag{1}$$

$$FC_{iF} = PC_{iP} + WC_{iW} \ , \ \mathrm{i} = 1,2,...,\mathrm{m}. \tag{2}$$

The internal parameters of the stage with the number s (L_s, L'_s, L''_s, $C_{is}, C'_{is}, C''_{is}$) in the stationary mode of the cascade, in the absence of losses of the working substance at the stages of the cascade, are related by the equations of the balance of matter and each component:
for the enrichment part of the cascade

$$\theta_s L_s - (1-\theta_{s+1})L_{s+1} = P, \quad (3)$$

$$G'_{i,s} - G''_{i,s+1} = PC_{iP}, \quad (4)$$

for the depletion part of the cascade

$$\theta_s L_s - (1-\theta_{s+1})L_{s+1} = -W, \quad (5)$$

$$G'_{i,s} - G''_{i,s+1} = -WC_{iW}. \quad (6)$$

The external and internal parameters of the cascade are connected by the following boundary conditions

$$L'_N = \theta_N L_N = P, \quad (7)$$

$$L''_1 = (1-\theta_1)L_1 = W, \quad (8)$$

$$C'_N = C_{iP}, \quad (9)$$

$$C''_1 = C_{iW}, \; i = 1,2,..,m. \quad (10)$$

To calculate the cascade parameters, it is necessary to specify the separation characteristics of a stages. As a rule, they are expressed in the form of dependencies:

$$q_s = q_s(\theta_s, g_s), \quad (11)$$

where q_s is the overall separation factor of a stage with the number s, g_s is the a feed flow to a single separation element in a stage s, and θ_s is a cut at a cascade stage with the number s ($\theta_s = L'_s/L_s$). A similar dependence of the separation factor is due to the independence of isotopic composition observed for kinetic methods of separation [8-9].
In the case of a binary mixture (m = 2), for each stage with the number s one can find the concentrations C'_s, C''_s, according to formulas

$$C'_{is} = \frac{q_s C''_s}{1+(q_s-1)C''_s}, \quad (12)$$

$$C''_1 = C_W; \quad C''_s = C'_{s-1}\left(1-\frac{\tau'_s}{G''_s}\right)+\frac{\tau'_{us}}{G''_s}, \quad (13)$$

where τ'_s and τ'_{us} are the transit flows of a separated mixture and a target isotope in the direction of the cascade waste flow, that are equal to $\tau'_i = W$, $\tau'_{ui} = WC_W$ for s <f, $\tau'_i = -P$, $\tau'_{ui} = -PC_P$ for f <s <N.

The task of cascade optimization is formulated as the definition of the internal parameters that minimize the selected effectiveness criterion on a set of acceptable values, with the selected external parameters satisfying the condition $C'_N = C_P$, $C''_1 = C_W$. As an example it can be the total number of separating elements (in the case under review these are gas centrifuges), $\gamma = \sum Z_i \to \min$.

To obtain the desired equations, it is assumed that the number of gas centrifuges in the first stage Z_1 is defined as an implicit function of the quantities Z_2, …, $L'_2,\ldots L''_N$, and then the extreme condition of the function is written γ.

For the closure of the system of equations (1) - (13), it is necessary to know the dependence of q_s on a feed flow g_s to a centrifuge and on the cut θ_s, which can differ from one to another. The present paper uses the dependence $q_s = q_s(\theta_s, g_s)$ proposed in [9]:

$$q = \exp(a_0 + a_1\theta - a_2\theta^2)\, g^{-a_3}, \tag{14}$$

where a_0=1.2, a_1=1.8, a_2=2.2 and a_3=0.4;, g (mg / s) is the flow of UF_6 to single gas centrifuge. Since the case of obtaining low-enriched uranium from natural raw materials was considered, a formula corresponding to the condition C << 1 [10] was used to calculate the centrifuge separation capacity:

$$\delta U = g \cdot \left[\ln(1 + \theta(q-1)) - \theta \ln q\right]. \tag{15}$$

3. Method of optimization

In view of the fact that the optimization problem being solved is a task of multi-parametric optimization, since a large number of quantities act as parameters in its solution. The latter increases in proportion to the number of stages, what make us carefully choose the optimization method. Within the framework of the work, the genetic algorithm (GA) has been chosen as the optimization method. The main reason for that is as follows. A GA is sufficiently universal and can be suitable for solving a wide class of problems, and also allows us to reduce the calculation time with respect to traditional methods based on quadratic programming. As is known, a GA is one of the methods for solving optimization problems, using an abstract version of evolutionary processes to develop solutions of such problems. Using the terminology adapted to the theory of a GA, let us briefly explain the principle of their realization. Each GA works on a 'population" of the artificial "chromosomes." These are strings in the final alphabet (usually binary). Each a chromosome is a solution to the problem and has "suitability" is a number that is a measure of how close the solution is to the optimum [11]. Below is a brief review of the methodology for solving the problem, implemented on the basis of a GA.

The solution is encoded in the form of a vector ("genome") (16), where the first *N* numbers are the cuts of the flow at each cascade stage, and the next *N* numbers are the feed flows to the separation elements at each stage (where *N* is the number of stages in the cascade)

$$Ch = \{\theta_1, \theta_2, ..., \theta_N, g_1, g_2, ..., g_N\}. \tag{16}$$

At the first step, a set of vectors ("population") is created randomly. Then, the cascade parameters are calculated and the value by which the cascade is optimized for each set of vectors. The vectors that do not correspond to the boundary conditions are removed and replaced by new ones. Further, all vectors are sorted according to the magnitude of the increase of the optimized value. Then, using the crossovers function, all genomes are sequentially crossed, while the same genomes do not cross.

3.1. Crossover function:

The work uses a modification of the genetic algorithm, in which the crossovers function is combined with the mutation function. In other words, each gene is assigned a number from 0 to 1. If the number is in the range from 0 to 0.2, the new genome receives the gene from the first parent unchanged (17)

$$T_{i+1}^{cross1} = T_i^p, \tag{17}$$

in the range from 0.2 to 0.4 from the second parent (18).

$$T_{i+1}^{cross2} = T_i^q. \tag{18}$$

In the case when the number is in the interval from 0.4 to 0.8, the crossing of the genomes takes place according to formula (19),

$$T_{i+1}^{cross3} = \frac{T_i^p + T_i^q}{2}, \tag{19}$$

0.8 to 1 new gene is selected randomly (20)

IOP Conf. Series: Journal of Physics: Conf. Series **1099** (2018) 012009 doi:10.1088/1742-6596/1099/1/012009

$$T_{i+1}^{cross4} = Rand \ . \tag{20}$$

Then for the genomes obtained, the internal parameters of the cascade and the value of the "fitness function" (suitability function), on which optimization is performed, are calculated. Further, all genomes are sorted according to the magnitude of the optimized value increase and the destruction of genomes with the least suitable value of the optimized value to the amount of the original population. This process is repeated a certain number of times. The more the number of iterations completed, the more accurate the data are obtained. In the case where the difference between genomes is less than a given value, 90% of the population with the worst value of the optimized value is destroyed, and new ones are created randomly instead of the destroyed genomes. Figure 2 schematically shows all the main steps of the genetic algorithm.

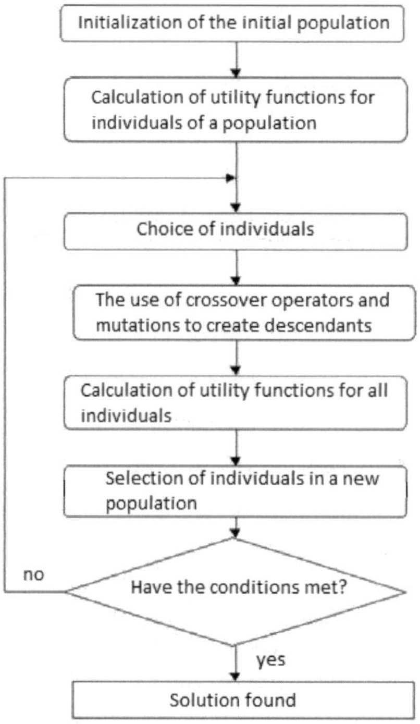

Figure 2. Diagram explaining how the genetic algorithm works

4. Results and discussion

The described algorithm was applied to the above optimization problem of a cascade with variable overall separation factors. Approbation of the algorithm was carried out using cascade optimization for enrichment of natural uranium in the form of uranium hexafluoride (UF_6). The concentration of the target isotope ^{235}U in the product flow is assumed equal to 3%, in the waste flow concentration of this component is taken equal to 0.3%. The product flow is 1 g/s. The number of stages in cascade is five, the feed flow is entered into the second stage. For this set of parameters, the data on the optimal parameters of such a cascade are available in [4, 9], which made it possible to check the developed algorithm. As can be seen, from Table 1, the results showed a satisfactory match in optimizing such a separation stage by the criterion of the minimum number of separation elements at variable coefficients of cascade stage, since the difference in the values given in the Table is less than 0.2%.

Table 1. Comparison of cascade optimization results by the total number of separation elements for different optimization methods

	Optimization by the method 1 from [12]	Optimization by the SQP method from [12]	Optimization by a genetic algorithm
Total number of separation elements	6352	6355	6363

Then, optimization was performed using several optimization criteria for a cascade of the same configuration. Namely, the total number of separation elements:

1) $Z=\sum_{s}^{N} Z_s$ is the criterion characterizing the amount of unit costs for production of isotope product of necessary quality

Optimization of the cascade parameters by the total flow:

2) $L=\sum_{s}^{N} L_s$ is the criterion for the case of the constant overall separation factors at cascade stages., This criterion is equivalent to the total number of separation elements.

Deviation from the maximum separation capacity:

3) $\Psi=\dfrac{\sum_{s}^{N} Z_s*(\delta U_{max}-\delta U_s)}{\sum_{s}^{N} Z_s*\delta U_{max}}$ is the criterion characterizing the efficiency of the separation element taking into account the deviation of the relative sum of separation capacities stages from its maximum possible value [5].

The total relative deviation from the maximum separation ability:

4) $\chi=\dfrac{\sum_{s}^{N} Z_s*(\delta U_{max}-\delta U_s)}{\delta U_{max}}$ is the criterion which characterizes vicinity of the separation capacity of centrifuges at each cascade stage to their maximum possible value. This takes into account the number of elements at each stage, thus excluding the effect of the number of separation elements on the absolute values of the deviation of the separation capacity from the maximum value [12].

The results of optimization are presented in Table 2 (the parameters corresponding Ψ to are multiplied into 10^{-5} by the order of magnitude of the remaining parameters).

Table 2. Results of optimization by various efficiency criteria

Basic parameters	Criteria of optimization			
	$\sum Z_s \rightarrow \min$	$\Psi \rightarrow \min$	$\sum L \rightarrow \min$	$\chi \rightarrow \min$
$\sum Z_s$	**6363**	6385	6776	6378
ΔU mg/s	**3429**	3449	3440	3446
$\sum L_s$, g/s	**43.71**	44.56	34.13	44.79
$\Psi \cdot 10^{-5}$	**272**	23	6051	26

ΔU is calculated as the difference between δU_{max} and the value of δU, calculated for each step

From the analysis of the obtained data it follows that for the criteria of the minimum of separation elements and the minimum of the total deviation from the maximum separation capacity, the difference in the total number of centrifuges is less than 0.4%, for a total separation capacity of less than 0.8%, and for a total flow of about 5%. It can be explained by the fact that all three criteria actually characterize the efficiency of using a single separation element in a cascade, so the optimal parameters for these cases turned out to be quite close. However, it is important to note the fact that in the case of optimization of the cascade by the smallest number of separation elements, the total deviation of the separation capabilities of the stages was not minimal. This means that in terms of cascade optimization, it is not necessary to work at maximum separating ability for each single separation elements in the cascade. As for optimization in the minimum of the total cascade flow in a cascade, it can be seen that the difference in the total number of centrifuges is much larger than for all others optimization criteria, what is more than 6%. Consequently, as a criterion of optimization of the cascade within the framework of the task under investigation, it cannot be used, since the obtained cascade parameters will not provide the minimum unit costs for obtaining the desired isotope product. To verify the "generality" of the above conclusions, an additional series of calculation of the optimum parameters of the cascade for enrichment of ^{235}U to 5% from the natural abundance with a 0.3% in a waste flow was carried out. The product flow is 1 g / s. The total number of stages in the cascade is equal eight. A feed flow goes to the third stage. The number of stages in the cascade and the number for a feed flow stage were estimated using the ideal cascade model. The optimization results are presented in Table 3.

Table 3. Results of optimization by various criteria, obtained using the genetic algorithm

Basic parameters	Criteria of optimization			
	$\sum Z_s \rightarrow \min$	$\Psi \rightarrow \min$	$\sum L \rightarrow \min$	$\chi \rightarrow \min$
$\sum Z_s$	**14006**	15136	15250	14111
ΔU mg/s	**7254**	7852	7564	7289
$\sum L_s$, g/s	**134.35**	126.39	110.40	131.31
$\Psi \cdot 10^{-5}$	**4154**	3997	8205	4410

As one can see the exact total number of separation elements, the total separation capacity, the deviation from the total separation capacity, the total flow and the total relative deviation from the possible maximum separation ability, for each criterion the deviation in the number of separation elements from the number of separation elements obtained by the criterion of the total number of separation elements is 21.19%, 7.47%, 8.16%, 0.74%, respectively. Thus, we can conclude the regularities inherent in the previous case are also fulfilled qualitatively in this example as well. However, the absolute values of the discrepancy in the set of parameters obtained my means of various criteria have increased. The latter can be explained by an increase in the total number of stages in the cascade.

5. Conclusion

In this paper, we propose a method of optimization of a symmetric countercurrent cascade with variable separation factors at its stages obtained by various efficiency criteria, i.e. the minimum number of separation elements, the minimum of the total flow and the minimum deviation from the maximum of the separation ability based on the genetic algorithm. With the use of the developed technique, computational experiments were carried out on optimization of the cascade parameters for enriching natural uranium under the condition of the variable overall separation factors at cascade stages and usage of various efficiency criteria. The analysis of the obtained results showed that the difference in the number of separation elements (and, consequently, unit costs) for a given task when using the number of separation elements and deviation from the maximum of the separation ability as a criterion is relatively small and does not exceed fractions of a percent. Wherein, the difference in the case of optimization in the total flow reaches several percent. Moreover, the minimum value of the total flow does not correspond to the case of the minimum number of separation elements, which indicates that the total flow cannot be used as an optimization criterion in cascades with variable separation coefficients.

Acknowledgments

This research is supported by Natural Science Foundation of China (Grant No. 11575097), the National Key Basic Research and Development Program of China (Grant No. 2014CB744100), and the Sino-Russian joint project from National Natural Science Foundation of China (Grant No. 1151101155) and Russian Fund for Basic Research (Grant No. 16-58-53058 GFEN_a).

6. References

[1] Borisevich V.D. et al. Physical backgrounds of isotope separation by gas centrifuge. *MEI Publishing House*, 2011.

[2] Cohen K., The theory of isotope separation as applied to the large-scale production of U-235, N.Y., McGraw-Hill, 1951.

[3] Smirnov A.Yu., Borisevich V.D., Sulaberidze G.A., Evaluation of specific cost of obtainment of lead-208 isotope by gas centrifuges using various raw materials, *Theor. Found. Chem. Eng*. 2012. 46 (4). P. 373-378.

[4] Sulaberidze G.A., Zeng S., Smirnov A.Yu., Bonarev A.K., Borisevich V.D., Jiang D. Efficiency criteria for optimization of separation cascades for uranium enrichment. *Nuclear Engineering and Technology*. 2018. V. 50. P. 126-131.

[5] Delbeke J.F.A. Theoretical analysis to assess the separative power of reconfigured cascades of predesigned gas centrifuges. *Ind. Eng. Chem. Res*., 2009, v.48, pp.4960-4965.

[6] Orlov A.A., Tsimbalyuk A.F., Malyugin R.V. Desublimation for Purification and Transporting UF6: Process Description and Modeling, *Separation and Purification Reviews*. 2017. 46(1). P. 81-89.

[7] Orlov A.A., Tsimbalyuk A.F., Malyugin R.V., Glazunov A.A. Dynamics of UF 6 Desublimation with the Influence of Tank Geometry for Various Coolant Temperature, *MATEC Web of Conferences*. 2016. 72:01079

[8] Palkin V.A. Optimization of a cascade with arbitrary specified separation coefficients of the stages, *Atomic energy*. 1997. V. 82. No. 4. P. 190-195.

[9] Palkin V.A. Determination of the optimal parameters of a cascade of gas centrifuges, *Atomic energy*. 1998. V. 84. No. 3. P. 246-253.
[10] Sulaberidze G.A, Borisevich V.D. Application limits of the classical concepts "separation potential" and "separative power" *Atomic Energy*. 2013. 114(6). P.412-420.
[11] M. Mitchell, An Introduction to Genetic Algorithms, *MIT Press*, Cambridge, MA, 1998.

A homotopy algorithm to solve the problems of flows under strong rotation

Z Y GU[1], D J JIANG[1], S ZENG[1] and V D Borisevich[2]

[1] Tsinghua University, Beijing, China
[2] National Research Nuclear University MEPhI, Moscow, Russia

jiangdj@tsinghua.edu.cn

Abstract. A finite difference method for numerical simulation of a high speed rotating flow is improved by applying a homotopy algorithm. As a result a large-scale convergence algorithm is constructed to make numerical simulation of a rotating flow in the 2D axisymmetric model of a gas centrifuge for uranium enrichment. The computing code is verified and then applied for calculation of the flow field in a 2D GC operating in a total reflux mode. The peculiarities of an axial circulation flow in a working chamber of a GC are studied. The conclusion about the rate of attenuation of an axial circulation flow is made on the basis of the results obtained.

Keywords: Gas centrifuge; Homotopy algorithm; Finite difference method; Decay of circulation

1. Introduction

An extremely strong rotating flows are realized inside the modern gas centrifuge (GC) for uranium isotope separation. The pressure and density gradients of a process gas (UF_6) over a radial direction are very large and changing by 5-6 orders on 1cm near the rotor side wall. It does not allow us to study properties of such flow fields experimentally. That is the reason why the flow field inside a GC is explored already more than 40 years by numerical simulation [1, 2].

The presence of static devices inside a GC, i.e. scoops for extraction of enriched and depleted in the target component fractions, demands the use of 3D modeling in the numerical study of a rotating flow in this apparatus. However, the 2D axisymmetric model is usually taken to study the general regularities of an internal flow structure in a GC to simplify the problem. Kai [2] has applied a staggered mesh finite difference scheme for numerical simulation of nonlinear flow field in a GC when a linear speed of GC rotation was about 500 m/s, whereas the modern GCs have the speed of rotation more than 800 m/s [3]. The studies [3-7] have demonstrated that with increase of a rotation velocity in a GC, the problems associated with the convergence of the obtained numerical solution is also increased.

In this paper it is presented the way to improve the traditional finite difference method (FDM) regularly applied to explore the 2D rotating flow in a GC by its combination with a homotopy method what results to a large-scale convergence algorithm suitable for numerical simulation of nonlinear flow field under strong perturbation.

IOP Conf. Series: Journal of Physics: Conf. Series **1099** (2018) 012010 doi:10.1088/1742-6596/1099/1/012010

2. Homotopy algorithm

A homotopy algorithm allows us to take into account the topology in the algebra fields.This makes it a large-scale convergence algorithm [6]. With the development of theory a homotopy algorithm has been applied in such areas of human activities as economy, electronics, machinery, and many others. Now it became a powerful tool to solve various sets of nonlinear equations, fixed point problem, optimization problem, and so on. The main idea of this algorithm is to construct a homotopy mapping.

Assume that there is an n order system of equations $F(\boldsymbol{x})=0$ in the real number field $\boldsymbol{R}$, and the solutions of this system $\boldsymbol{x}=\boldsymbol{x}^* \in \boldsymbol{R}^n$ is unknown and to be solved. Then we have a mapping which is

$$F:\boldsymbol{R}^n \to \boldsymbol{R}^n.$$

If there is another n order system of equations $G(\boldsymbol{x})=0$ in the field $\boldsymbol{R}$, and the solutions of this system $\boldsymbol{x}=\boldsymbol{x}_0 \in \boldsymbol{R}^n$ is already known or easy to get, then we can construct another one mapping.

$$H:\boldsymbol{R}^n \times [0,1] \to \boldsymbol{R}^n$$

Additionally, we can make the new mapping meet the following conditions

$$\begin{cases} H(\boldsymbol{x},0)=G(\boldsymbol{x}) \\ H(\boldsymbol{x},1)=F(\boldsymbol{x}) \end{cases} \tag{1}$$

We call the mapping $H:\boldsymbol{R}^n \times [0,1] \to \boldsymbol{R}^n$ as a homotopy mapping and the problem of solving the system of equations $F(\boldsymbol{x})=0$ can be converted to solving the homotopy system of equations $H(\boldsymbol{x},\alpha)=0, \alpha \in [0,1]$.

In fact, it is easy to construct a mapping meeting conditions (1). For example, we can construct the following kinds of mapping only from the mathematical point of view ignoring the physics.

$$\begin{cases} H(\boldsymbol{x},\alpha)=(1-\alpha)F(\boldsymbol{x})+\alpha G(\boldsymbol{x}) \\ H(\boldsymbol{x},\alpha)=\alpha F(\boldsymbol{x})+G(\boldsymbol{x}) \\ \qquad \vdots \end{cases}$$

It is also easy to find that the solution of a homotopy system of equations is a function of a parameter α, so a parameter α and the corresponding solution $\boldsymbol{x}(\alpha)$ define a curve in the n dimensional space $\boldsymbol{R}^n$, which is called a homotopy curve. Additionally, the start and end of a homotopy curve are $\boldsymbol{x}(0)$ and $\boldsymbol{x}(1)$ respectively. They will meet the following conditions

$$\begin{cases} G(\boldsymbol{x}(0))=H(\boldsymbol{x}(0),0)=0 \\ F(\boldsymbol{x}(1))=H(\boldsymbol{x}(1),1)=0 \end{cases} \tag{2}$$

The conditions (2) show us that the start of a homotopy curve is the solution of $G(\boldsymbol{x})=0$, and the end of a homotopy curve is the solutions of $F(\boldsymbol{x})=0$. As we assumed above, the solution of $G(\boldsymbol{x})=0$ is known or easy to get, which means that we will have known the start of homotopy curve once the homotopy mapping constructed. Therefore, if we can follow the curve from start to end, we will find the solution of $F(\boldsymbol{x})=0$.

In sum, the main procedure of carry out a homotopy algorithm is to construct a proper homotopy mapping based on the problem and then to follow a homotopy curve to the end.

3. Homotopy algorithm based on the FDM

3.1. Kai's FDM

Kai gave us a kind of a FDM with technique of the staggered meshes to solve the 2D axisymmetric Navier_Stokes equations in the paper [2]. The dimensionless 2D axisymmetric Navier-Stokes equations are given as follows

$$\frac{1}{r}\cdot\frac{\partial}{\partial r}(r\rho\rho_0 u)+\frac{\partial}{\partial z}(\rho\rho_0 w)=0, \tag{3}$$

$$\begin{aligned}\rho[u\frac{\partial u}{\partial r}+w\frac{\partial u}{\partial z}-(2+\frac{v}{r})v]&=-[\rho r(T-1)+\frac{1}{2A}(\rho\frac{\partial T}{\partial r}+T\frac{\partial \rho}{\partial r})]\\&+\frac{E}{\rho_0}[\frac{4}{3}\cdot\frac{\partial^2 u}{\partial r^2}+\frac{4}{3}\cdot\frac{1}{r}\cdot\frac{\partial u}{\partial r}-\frac{4}{3}\cdot\frac{u}{r^2}\\&+\frac{\partial^2 u}{\partial z^2}+\frac{1}{3}\cdot\frac{\partial^2 w}{\partial r\partial z}],\end{aligned} \tag{4}$$

$$\rho[u\frac{\partial v}{\partial r}+w\frac{\partial v}{\partial z}+(2+\frac{v}{r})u]=\frac{E}{\rho_0}[\frac{\partial^2 v}{\partial r^2}+\frac{1}{r}\cdot\frac{\partial v}{\partial r}-\frac{v}{r^2}+\frac{\partial^2 v}{\partial z^2}], \tag{5}$$

$$\begin{aligned}\rho[u\frac{\partial w}{\partial r}+w\frac{\partial w}{\partial z}]&=-\frac{1}{2A}[\rho\frac{\partial T}{\partial z}+T\frac{\partial \rho}{\partial z}]\\&+\frac{E}{\rho_0}[\frac{\partial^2 w}{\partial r^2}+\frac{1}{r}\cdot\frac{\partial w}{\partial r}+\frac{4}{3}\cdot\frac{\partial^2 w}{\partial z^2}\\&+\frac{1}{3}\cdot\frac{1}{r}\cdot\frac{\partial u}{\partial z}+\frac{1}{3}\cdot\frac{\partial^2 u}{\partial r\partial z}],\end{aligned} \tag{6}$$

$$\begin{aligned}\rho[u\frac{\partial T}{\partial r}+w\frac{\partial T}{\partial z}]&=\frac{\gamma E}{\rho_0 Pr}[\frac{\partial^2 T}{\partial r^2}+\frac{1}{r}\cdot\frac{\partial T}{\partial r}+\frac{\partial^2 T}{\partial z^2}]\\&-(\gamma-1)\rho T(\frac{\partial u}{\partial r}+\frac{u}{r}+\frac{\partial w}{\partial z})+(\gamma-1)(2AE\Phi),\end{aligned} \tag{7}$$

where $\rho_0=\exp(A(r^2-1))$ is the exponential distribution term, and Φ in eq. (7) is the viscous dissipation term in form of

$$\begin{aligned}\Phi=&-\frac{2}{3}(\frac{\partial u}{\partial r}+\frac{u}{r}+\frac{\partial w}{\partial z})^2+2[(\frac{\partial u}{\partial r})^2+(\frac{\partial w}{\partial z})^2+(\frac{u}{r})^2]\\&+[(\frac{\partial u}{\partial z}+\frac{\partial w}{\partial r})^2+(\frac{\partial v}{\partial z})^2+(\frac{\partial v}{\partial r}-\frac{v}{r})^2].\end{aligned} \tag{8}$$

The dimensionless numbers in the above equations are defined below.

IOP Conf. Series: Journal of Physics: Conf. Series **1099** (2018) 012010 doi:10.1088/1742-6596/1099/1/012010

$$A \equiv \frac{M\Omega^2 r_a^2}{2R_0T_0},\ E \equiv \frac{\mu}{\rho_w r_a^2 \Omega},\ Pr \equiv \frac{C_p\mu}{k},$$

where R_0 is the universal gas constant, T_0 is the mean temperature in rotor which we can get from the preset conditions, ρ_w is a gas density near a rotor wall, r_a is a rotor radius, and Ω is the angular velocity of a rotor. The ideal gas state equation should also be added to the Navier-Stokes equations to make the system be able to solve.

$$pM = \rho R_0 T \tag{9}$$

The reference values to do nondimensionalization are a radius of rotor r_a, a linear speed of rotor rotation ωr_a, the density near a side wall ρ_w and the mean temperature in rotor T_0. The procedure of nondimensionalization is shown below. The variables with a hat symbol "$\wedge$" represent the dimensional quantities.

$$r = \hat{r}/r_a,\ z = \hat{z}/r_a,\ u = \hat{v}_r/(\omega r_a),\ v = (\hat{v}_\theta - \omega r)/(\omega r_a),\ w = \hat{v}_z/(\omega r_a)\ \ \rho = \hat{\rho}/(\rho_w\rho_0) - 1,\ T = \hat{T}/T_0 - 1$$

The technique of a staggered mesh is used to discretize equations, and the position of original variables in cell (i, j) is shown in Figure 1. Density ρ and temperature T are located in the center of cell, radial velocity u and angular velocity v are located in the center of left side of cell, and axial velocity w is located in the center of down side of cell.

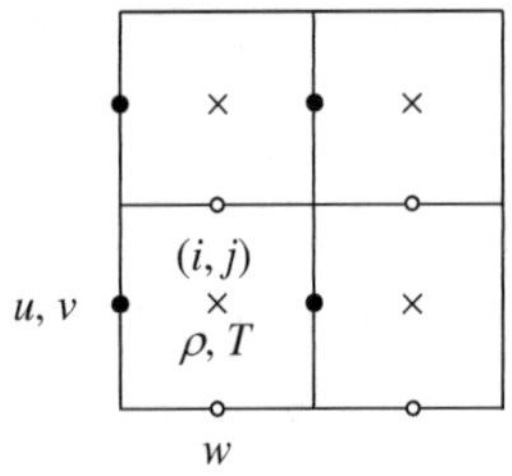

Figure 1. Schematic diagram of a staggered mesh

The central difference scheme is used to discretize the equations (3-9), and to avoid errors caused by the large density gradient in the equation (3). This equation should be integrated in a cell.

$$\int_i^{i+1}\int_j^{j+1}[\frac{1}{r}\frac{\partial}{\partial r}(r\rho\rho_0 u) + \frac{\partial}{\partial z}(\rho\rho_0 w)]r\mathrm{d}r\mathrm{d}z = 0.$$

The density and temperature in the center of a cell, the radial and angular velocities in a center of the left side of a cell, and the axial velocity in a center of bottom side of a cell can be directly used during the discretization, while the variables in other location will be calculated by a linear interpolation procedure. If the discrete left hand side in the cell (i, j) are represented by $Y_{\rho_{i,j}}$, $Y_{u_{i,j}}$, $Y_{v_{i,j}}$, $Y_{w_{i,j}}$, $Y_{T_{i,j}}$, respectively, then one can get the discrete algebra system of the equations below.

$$F(\boldsymbol{x})=\begin{pmatrix} Y_{\rho_{1,1}} \\ Y_{u_{1,1}} \\ Y_{v_{1,1}} \\ Y_{w_{i,j}} \\ Y_{T_{i,j}} \\ \vdots \\ Y_{\rho_{i\max,1}} \\ \vdots \\ Y_{T_{i\max,j\max}} \end{pmatrix}=0\,,\ \boldsymbol{x}=\begin{pmatrix} \rho_{1,1} \\ u_{1,1} \\ v_{1,1} \\ w_{1,1} \\ T_{1,1} \\ \vdots \\ \rho_{i\max,1} \\ \vdots \\ T_{i\max,j\max} \end{pmatrix}$$

The system of equations above is a nonlinear one, and the Newton iteration method can be applied to solve it. If the Jacobian matrix is represented by $\boldsymbol{J}$, the iteration procedure can be written in a such way as follows.

$$\begin{cases} \boldsymbol{x}_0 = \text{Initial Value} \\ \boldsymbol{x}_{i+1} = \boldsymbol{x}_i - \boldsymbol{J}^{-1}F(\boldsymbol{x}_i),\ i = 0,1,2,\cdots \end{cases}$$

Consider a linear system of equations $\boldsymbol{J}\Delta\boldsymbol{x} = F(\boldsymbol{x})$, the term $\boldsymbol{J}^{-1}F(\boldsymbol{x})$ in the procedure of the Newton iteration can be calculated by solving this system of equations, so there is no need to calculate the matrix which is reversed to the Jacobian one.

3.2. A homotopy algorithm based on FDM

The procedure of the Newton iteration in section 3.1 is extremely sensitive to the initial value and the iteration will be diverging when the initial value is much different from the solution. In order to solve this problem, one can improve the Kai's FDM by using the idea of a homotopy algorithm to get a large-scale convergence tool.

For example, let us we take a total reflux GC without upper and lower baffle, all disturbances are spread from the boundary. The disturbance term in the discrete boundary equations should appear. We use $\Delta(\boldsymbol{x})$ to represent these disturbance, and the flow field in a GC can be regarded as a superposition of the rigid body solution $\boldsymbol{x}_0$ and the disturbance solution $\boldsymbol{x}_\Delta$. Then we can construct a homotopy mapping as indicated below.

$$H(\boldsymbol{x},\alpha) = F(\boldsymbol{x}) + \alpha\Delta(\boldsymbol{x})\,,\ \alpha \in [0,1]$$

Besides, the homotopy mapping will meet a condition that when $\alpha = 0$, the solution of the homotopy equation is $\boldsymbol{x} = \boldsymbol{x}_0$, which is equal to the rigid body solution as is known, and when $\alpha = 1$, the solution of the homotopy equation is $\boldsymbol{x} = \boldsymbol{x}^*$, which is exactly the solution of the flow field we need. Then, we can follow the homotopy curve from $\boldsymbol{x} = \boldsymbol{x}_0$ to the end and get the solution we need.

In order to follow the homotopy curve, we need to calculate the slope of the curve so that we can predict the next point on the curve from the point which is already known. After calculation of the total derivative of the homotopy equation one can get

$$\frac{\partial H}{\partial \alpha} = \frac{\partial H}{\partial \boldsymbol{x}}\frac{\partial \boldsymbol{x}}{\partial \alpha} \tag{10}$$

The term $\partial \boldsymbol{x} / \partial \alpha$ is the slope we need and the equation (10) can be regarded as a linear system of equations in which the coefficient $A = \partial H / \partial x = \boldsymbol{J}$, the right-hand side is $b = \partial H / \partial \alpha$ and the unknown number is $z = \partial \boldsymbol{x} / \partial \alpha$.

Now we use the forward Euler interpolation method based on the slope to predict the next point, and after that we use the Newton iteration method to correct the point to the homotopy curve. The step length for the Euler interpolation is adaptable: if the predicted point can be corrected to the curve with the last step length $\mathrm{d}\alpha_{i-1}$, then the next step length will be $\mathrm{d}\alpha_i = 2\mathrm{d}\alpha_{i-1}$, else the last step length will be reduced by half $\mathrm{d}\alpha_{i-1} = \mathrm{d}\alpha_{i-1} / 2$.

The pseudocode description of a homotopy algorithm based on a FDM is shown below.

(1) Input the initial value $\boldsymbol{x}_0$, the initial step of the length $\mathrm{d}\alpha_0$ and set the parameter $\alpha_0 = 0$;

(2) Continue while $\alpha < 1$.

Prediction step

Predict the next point by the forward Euler interpolation: $\boldsymbol{x}'_{i+1} = \boldsymbol{x}_i + \mathrm{d}\alpha_i \dfrac{\partial \boldsymbol{x}}{\partial \alpha}$, and set the parameter $\alpha_{i+1} = \alpha_i + \mathrm{d}\alpha_i$.

Correction step

Solving the system of equations $H(\boldsymbol{x}, \alpha_{i+1}) = 0$ by the Newton iteration method with the initial value $\boldsymbol{x}'_{i+1}$, and the upper limit of a number of iterations is n_i, and the upper limit of an error is ε_i.

If the iteration converged within the n_i iterations and get the point on the curve $\mathbf{x}_{i+1}$,

calculate the next step length $\mathrm{d}\alpha_{i+1} = \min(2\mathrm{d}\alpha_i, 1 - \alpha_{i+1})$, and do the next loop,

else

decrease the step length $\mathrm{d}\alpha_i = \mathrm{d}\alpha_i / 2$, and do the loop again,

end.

(3) Output the result $\boldsymbol{x}^*$.

4. Verification of a homotopy algorithm based on the FDM

For verification of a created computing code we consider a total reflux GC with a length of a rotor 1m and other parameters corresponding to the Iguassu GC model. The parameters of the model GC and the process gas are shown in Table 1 and Table 2.

The comparisons of results calculated in this paper and the data obtained by Bogovalov[1] playing a role of a benchmark for our code are shown in Figures from 2*a*, *b* to 6. It can be seen from the comparison that the constructed homotopy algorithm based on the FDM in this research can be used to simulate the flow field in GC.

Table 1. Parameters of the Iguassu model GC

Parameter	Value
V, linear speed of rotor wall	600 m/s
P_w, pressure near rotor wall	80 mmHg
T_0, mean temperature in rotor	300 K
R_{out}, rotor radius	0.06 m
R_{in}, inner boundary radius	0.048 m
ΔT, maximum of temperature disturbance	0.3 K

Table 2. Parameters of the process gas

Parameter	Value
M , molecular weight	0.352 kg/mol
μ , dynamic viscosity	1.83×10^{-5} Pa·s
k , heat conductivity	6.1×10^{-3} W/m/K
C_p , specific heat	372.84 J/kg/K
γ , specific heat ratio	1.068

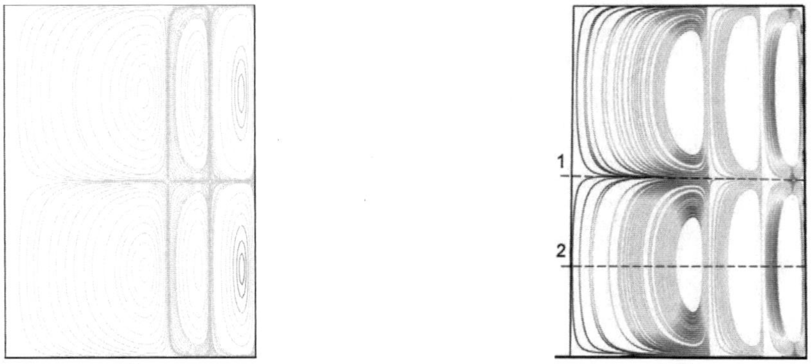

a) *b*)

Figure 2. Comparison of the flow fields
a) calculated in this paper. *b*) calculated in the paper[1]

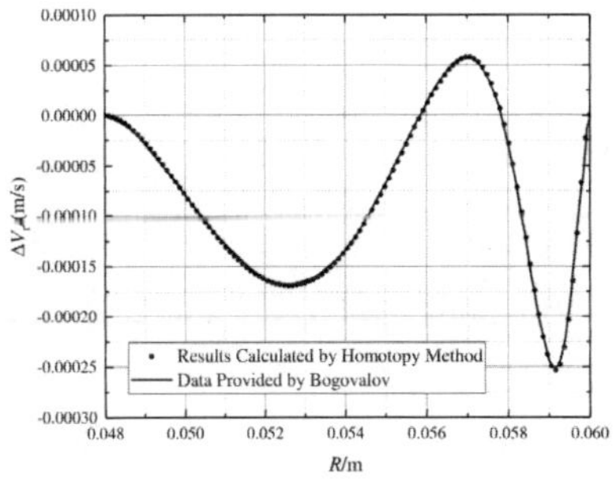

Figure 3. Comparison of disturbance for the radial velocity in the cross-section 1 in Figure 2*b*.

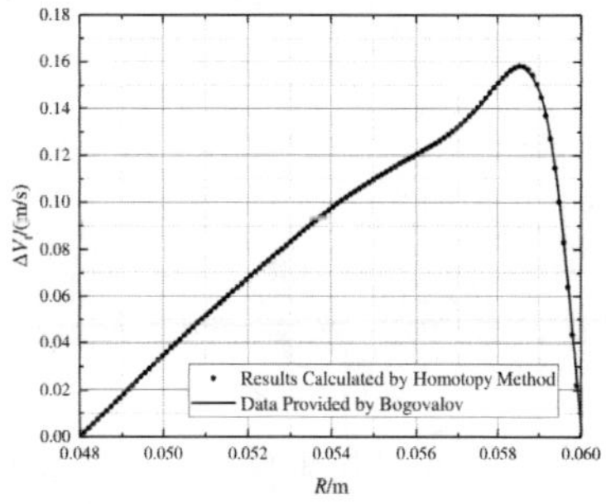

Figure 4. Comparison of the angular velocity disturbance in the cross-section 1 in Figure 2*b*.

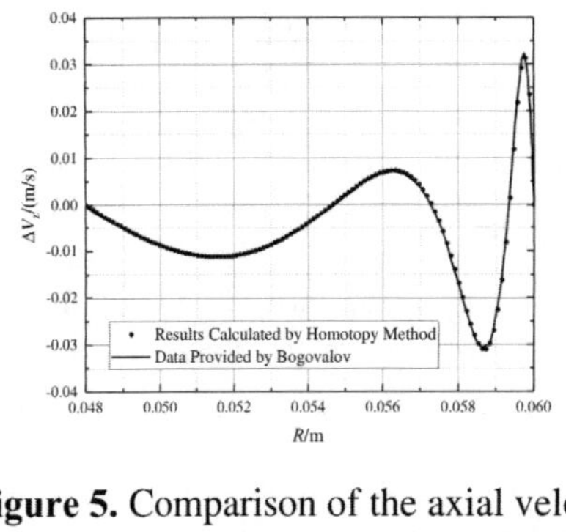

Figure 5. Comparison of the axial velocity disturbance in the cross-section 2 in Figure 2.

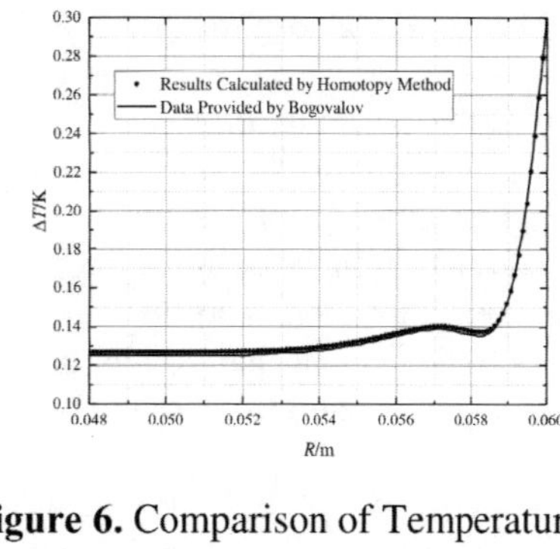

Figure 6. Comparison of Temperature disturbance in the cross-section 1 in Figure 2.

5. Simulation of flow field in GC under strong disturbance

5.1. Introduction to the case

In this case, the total reflux GC without upper and lower baffle is used for research and all parameters of a GC correspond to the Iguassu model GC. The mechanical drive disturbance is located in the lower boundary, the end cap thermal drive disturbance is applied both in the upper and lower boundaries, and the side wall thermal drive disturbance is placed on a rotor wall boundary. The schematic diagram of the boundaries mentioned above is shown in Figure 7 and listed in Table 3.

The parameters of a process gas are the same as in Table 2. The inner boundary is free one and the boundary conditions on it are as follows

$$\frac{\partial \rho}{\partial r}=0,\ u=0,\ \frac{\partial}{\partial r}\left(\frac{v}{r}\right)=0,\ \frac{\partial w}{\partial r}=0,\ \frac{\partial T}{\partial r}=0$$

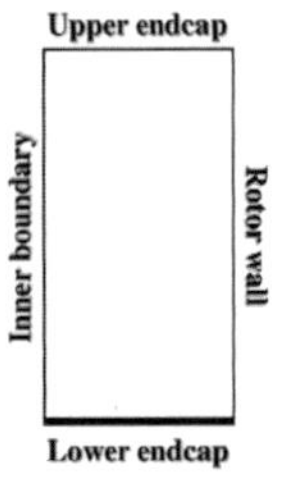

Figure 7. Schematic diagram of the model GC

Table 3. Parameters of a case under strong disturbance condition

Parameter	Value
L, length of rotor	0.48 m
r_a, radius of rotor	0.06 m
r_{in}, radius of inner boundary	0.048 m
V, linear speed of rotor wall	600m/s
P_w, pressure near rotor wall	100 mmHg
T_0, mean temperature in rotor	300 K
ΔT_{up}, temperature disturbance on upper endcap	-30 K
ΔT_{down}, temperature disturbance on lower endcap	30 K
ΔT_w, temperature disturbance on rotor wall	-30 K – 30 K linear distribution
$\Delta\Omega / \Omega$, mechanical disturbance on lower endcap	40%

Note that the thermal and mechanical drives mentioned above are strong enough. The flow field is simulated by a homotopy algorithm based on the FDM and the iteration is convergent.

5.2. Results of numerical simulation results

Results of the density, velocity, temperature and stream lines distributions found in this research are presented in Figures from 8 to 13. It can be seen that the flow field in a total reflux GC without baffle under such disturbances is much different from the simple rigid body solution, the radial and axial velocities reach tens of meters, and this cannot be ignored comparing to the linear speed of the rotor wall. Because of the strong mechanical disturbance, the distribution of the angular velocity is also much different from the linear distribution in the rigid body solution.

IOP Conf. Series: Journal of Physics: Conf. Series **1099** (2018) 012010 doi:10.1088/1742-6596/1099/1/012010

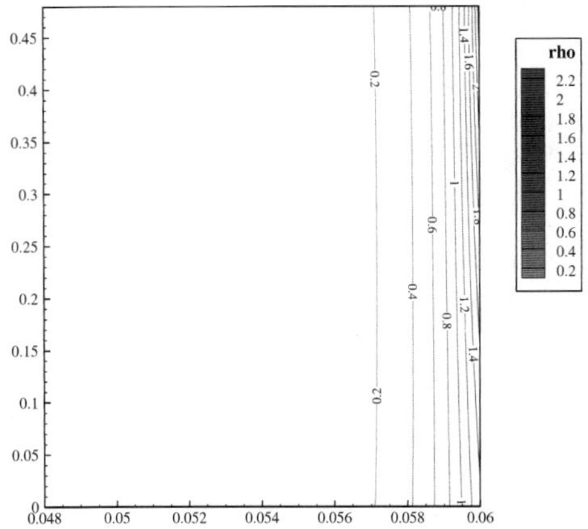

Figure 8. Distribution of a density

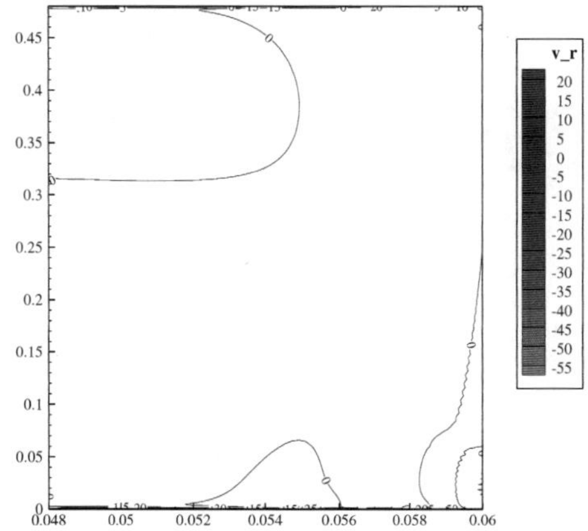

Figure 9. Distribution of the radial velocity

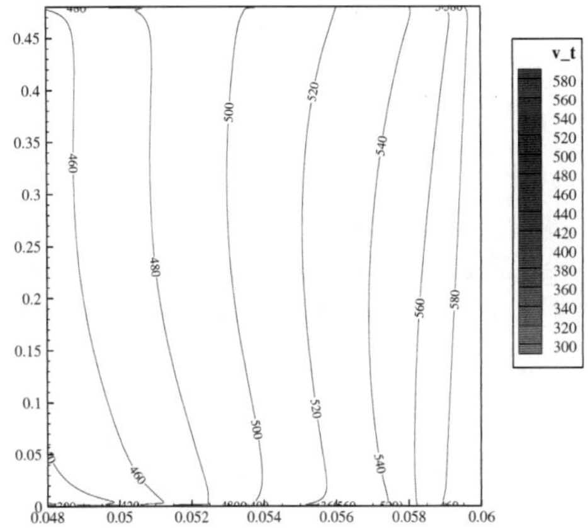

Figure 10. Distribution of the angular velocity

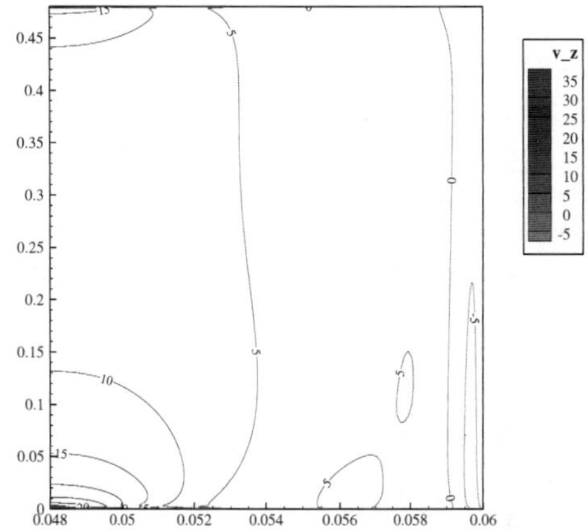

Figure 11. Distribution of the axial velocity

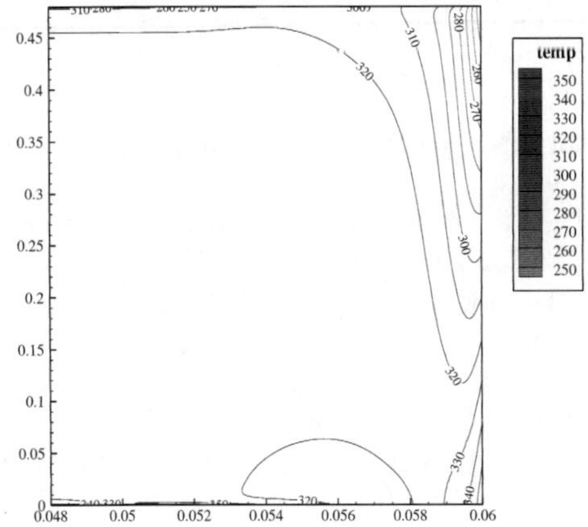

Figure 12. Distribution of the temperature

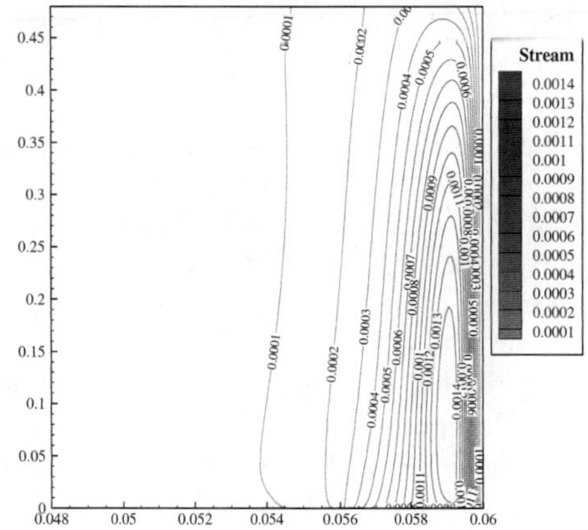

Figure 13. Distribution of the stream lines

Also, we use the maxima of the stream function in various cross-sections of a GC as the metrics of the axial circulation intensity. One can see in Figure 14, the dependence of the axial circulation intensity in the following cross-sections of a GC: 1/8, 2/8, 3/8, 4/8, 5/8, 6/8 and 7/8. The behavior of the axial circulation is shown in Figure 14. It can be seen that the circulation intensity under the condition of strong mechanical and thermal drives has a parabolic character.

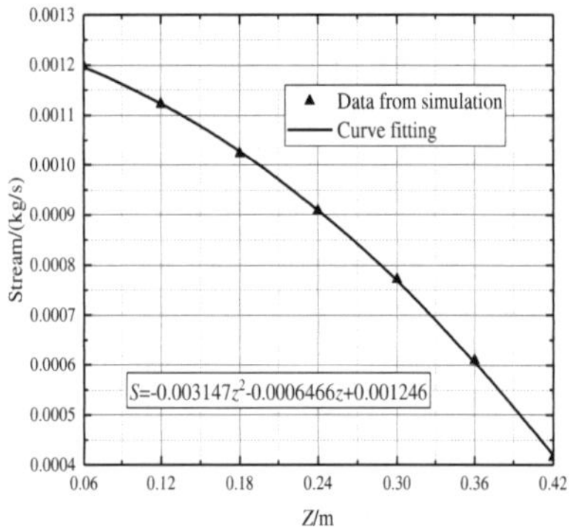

Figure 14. Damping of the axial circulation intensity in the Iguassu GC.

6. Conclusions

1. It is demonstrated that the FDM can be radically improved by a homotopy algorithm what allows us to construct a large-scale convergence algorithm and on its basis a computing code.
2. The new algorithm is verified and proved to be valid using as a benchmark the data obtained for the model GC Iguassu by means of the independent algorithm.
3. Using the maxima of the stream function in various cross-sections of a total reflux GC as the metrics for the axial circulation intensity it is found that its damping along the axial coordinate has a parabolic character.

Acknowledgement

Project 11475096 supported by National Natural Science Foundation of China.
Project 11611130021 supported by NSFC-RFBR.

References

[1] Bogovalov S V, Borisevich V D, Borman V D, et al. Verification of numerical codes for modeling of the flow and isotope separation in gas centrifuges [J]. Computers & Fluids, 2013, 86: 177-184.

[2] Kai T Basic characteristics of centrifuges, (III): Analysis of separation performance of centrifuges [J]. J. Nucl. Sci. Technol, 1977, 14(7).

[3] Borisevich V D, Borman V D, Bogovalov S V et al. On a formula to evaluate the separative power of long gas centrifuges [J] Separation Science and Technology, 2014, 49: 329–334.

[4] Omnes P Numerical and physical comparisons of two models of a gas centrifuge [J]. Computers & Fluids, 2007, 36(6): 1028-1039.

[5] Borisevich V D, Levin E V, Naumochkin V V. Numerical investigation of viscous gas secondary flows in a rotating cylinder with sources and sinks [J]. Fluid Dynamics, 1989, 24(4): 520-524.

[6] Xu L, Zeng S, Chen N et al. Study on calculation of nonlinear flow field in gas centrifuge and its separation performance [J]. Atomic energy science and technology, 2016, 50(03):385-389

[7] Zhang L, Zhang X Homotopy algorithm of nonlinear equation set based on MATLAB [J]. Journal of Baicheng Normal College, 2013, 27(03):9-12.

[8] Bogovalov S V, Borisevich V D, Borman V D et al. Numerical modeling of the flow and isotope separation in centrifuge Iguasu for different lengths of the rotor [J]. AIP Conference Proc., 2016, 1738, 030008.

IOP Conf. Series: Journal of Physics: Conf. Series **1099** (2018) 012011 doi:10.1088/1742-6596/1099/1/012011

Plasma centrifuge for isotope separation with axial circulation caused by a traveling magnetic field

V D Borisevich[1], **E P Potanin**[1, 2]

[1]National Research Nuclear University MEPhI, 31 Kashirskoe Shosse, Moscow 115409. Russian Federation

[2]National Research Center "Kurchatov Institute", 1 Academician Kurchatov Square, Moscow 123098, Russian Federation

E-mail: VDBorisevich@mephi.ru

Abstract. The isotope separation processes in the high-frequency plasma centrifuge operating in the withdrawal mode are studied. The rotation of weakly ionized plasma is provided by a rotating magnetic field, and the axial countercurrent flow (circulation) is caused by a traveling magnetic field. The dependences of the overall enrichment factor and the separation capacity of a plasma centrifuge on a product flow rate are calculated. The case when a separation column is operated in a symmetrical mode is considered. The optimum characteristics of the separation device are discussed. It is shown that there is the optimal value of a product flow rate at which the maximum value of the separative power of a plasma centrifuge is reached.

1. Introduction

Separation of stable isotopes is an important branch of the country's economy, serving nuclear industry power, basic and life sciences, medicine, agriculture, and many others. Without a number of stable isotopes, it is difficult to imagine the further development of nuclear power and its efficiency and safety, basic research in nuclear physics, diagnosis and treatment of many diseases. Interest in the separation of stable isotopes in the plasma phase is associated with inability to use gas centrifuges to enrich isotopes, if there are no suitable volatile compounds at room temperature which can be used as process gases [1-3]. In addition, the problem of reprocessing spent nuclear fuel from nuclear power plants can be solved, as suggested, with using rapidly rotating plasma [4, 5].

A new concept of a circulating plasma centrifuge (PC) is proposed in [6], in which rotation of a weakly ionized ^{20}Ne-^{22}Ne isotope mixture is provided by a rotating magnetic field, and the axial circulation flow multiplying the radial separation effect is created by a traveling magnetic wave. There were determined the hydrodynamic characteristics of a PC and the overall enrichment factor of the apparatus for the so-called total reflux operating mode. In practice, the separation unit operates in a mode when the initial isotope mixture enters the device and the enriched in a target component fraction is withdrawn.

In this paper, basing on the results of an analytical approach for calculation of the azimuthal $V_\varphi(r)$ and axial $V_r(r)$ components of plasma velocities, the overall enrichment factor and the separation capacity of a PC operating in the working regime with a product flow were calculated. Notice that the analytical solution obtained in this research allows us not only to estimate the optimum parameters of a single separation device, but also to trace, using the found theoretical dependences, the ways of intensification the separation processes in a PC.

IOP Conf. Series: Journal of Physics: Conf. Series **1099** (2018) 012011 doi:10.1088/1742-6596/1099/1/012011

2. Design principles

A PC with an axial circulation flow excited by a traveling magnetic field is presented in Fig.1. The rotation of weakly ionized plasma is produced by a transverse HF magnetic field having the angular velocity ω_0, generated by the axial current carrying rods, two of which are shown in the front view, and all six rods in the top view in Fig.1.

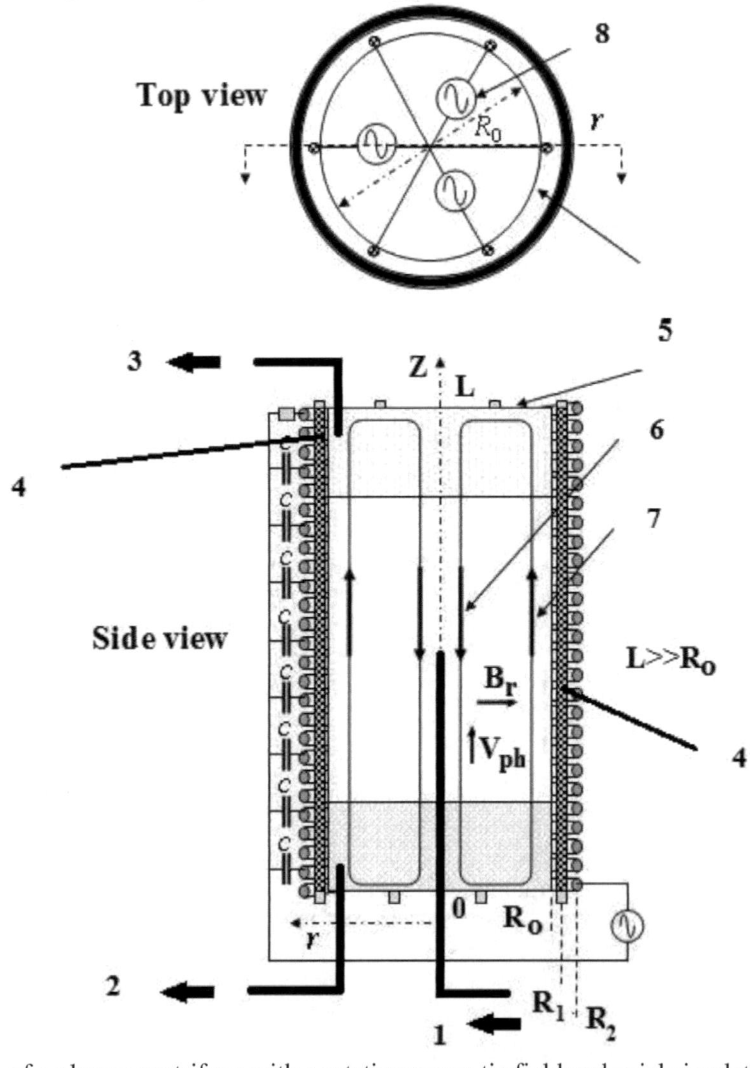

Figure 1. Design of a plasma centrifuge with a rotating magnetic field and axial circulation stimulated by a traveling magnetic field. 1, 2, 3 - the feed, waste, and product flows in a PC, respectively; 4 – current carrying rods; 5 – a quartz camera; 6 – a flow of a light fraction; 7 – a flow of a heavy fraction; 8 - 3-phase HF power supply.

The circulation flow is excited by a travelling magnetic field caused by a delay line which is a system of capacitors with a capacitance *C*. The radial magnetic field propagates along the axis of the discharge chamber with a phase velocity V_{ph} and varies in space and time according to the law $B_r = B_1 \cos\left[\omega_1\left(t - z/V_{ph}\right)\right]$, where ω_1 is the circular frequency of a traveling magnetic wave, B_1 is the amplitude value of the magnetic induction of the traveling magnetic field, z is the longitudinal coordinate.

A detailed description of the separation installation and the procedure to calculate the hydrodynamic characteristics of a PC with separate excitement of a centrifugal field of force forces and an axial countercurrent flow are given in [6]. The case was considered when the length of a PC is much larger than the radius of the discharge chamber. It was demonstrated that in the total reflux mode of operation of A PC, when the highest degree of separation is achieved, it is possible to obtain significant multiplication of the primary (radial) separation effect along the length of a PC due to the created an axial countercurrent flow.

A schematic drawing of a separating element is shown in Fig.2. A separated mixture with a given flow rate G and concentration of a target component C_0 enters a continuously working separation element. Withdrawals of heavy W, C_W and light P, C_P fractions occurs near the upper and lower ends of the working chamber (Fig.1), respectively. In this case, the longitudinal enrichment factors become lower in contrast to the case of the total reflux regime considered in [6].

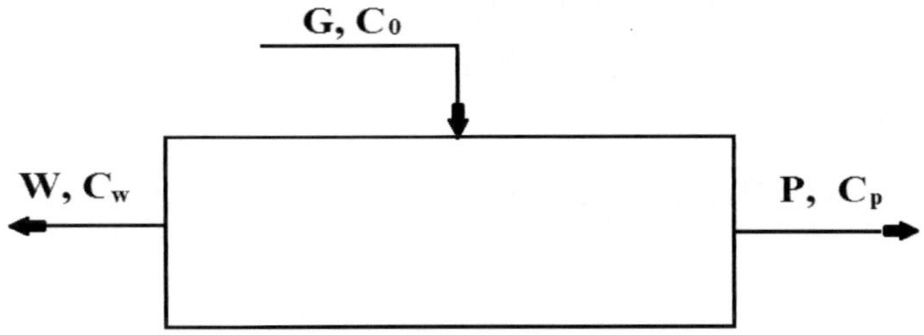

Figure 2. Schematic drawing of a separating element

As is known, the separation device, in addition to the overall separation factor determining a degree of enrichment of a separated mixture, is characterized by the so-called "the separative power" or "separation capacity". This value is related to a speed of a diffusion process and ultimately determines the cost of separation.

3. Estimation of separation performance

In the absence of circulation ($B_1 = 0,\ V_z = 0$), the radial enrichment factor for the binary isotope mixture is determined by the expression [6]

$$\varepsilon_0 = \left(\frac{C}{1-C}\right)_{r=R_0} \Bigg/ \left(\frac{C}{1-C}\right)_{r=0} - 1 = \int_0^{R_0} \varepsilon_r(r)dr\,, \tag{1}$$

where $\varepsilon_r(r) = \Delta M V_\varphi^2(r) / \Re T r$, C is the averaged over a cross-section of a PC concentration of a heavy component, ΔM is a difference in the molecular weights of the separated isotopes, T is a temperature of plasma, $\Re$ is the universal gas constant, $V_\varphi(r)$ is the azimuthal velocity component.

The latter in the regime of moderate induced magnetic fields [5] and under condition when braking the medium in the Ekman layers on the lids of a working chamber is neglected [7, 8], is determined by the expression

$$V_\varphi = \frac{B_0^2 R_0 \operatorname{Re}_m}{16\mu_0 \langle\eta\rangle \chi} y\left(1-y^2\right), \tag{2}$$

where $B_0 = I_{z0}\mu_0/2$ is the amplitude value of induction for the vacuum rotating magnetic field, I_{z0} is the amplitude value of the surface density of an axial current induced a rotating magnetic field, μ_0 is a magnet constant, $y = r/R_0$, $\langle \eta \rangle$ is an averaged value of the coefficient of dynamic viscosity, χ is an dimensionless parameter depending on the magnet Reinolds number $\mathrm{Re}_m = \mu_0 \sigma \omega_0 R_0^2$, σ is a plasma conductivity.

According to [1], the transport equation for a stripping part of a column can be written in the form

$$(K_1 + K_2)\frac{dC}{dz} = K_3 C(1-C) + P(C - C_p) \tag{3}$$

Eq. (3) differs from the transport equation used in [6] for the total reflux regime by the presence of an additional term characterizing the flow of a target component in a product flow. Here K_1, K_2, and K_3 are the coefficients defined by the relations as follows

$$K_1 = 2\pi \int_0^{R_0} \psi^2(r)/\rho D r\, dr\,,\ K_2 = 2\pi \int_0^{R_0} \rho D r dr \text{ и } K_3 = 2\pi \int_0^{R_0} \psi(r)\varepsilon_r(r) dr\,,$$

where $\psi(r) = \int_0^r \rho V_z(r) r dr$ is a flow function, ρ is a density of a separated mixture, D is a coefficient of radial diffusion.

Following [6], one can determine a plasma density distribution a along the radius of a working chamber as

$$\rho = \rho_0 \exp(Ay), \tag{4}$$

where $A = \dfrac{\langle M \rangle B_0^4 R_0^2 \mathrm{Re}_m^2}{2 \cdot 512 \Re T_0 \mu_0^2 \langle \eta \rangle^2 \chi^2}$, ρ_0 is a density on a working chamber , T_0 is a temperature on a PC axis.

Now we introduce gas-filling density value ρ^* and the parameter $V = \dfrac{\rho^* V_{z0} R_0}{2\langle \rho D \rangle} \dfrac{A^2}{(1+(A-1)\exp(A))}$, where $V_{z0} = \sigma B_1^2 R_0^4 k_z \omega_1 \mathrm{Re}_{m1} / 128 \langle \eta \rangle (f_1^2 + f_2^2)$, $k_z = \omega_1 / V_{ph}$ [6]. Whereupon, one can demonstrate that the dependence of the dimensionless optimal circulation flow $U = \dfrac{2\langle \rho D \rangle (V_{z0})_m}{\rho^* R_0} = \dfrac{2V_m \langle \rho D \rangle (1+(A-1)\exp(A))}{\rho^* R_0 A^2}$ on the compressibility parameter A in the total reflux mode of PC operation has the form shown in Fig.3. As expected, increase of the compressibility parameter A leads to increase of the optimal circulation flow rate.

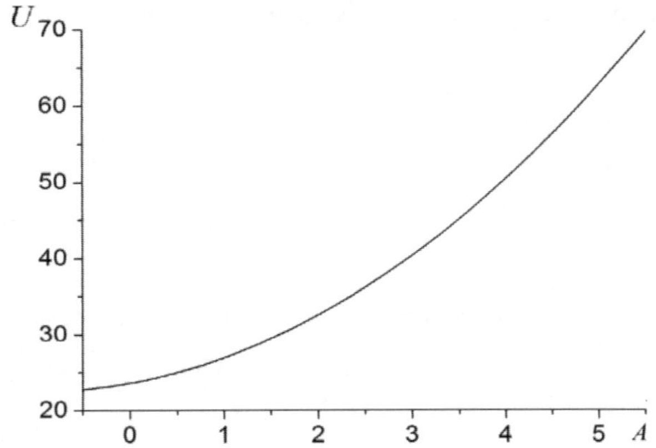

Figure 3. Dependence of the dimensionless optimal velocity of circulation U on the compressibility parameter A.

Integration of Eq. (3) under the assumption of small enrichment factors and the condition $C(1-C)\approx const$ leads to the following dependence of the enrichment factor over the length of the stripping part of a PC [2]

$$\varepsilon_z = \frac{4}{\sqrt{2}}\varepsilon_0 \frac{L_p}{R_0}\frac{m}{1+m^2} N \frac{1-\exp(-a)}{a} \qquad (5)$$

where L_p is a length of a stripping part of a column, $m = F/F_0$ is a dimensionless twofold circulation flow, $F = 2\pi\int_0^{R_0}|\rho V_z| r dr$, $F_0 = \sqrt{K_2/k_1}$, $k_1 = K_1/F$, $a = \frac{\theta(1-\theta)}{K_1+K_2}L_p$,

$\theta = P/G$ is a cut, $N = 2\int_0^1 y(1-y^2)\psi_0 dy \Big/ \sqrt{\int_0^1 \frac{\psi_0^2}{y}dy}$,

$\psi_0 = \psi/V_{z0}\rho_0 R_0^2 = \exp(-Ay)\left[\lambda(y^2-1)+1-y^4\right]$, $\varepsilon_0 = \Delta M B_0^4 R_0^2 \mathrm{Re}_m^2 / 4\cdot 256\langle\eta\rangle^2 \mu_0^2 \Re T_0 \chi^2$,

$\mathrm{Re}_{m1} = \mu_0\sigma\omega_1 R_0^2$, λ is a dimensionless coefficient defined in [6].

The obtained analytical dependence of the overall enrichment factor on the circulation value demonstrates the existence of the optimal countercurrent flow. It also confirms that the efficiency of multiplication of the primary (radial) separation effect depends to a significant extend on a circulation profile described by the dependence of the N parameter on the shape of a stream function ψ (y). The approach developed in the paper allows us to define the circulation profile at which the maximum overall enrichment factor is reached.

In the regime of the optimal by enrichment factor circulation flow (m=1) and a symmetrical scheme of working of a separation element ($\theta = 1/2$), the expression for the enrichment factor (5) takes the following form

$$\varepsilon_z = \varepsilon_0 \frac{4L_1}{\sqrt{2}} N\left(1-\exp(-L_1 g/8\pi)\right)/(L_1 g/8\pi), \qquad (6)$$

where $L_1 = L_p/R_0$, $g = G/\langle\rho D\rangle R_0$. Note that in the case when $\theta = 1/2$ (the symmetric working regime of a PC, the value of a dimensionless product flow will be equal to $p = g/2$.

Then the separation performance of a PC at moderate enrichments is defined as [9]

$$\delta U = G\varepsilon_z^2 / 2 . \tag{7}$$

It is quite obvious that the separative power of a PC calculated by this way has the dimension of a flow.

The dependences of the normalized enrichment factor $\varepsilon=\varepsilon_z/\varepsilon_0$ and the separative power $\delta u = \delta U / \langle \rho D \rangle R_0 \varepsilon_0^2$ on the dimensionless feed flow g for the case of neon isotope separation are presented in Fig.4.

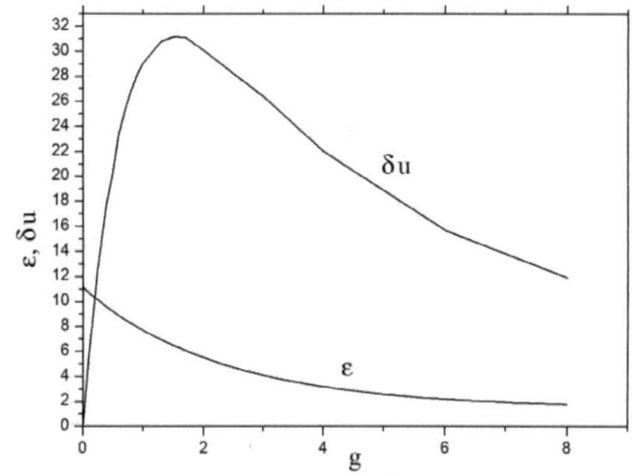

Figure 4. Dependences of the enrichment factor ε and the dimensionless separative power of a PC δu on the dimensionless feed flow rate *g*=2*p*.

As would be expected, the enrichment factor of a PC operating in withdrawal regime decreases with the increase in a product flow. Unlike the enrichment factor, the dependence of the separative power of a PC on a dimensionless feed flow has an optimum, the value of which for the parameters specified in this research is equal to $g_{opt} \approx 2$.

Estimating the optimal feed flow as $G_{opt} \approx 2\langle \rho D \rangle R_0$, one can find for natural abundance of neon isotopes at the characteristic temperature of a neutral gas $T \approx 1000K$ ($\langle \eta \rangle \approx 10^{-4}$ kg/m·s) the optimum value of a feed flow rate to a PC $G_{opt} \approx 10^{-2}$ g/s. For $B_0 \approx 1.3 \cdot 10^{-2}$ T, $B_1 = 6.10^{-3}$ T, $\omega_0 = 2.10^5$ 1/s, $\omega_1 = 2.10^6$ 1/s, $\sigma = 2.10^3$ 1/Ω·m, $\mathrm{Re}_m = 0.6$, $\mathrm{Re}_{m1} = 12.6$, $\chi \approx 2$, $R_0 = 0.05$ m, $\Delta M = 2$ kg/Kmol, $L_p = 1$ m, it will be obtained that $\varepsilon_0 \approx 0.06$ and $\varepsilon(g=0) \approx 11$. In this case, the maximum separative power of a PC will be equal to $\delta U_{\max} \approx 5\,10^{-4}$ g/s.

4. Conclusions

1. Theoretical study of neon isotope separation in a PC with operating in a mode with a product flow is carried out. There were calculated the overall enrichment factor and the separative capacity of the apparatus in which the primary (radial) separation effect and its multiplication in axial direction are achieved by rotating and traveling magnetic fields, respectively.
2. The dependence of the characteristics under investigation on a feed flow rate is obtained for the symmetric mode of PC operation. The separative power of the installation having the

maximum separative power δU_{max} at the value G_{opt} of a feed flow rate for the given working parameters of a PC and the characteristics of the weakly ionized plasma are estimated.

Acknowledgement. This research is funded partly by the MEPhI Academic Excellence Project (contracts No. 02.a03.21.0005, 27.08.2013) and The Russian Science Foundation (RSF), the grant number 18-19-00447.
Also, the authors also would like to express their gratitude to Mr. Julian Whichello, Applied and Plasma Physics, University of Sydney for useful discussion of the paper.

References

[1] Villani S Uranium enrichment 1979 Springer – Verlag Berlin Heidelberg New York.
[2] Borisevich V D, Borman V D, Sulaberidze G A et al. Physical fundamentals of isotope separation by gas centrifuge 2011 Moscow MEI Publ. House (in Russian).
[3] Karchevskiy A I, Potanin E P ISOTOPES Properties Production Application Ed. Baranov V Y 2005 Moscow IzdAT 1 326 (in Russian).
[4] Zhil'tsov V A Kulygin V M Semashko N N et al 2006 *At. Energy* **101** (4) 755.
[5] Fetterman, A J Fisch N J 2009 *Plasma Sources Sci. Technol.* **18** 045003.
[6] Borisevich V D Potanin E P 2017 *Phys. Sripta* **92** 07560.
[7] Borisevich V D, Potanin E P, Whichello J J 2017 *Fluid Mech.* **829** 328.
[8] Gorbachev L P, Nikitin N V, Ustinov A P 1974 *Magnetohydrodyn.* **4** 32.
[9] Rosen A M Theory of isotope separation in columns 1960 Moscow Atomizdat (in Russian).

On the scoop heating effect of a gas centrifuge in numerical simulation

Y N Zhang[1], S Zeng[1,3], D J Jiang[1] and V D Borisevich[2]

[1] Department of Engineering Physics, Tsinghua University, Beijing 100084, China

[2] Department of Molecular Physics, National Research Nuclear University (MEPhI), Moscow 115409, 31 Kashirskoe Shosse, Russia

[3] Corresponding author

Email: zengs@tsinghua.edu.cn

Abstract. The withdrawal scoops in a gas centrifuge generate heat due to frictions and shock waves from the interaction between the scoops and the gas at a high speed motion. The heat changes the flow field in the centrifuge and influences the separation performance, giving rise to the heating effect. In the study of the flow field in a gas centrifuge with the two-dimensional axisymmetric approximation, the effects of a scoops are modelled by momentum and energy sources/sinks, or by a rotating disk with an angular speed different from that of the centrifuge rotor. The scoop-gas interaction is very complicated, and it is important to ensure to have the heating effect appropriately considered. Numerical simulation shows that, if the heating effect is not properly taken into account, the gas temperature around the scoops decreases, which is inconsistent with the real physical process. Some approaches dealing with the heating effect are discussed to avoid making incorrect modelling. Simulation results reveal that the scoop heating effect has a significant influence on the separation performances of the gas centrifuge.

1. Introduction

In the numerical study of the flow field of a gas centrifuge, the two-dimensional axisymmetric approximation is widely adopted to simplify the N-S equations, because there exist many difficulties in simulating directly the strongly rotating three-dimensional flow field in the centrifuge with a very complicated geometry, such as accurate capture of the shock waves in front of the withdrawals under a strong pressure gradient, computation of non-stationary flow, computation of the feed jet flow, and consumption of a significant amount of computing resource. Under this approximation, the true three-dimensional flow situation in an actual gas centrifuge, for instance, resulting from the interaction of the stationary withdrawal scoops with the rotating gas at a very high speed, cannot be reflected and, therefore, models with great simplifications must be introduced to simulate the mechanical driving effects of the scoops on the flow field. Currently, there are two main types of models often used: one is the model of rotating disk proposed by Cenedese and Cunsolo [1], which uses a disk rotating at an angular speed slower than the centrifuge rotor to simulate the dragging effect of a scoop on the gas; another is the model of sources and sinks employed by Wood and Sanders [2], which introduces momentum and energy sources and sinks to simulate the interaction between the scoop and the gas. Of course, the feed or withdrawals can also be simulated with this model, with mass sources of sinks in addition to the momentum and energy sources and sinks.

The two models both replace the interaction between a scoop and the gas by means of introducing external resistances of some kind to the high speed gas flow. The interaction converts the kinetic energy of the gas into heat and produces heating effect, which heats up the gas around the scoop. In fact, at the normal operational state, the heat produced by the scoops is one of the primary heat sources inside a gas centrifuge. The heating effect increases the temperature of the gas. On one hand, the heated gas affects the temperature distribution of the centrifuge rotor, and hence affects the fluid flow. On the other hand, the heated gas changes the pressure distribution in the gas, which produces an effect similar to the thermal drive and also affects the fluid flow. Therefore, the heating effect of the scoops would eventually influence the separative performance. To simulate the heating effect appropriately, some cares should be taken when using either the rotating disk model or the sources and sinks model, or otherwise some problems, which have not been explicitly pointed out in literature or have been mentioned but not paid much attention and analyzed systematically, can lead to erroneous results deviating from the situation of flow in reality. This paper is meant to make clear these problems so that a simulation can be performed correctly.

2. A scoop model with two-dimensional axisymmetrical approximation

2.1. Model of sources and sinks

The N-S equation system describing the law of gas motion in a gas centrifuge can be presented as follows.

The equation of mass conservation:

$$\frac{D\rho}{Dt} + \rho\nabla \cdot \mathbf{V} = \dot{m} \tag{1}$$

the momentum equation:

$$\rho\frac{D\mathbf{V}}{Dt} = \mathbf{F} - \nabla p + \frac{\mu}{3}\nabla(\nabla \cdot \mathbf{V}) + \mu\Delta\mathbf{V} + \dot{m}(\mathbf{V}_m - \mathbf{V}) \tag{2}$$

and the energy equation:

$$\rho\frac{De}{Dt} = \mathbf{F} \cdot \mathbf{V} + \frac{p}{\rho}\frac{D\rho}{Dt} + \Phi + \kappa\Delta T + \dot{q} + \dot{m}\left[\left(e_m - e - \frac{p}{\rho}\right) + \frac{1}{2}|\mathbf{V}_m - \mathbf{V}|^2\right] \tag{3}$$

Here, $\mathbf{V}$ is the gas velocity; ρ, p and T are, respectively, the gas density, pressure, and temperature; $e = c_v T$ is the internal energy (c_v is the specific heat at constant volume); Φ is the viscous dissipation; μ and κ are, respectively, the kinetic viscosity and heat conductivity coefficients. The terms for simulating the feed, withdrawals, mechanical, and thermal drives are: mass, momentum and energy sources/sinks $\dot{m}$, $\mathbf{F}$, and $\dot{q}$, respectively. The quantities e_m and $\mathbf{V}_m$ with a subscript 'm' represent the internal energy and momentum of gas introduced into or extracted from the flow field by the sources or sinks. In addition to the above N-S equation system, the perfect gas law is employed in the solution of the system.

For the convenience of analysis, the N-S equations uses a cylindrical coordinate system that is attached to the rotor, as shown in figure 1. Ω is the angular speed of the rotor, a and Z are the radius and height of the rotor, respectively. Since the drag of a scoop to the gas take place mainly in the direction of the incoming flow (i.e., the θ-direction), a mementum sink in the θ-direction $\mathbf{F} = F_\theta \mathbf{e}_\theta$, distributed in the influential area of the scoop, can be used to model the scoop drag.

A momentum source/sink $\mathbf{F}$ in the N-S equations is equivalent to a field of an external force. Considering the force only in the momentum equation is not enough. Because of the resistance of the scoop to the flow in the θ-direction, the direction of the external force is against the direction of the incoming flow and, therefore, it does negative work $\mathbf{F} \cdot \mathbf{V}$ to the incoming flow. In other words, the gas does work to the scoop and outputs energy, and the total energy of the gas reduces. However, a scoop is stationary in reality, and the gas interacting with the scoop has no velocity, $\mathbf{V} = \mathbf{0}$, which means that the scoop does not do any work to the gas and does not consume energy, that is, there is no loss of the total energy of the gas incurred from the scoop resistance. As such, from one point of view, there is a loss of total energy, but from another point of view, there is no loss of total energy. This seemingly contradiction result comes from the essential difference of consideration between adopting a model of sources and sinks and adopting the actual three-dimensional flow field. The truth is that there is no loss of total energy, even from the first point of view, which is explained in the following.

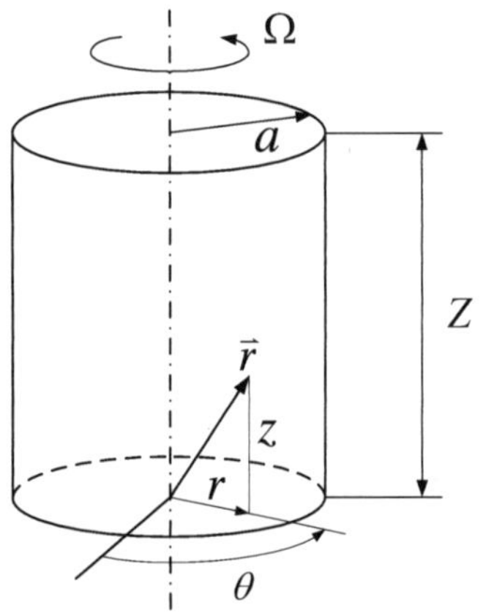

Figure 1. The cylindrical coordinate system used in the analysis

The total energy of the gas should not change, now matter whether or not a scoop does do work to the gas of the incoming flow. The resistance of the scoop to the gas moving at a very high speed produces complicated shock waves, through which the gas entropy increases. Meanwhile, near the surface of the scoop a large velocity gradient is produced when the gas moves around. All these convert the mechanical energy of the gas into the internal energy, causing both the temperature and pressure to increase. But with the model of momentum sources and sinks under the two-dimensional axisymmetric approximation, the resistance of the scoop is simulated by a momentum sink, and the local velocity $\mathbf{V}$ is not zero, producing an energy sink $\mathbf{F}\cdot\mathbf{V}$ (<0), that is, the scoop does negative work to the gas, reducing the total energy of the gas. Furthermore, because the two-dimensional approximation is unable to simulate the phenomena of shock wave, the process of the energy conversion from mechanical energy into internal energy cannot be taken into account and, consequently, the heating effect of the scoop cannot be revealed by the change of the internal energy. Thus, to simulate the heating effect of the scoops, an energy source that is equal to the energy sink in absolute value must be introduced into equation (3), compensating the loss of mechanical energy due to the work done through the momentum sink $\mathbf{F}$ by the incoming flow. That is, when using the model of momentum sources and sinks, an energy source $\dot{q}$ should be provided in the enegy equation, whose relationsip with the momentum sink $\mathbf{F}$ satisfies the following:

$$\dot{q} = -\mathbf{F}\cdot\mathbf{V} \tag{4}$$

The above consideration was mentioned in some of the previous publications [3,4], but the reasons why so considering were not particularly emphasized or analyzed.

The N-S equation system is discretized on a staggered grid by means of the well-known hybrid scheme, which combines a central difference scheme of second order accuracy with an upwind scheme of first order accuracy. In the solution of the resulting nonlinear algebraic system, a homotopic method [5] is used to cope with convergence problems in presence of strong nonlinearity and achieve a better computational efficiency. Since the focus of this paper is on the study of the scoop heating effect, the numerical issues are not discussed here and can be found in [6].

2.2. Model of rotating disk

The model of rotating disk employs a disk in the area of a scoop, which is rotating at a lower angular speed than the rotor, to exert resistance to the surrounding gas and form the so-called Ekman layer on the surface of the disk. The mechanical drive is then generated. Assume that the rotating angular speed of the rotor is $\Omega\mathbf{e}_z$, and the lower angular speed of the rotating disk is $(\Omega-\Delta\Omega)\mathbf{e}_z$. The disk yields a moment of resistance $-M\mathbf{e}_z$ to the motion of the gas, and hence a power consumption (breaking power):

$$P_{\text{scoop}} = -M(\Omega - \Delta\Omega) \quad (< 0) \tag{5}$$

It is clear that, the same as with the model of sources and sinks, the work done by the disk is negative, causing the mechanical energy of the rotating gas to reduce and to lose its total energy. M is to be calculated through numerical integration, based on the friction exerted on the surface of the disk in the θ-direction.

If the loss of the total energy of the gas is not compensated, implying the internal energy of the gas surrounding the disk is consumed by the disk, both the temperature and pressure drop. Obviously, this is contradictory to the real situation, as in the case of the model of sources and sinks. In fact, the rotating disk is heated up by the consumed energy of the gas, and then the heated disk in turn returns the heat back to the gas. In this process the mechanical energy of the gas transforms into internal energy. Therefore, some compensation is necessary. The heating effect can be modelled by means of adjusting the temperature and its distribution on the disk. However, it is considerably less convenient to introduce the energy compensation in the model of rotating disk than in the model of sources and sinks, as the range and distribution of the temperature cannot be conveniently and appropriately set up in advance to precisely compensate the loss of the mechanical energy of the gas.

Furthermore, a drawback of using the model of rotating disk is that the flows above and below the disk is split into two areas, which is very different from the actual situation inside a real centrifuge, where the gas can flow freely up and down, except for the area occupied by the scoop (figure 2(a)). With the model of rotating disk, the upward and downward flows of the gas in the plane of the disk (shown as the grey area) is completely prohibited, as if the disk is treated as a solid wall without permeability (figure 2(b)). There is no any image of the gas flow available around the area of the disk and, consequently, it is very difficult to build up a model to simulate the upward and downward flow across the plane.

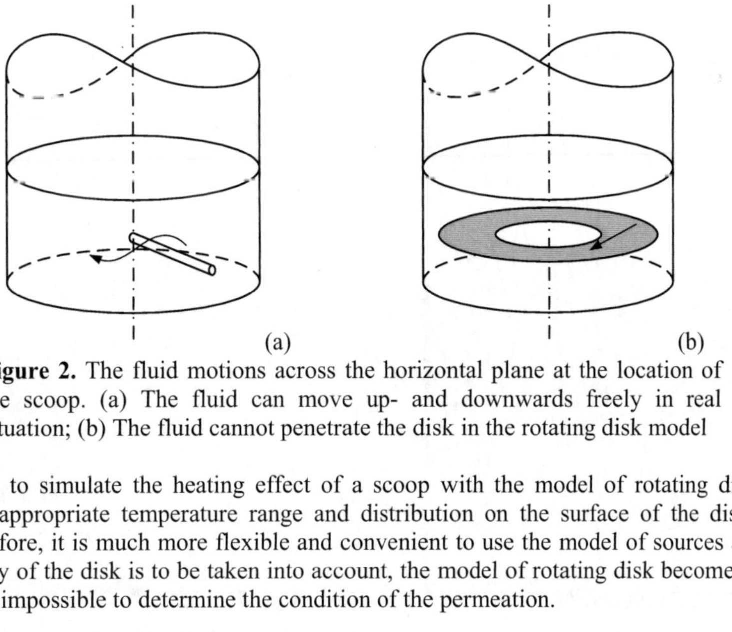

Figure 2. The fluid motions across the horizontal plane at the location of the scoop. (a) The fluid can move up- and downwards freely in real situation; (b) The fluid cannot penetrate the disk in the rotating disk model

To sum up, to simulate the heating effect of a scoop with the model of rotating disk needs to determine the appropriate temperature range and distribution on the surface of the disk, which is difficult. Therefore, it is much more flexible and convenient to use the model of sources and sinks. If the permeability of the disk is to be taken into account, the model of rotating disk becomes even more difficult, being impossible to determine the condition of the permeation.

3. Simulation results and discussions

To reveal the heat effect of a scoop on the flow field and separative performance of a centrifuge, the well-known Iguaçu centrifuge is chosen, using the above two different models of scoop simulation. Tab.1 presents the major parameters of the Iguaçu cemtrifuge. The process gas is natural UF_6.

Table 1. Major parameters of the Iguaçu centrifuge.

Parameter	Value
radius a (m)	0.06
length Z (m)	0.48
peripheral speed Ωa (m/s)	600
average gas temperature T_0 (K)	300
pressure at the rotor wall (Torr)	100
top-bottom temperature difference on the rotor wall ΔT (K)	6
feed F (mg/s)	12
cut θ	0.5
feed concentration C_F (%)	0.712

3.1. Results from the model of sources and sinks

The strength of a momentum source or sink is determined through a scoop drag coefficient to adjust the drag of the scoop [3,5,7]:

$$\iint_{\Xi} 2\pi r F_\theta drdz = -\frac{C_D}{2}\iint_{\Xi} \rho v^2 drdz \tag{6}$$

where v is the velocity of the incoming flow in the θ-direction, and Ξ denotes the area occupied by the momentum source or sink.

Figure 3 illustrates the streamlines for $C_D = 0.8$ in the two cases with the scoop heating effect ($\dot{q} = -\mathbf{F}\cdot\mathbf{V}$, Figure 3(a)) and without the heating effect ($\dot{q} = 0$, Figure 3(b)) taken into account. One clearly sees that for the same drag coefficient, the flow field with the heating effect distinguishes greatly from that without the heating effect. The circulation area in the former case is wider, which results from the additional drive caused by the gas heated by the scoop.

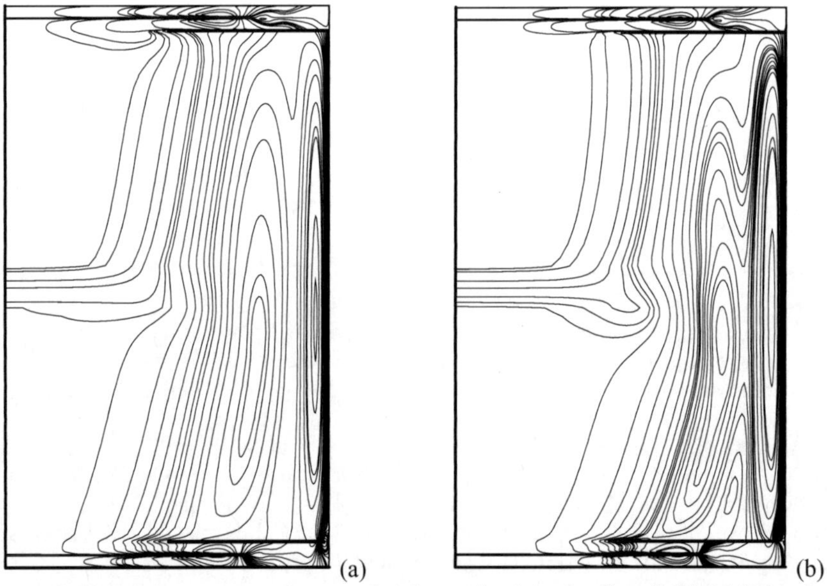

Figure 3. Using momentum sources and sinks to simulate the flow field. (a) With the scoop heating effect; (b) Without the scoop heating effect

Figure 4 shows the contour lines in the product and waste chambers with the heating effect; Figure

5 presents also the contour lines in the two chambers but without the heating effect. Clearly, in figure 4, the temperature in the adjacent areas of the sinks arises (the red areas), which is in accordance with the understanding of the physical process and verifies the provision of an energy source can truly reflect the scoop heating effect to the gas. However, in figure 5, the temperature in the adjacent areas of the sinks drops (the blue areas), which is due to the internal energy loss of the gas resulting from the work done by the gas, as pointed out in Section 2, and is not in conformity with the observation of the actual physical process inside a centrifuge.

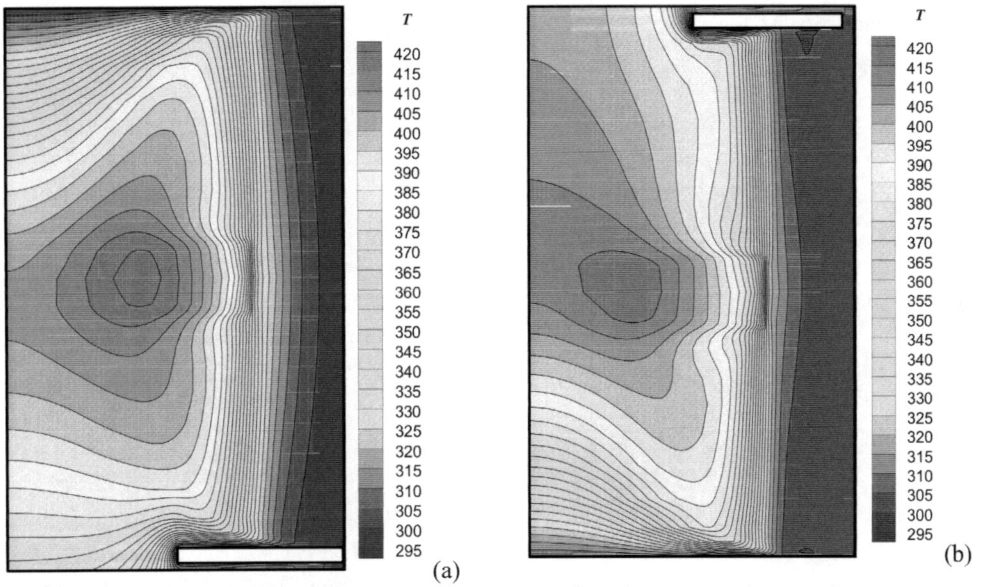

Figure 4. Gas temperature distributions with the scoop heating effect considered. (a) The product chamber; (b) The waste chamber

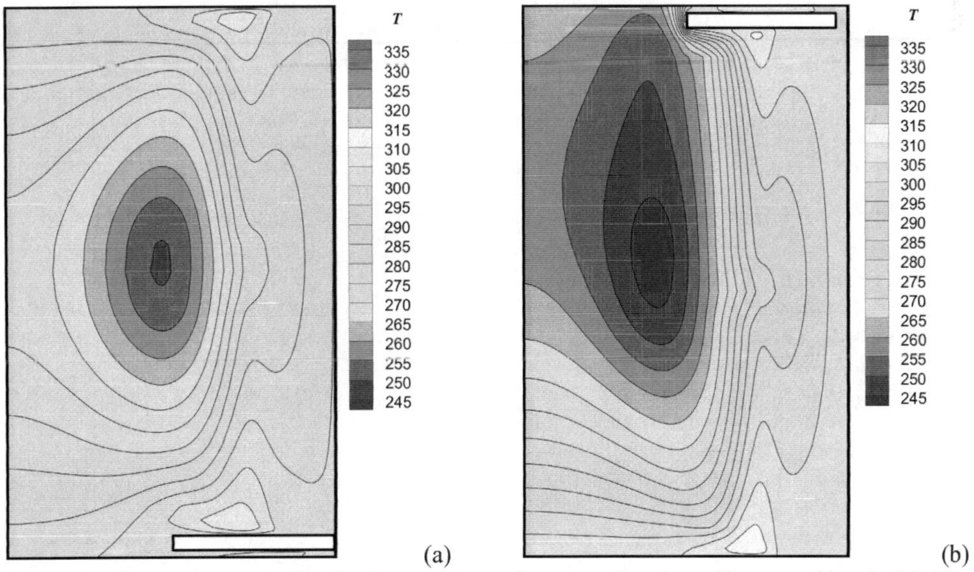

Figure 5. Gas temperature distributions without the scoop heating effect considered. (a) The product chamber; (b) The waste chamber

3.2. Results from the model of rotating disk

Now we use the model of rotating disk to replace the model of sources and sinks for the simulation of the mechanical drive. The strength of the drive is represented by the ratio of the angular speed difference $\Delta\Omega/\Omega$ between the rotor and the disk to the angular speed of the rotor. The temperature on the surface of the disk is set up to be the same as that at the rotor with the same axial location.

The temperature distributions in the product and waste chambers, respectively, are shown in Fig.6 for the driving strength $\Delta\Omega/\Omega = 20\%$.

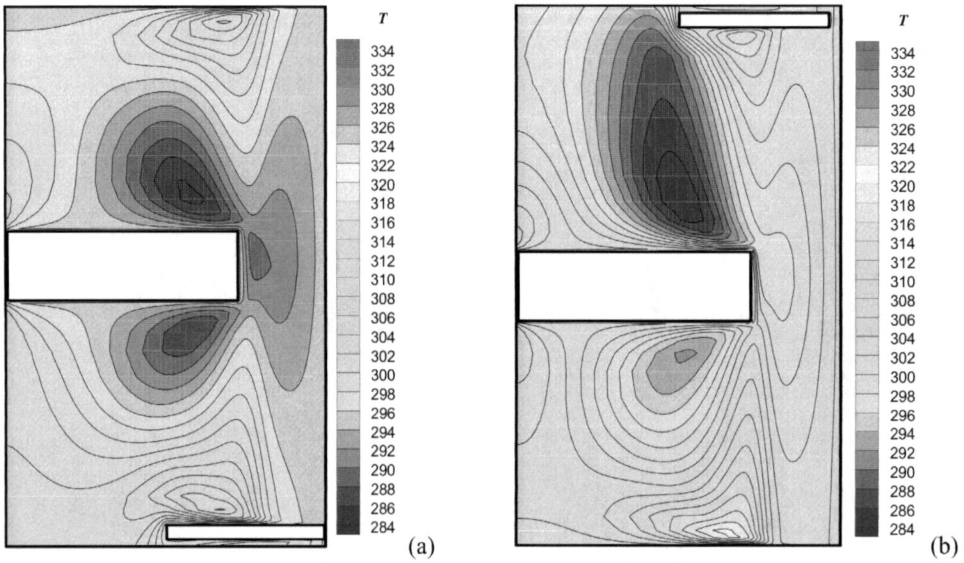

Figure 6. Gas temperature distributions when using the rotating disk model. (a) The product chamber; (b) The waste chamber

One can see that temperature drop happens in the gas surrounding the rotating disk in the withdrawal chambers (the blue areas), which conforms with the results in Section 2 and explains that the model of rotating disk can similarly result in the loss of the total energy of the gas and reduction of the internal energy. Therefore, this model is unable to simulate the scoop heating effect appropriately. The figures with the heating effect taken into account are not available, because, as pointed out previously, it is inconvenient to provide appropriate energy compensation.

3.3. Influences of models on separative performance

The additional drive from the scoop heating effect would yield different simulation results of separation performance if using different scoop models.

The variations of the separative power as a function of the scoop drag coefficient are compared in figure 7 for two cases corresponding to having the scoop heating effect considered or not in the model of sources and sinks. It is demonstrated that while maintaining other conditions unchanged, with the same drag coefficient C_D, taking into account the heating effect leads to a smaller scoop resistance at the maximum separative power. This is desirable in practice, because a smaller resistance means lower power consumption and also lower specific energy consumption. It is worth pointing out that, generally speaking, the same C_D does not mean the same resistance, the reason lying in the same C_D but different flow fields.

Figure 8 gives the breaking powers as a function of the drag coefficient in the model of sources and sinks. Here the breaking power P_{rotor} is calculated by integrating numerically the friction between the gas and the inner surface of the rotor, roughly indicating the strength of the scoop drags. In fact, the inner surface of the rotor drives the motion of the gas, whereas the scoops impede the motion. At the equilibrium state of the gas, the moment of driving the gas should be equal to that of impeding the gas, and has little to do with the temperature. As seen from figure 8, whether or not counting in the scoop heating effect does not have an obvious influence on the breaking powers. The simulation results presented in figure 7 and figure 8 give us a further proof that using the model of sources and sinks, with the scoop heating effect considered, the same drag coefficient would generate very different flow fields.

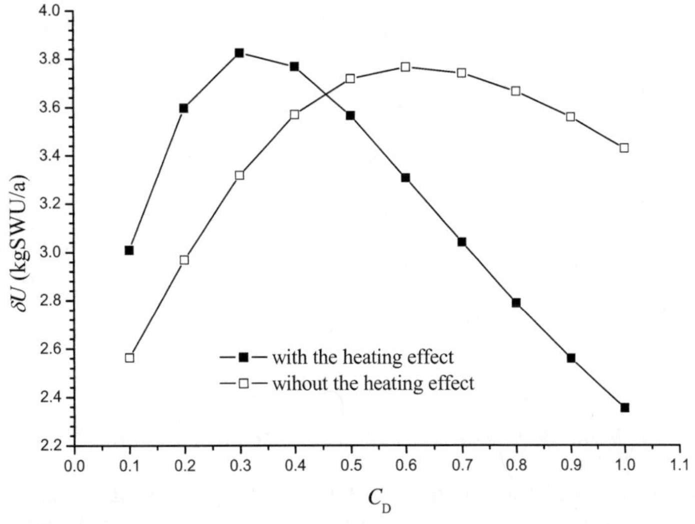

Figure 7. The frictional power consumption vs the drag coefficient

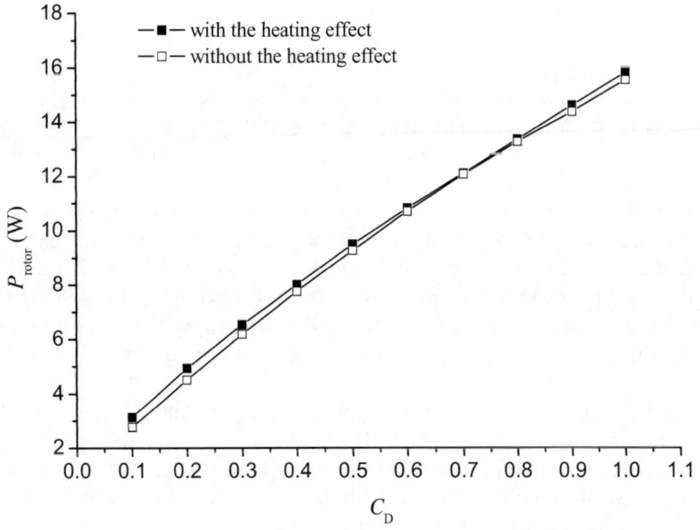

Figure 8. The separative power vs the drag coefficient

Figure 9 shows the variations of the separative power with the breaking power in the two cases using the model of sources and sinks and one case using the model of rotating disk. One sees that the variation tendency of δU with P_{rotor} in the model of rotating disk is close to that of the case without the heating effect in the model of sources and sinks. Therefore, with the same breaking power at a small drag, δU obtained from the former model is almost the same as that from the latter model. At a large drag, the values of δU obtained from the two models deviate from each other significantly.

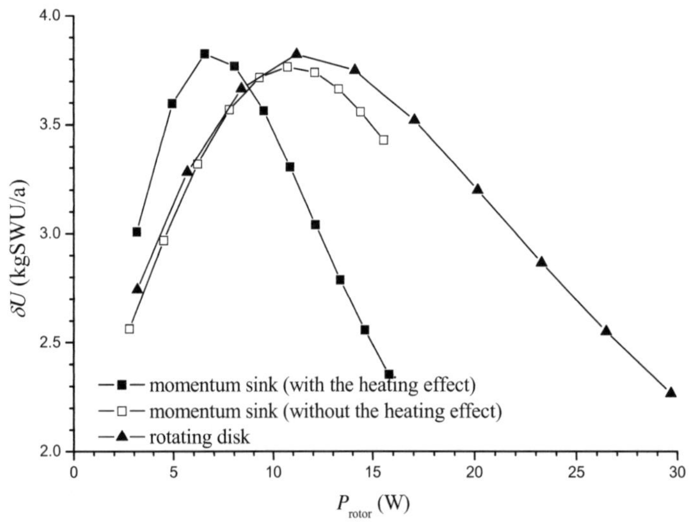

Figure 9. The separative power vs the breaking power

So far, the simulation results tell us that the heating effect has a significant influence on the separative power. Different treatments of the heating effect lead to different relationships of the measurable quantities of a centrifuge such as the separative power and breaking power with model parameters, which should be paid attention in determining model parameters in combination with experimental results.

4. Conclusions

The resistance of the withdrawal scoops causes the loss of the mechanical energy of the gas, which is transported to the gas in the form of heat energy. The overall effect is that the total energy does not lose in spite of the loss of the mechanical energy. In the investigation of the flow field inside the rotor of a gas centrifuge with the two-dimensional approximation, because differences exist between the simulation models of the scoops and the actual three-dimensional situation, the scoop heating effect resulting from the energy transformation taking place in the gas of a centrifuge cannot be accurately reflected, causing the temperature around the scoops to drop, which is not consistent with the phenomena in the actual physical process. In approximating the scoop effect with the two models, the model of sources and sinks and the model of rotating disk, appropriate source term and temperature distribution on the rotating disk, respectively, need be introduced to correctly simulate the heating effect.

Appropriately taking into account the scoop heating effect in simulation is crucial in the correct analysis of the separation performance of a centrifuge. The scoops is heated by the gas and then heat up the gas, which changes the temperature and pressure distributions of the gas and consequently generates a very different flow field from that without the heating effect considered. Therefore, the heating effect has a significant impact on the separative power.

Aknowledgement

This research is supported by Natural Science Foundation of China (Grant No. 11575097), the National Key Basic Research and Development Program of China (Grant No. 2014CB744100), and the Sino-Russian joint project from National Natural Science Foundation of China (Grant No. 1151101155) and Russian Fund for Basic Research (Grant No. 16-58-53058 GFEN_a).

IOP Conf. Series: Journal of Physics: Conf. Series **1099** (2018) 012012 doi:10.1088/1742-6596/1099/1/012012

References

[1] Cenedese A and Cunsolo D 1980 Scoop optimal shape in a gas ultracentrifuge. *Meccanica* **15**(1) 39–46
[2] Wood H G and Sanders G 1983 Rotating compressible flows with internal sources and sinks. *J. Fluid Mech.* **127** 299–313
[3] Omnes P 2007 Numerical and physical comparisons of two models of a gas centrifuge *Computers & Fluids* **36**(6) 1028–1039
[4] Zhang Y N and Zeng S 2016 Homotopic method applied to solving the flow field in a gas centrifuge *Journal of Physics: Conference Series* **751** (IOP Publishing) 012015
[5] Borisevich V D and Levin E V 2001 Separation of multicomponent isotope mixtures by gas centrifuge. *Sep. Sci. Technol.* **36**(8-9) 1697–1735
[6] Zhang Y N, Zeng S, Jiang D J, Borisevich V D, Smirnov A Yu and Sulaberidze G A 2018 On the static hydraulic properties of gas centrifuges *Proc. 14th Int. Workshop on Separation Phenomena in Liquids and Gases* (Stresa, Italy), to appear
[7] Kai T 1983 Numerical analysis of flow of binary gas mixture with large mass difference in rotating cylinder *J. Nucl. Sci. Technol.* **20**(4) 339–351

IOP Conf. Series: Journal of Physics: Conf. Series **1099** (2018) 012013 doi:10.1088/1742-6596/1099/1/012013

Analytical and numerical solutions of the shock tube problem in a channel with a pseudo-perforated wall

V A Shargatov[1], A P Chugainova[2], S V Gorkunov[1] and S I Sumskoi[1]

[1] National Research Nuclear University "MEPHI",Kashirskoye sh. 31,115409 Moscow, Russia

[2] Steklov Mathematical Institute, Gubkina str. 8, 119991 Moscow, Russia

E-mail: shargatov@mail.ru

Abstract. An approximate analytical solution is obtained for the shock tube problem in a rectangular channel with an array of rectangular grooves in the lower wall. The analytical solution is based on the approximate quasi-1D shock adiabat for a shock wave that propagates in a channel with periodically located barriers. This problem is also studied numerically. It is found that the approximate analytical solution correctly predicts the propagation velocity of the leading discontinuity and the flow parameters at this discontinuity. The leading gas-dynamic discontinuity is followed by a relaxation zone in which there are long-wave and short-wave flow oscillations. These oscillations are caused by waves arising from the interaction of the flow behind the leading shock wave with the side and bottom walls of the grooves. An approximate analytical solution allows one to obtain the values of the flow parameters at the end of the relaxation zone. These values are in good agreement with the numerical calculation.

1. Introduction

The problem of the propagation of a shock wave in a channel with obstacles is of great interest. This problem is important for the study of various processes in the chemical industry and nuclear power. Attenuation of a shock wave in a channel with obstacles is investigated in problems of industrial safety. A weak shock wave is generated when modern high-speed trains are entering in the tunnel [1]. Obstacles can be created specifically for attenuation of the shock wave or for rapid development of turbulence. To attenuate the shock wave, in addition, channels with perforated or pseudo-perforated walls are used. The attenuation of the shock wave in the channel filled with foams, grids and dust suspensions was investigated in recent decades. An overview of various ways of a shock wave attenuation in a channel is given in [2, 3].

Approximate analytical models for describing the process of attenuation of a shock wave in a channel with perforated (having through holes) walls were developed in [4, 5]. If the hole in the wall is closed from one end or has a closed cavity behind it, then such walls are called pseudo-perforated. The attenuation of a shock wave in a channel with pseudo-perforated walls was studied in [6]. In [7], an oblique interaction of a shock wave with a perforated plate was investigated. The methods of approximate calculation of the velocity of the leading shock wave in perforated channels and in channels with barriers and screens are described in [8, 9, 10, 11]. Obstacles can also arise as a result of instability of the flow, if the walls of the channel are deformable (see, for example, [12]).

The influence of the shape of the obstacles on the attenuation of the shock wave was studied in [13] by numerical simulation. The interaction of a shock wave with porous obstacles of various

IOP Conf. Series: Journal of Physics: Conf. Series **1099** (2018) 012013 doi:10.1088/1742-6596/1099/1/012013

shapes was investigated in [14]. In [15], experimental investigations of the influence of the shape and type of the barrier on the parameters of the flow behind the barrier during the propagation of the shock wave were performed. The interaction of a shock wave with rigid barriers was studied in [16] both numerically and experimentally.

In [17] the results of numerical and experimental studies of the interaction of a shock wave with a plane barrier are presented. In [1, 18, 19, 20, 21] studies of the interaction of shock waves with various types of obstacles have been carried out. In [18, 20] the barriers were equally spaced and parallelepipedic; the interaction of a wave with fourteen barriers was considered. Experiments on the interaction of a shock wave with obstacles of exactly the same "comb" type are described in [2, 22].

The results of the above studies dealt mainly with the specifics of the initial interaction of the shock wave with a small number of obstacles.

The present work is devoted to the study of the propagation of shock waves in a channel with grooves (a pseudo-perforated wall). The grooves have the same size and shape and are located at equal distances from each other. Such periodically located grooves are called regular or cascade [11]. We consider both the flow at a great distance from the leading gas-dynamic discontinuity, and the local flow structure in the region where the flow is determined primarily by the parameters of this gas-dynamic discontinuity.

The presence of a relatively narrow zone in which a nonmonotonic change in the flow parameters occurs to a certain constant value is characteristic of the structure of discontinuities in the presence of dispersion and dissipation [23, 27, 29, 24, 25, 26, 28, 30]. In this case, the problem of calculating the flow corresponding to the structure of the discontinuity may not have a stationary solution in the form of a traveling wave, but a solution that periodically depends on time, as shown in [25, 26, 28]. If the scale at which the flow parameters change as a whole is much larger than the width of the shock wave structure, then the average velocity and width of the structure can be considered constant during the time interval over which it travels over a distance equal to its width. In this case, it is possible to obtain relationships connecting the parameters behind the structure of the discontinuity with the structure velocity. Questions concerning the uniqueness of such a relationship are discussed in detail in [23, 27, 29, 24, 25, 26, 28, 30].

Note that discontinuities whose structure is not stationary (not a traveling wave), but time-dependent periodically, are considered, according to [28], to be admissible discontinuities.

In the previous paper, [32], we analyzed the possibility of the formation of slowly changing (with time) quasistationary wave structures, which consist of a gas-dynamic discontinuity, in which the parameters of the gas change abruptly, and the zone of the nonequilibrium flow following it, in which the pressure is relaxed to the new equilibrium value. In this case, the instantaneous velocity of the leading discontinuity may not be constant and vary periodically due to the periodic location of obstacles, but the average velocity changes on scales of a much larger scale of one obstacle, and uniquely determines the value of the pressure at the end of the relaxation zone. It was found in [32] that, from the point of view of large spatial scales, a complex consisting of a leading gas-dynamic discontinuity and a relaxation zone can be considered as a discontinuity with the structure, and an effective shock adiabat for such a discontinuity is obtained.

2. Statement of the problem

The computational model is shown in Figure 1. We consider a rectangular channel. The channel has a height of H. The channel consists of two sections. On the right side of the channel, for $x_0 < x < \infty$, there are grooves in the lower channel wall. The width of the rectangular grooves is l, the height is h, the distance between the grooves is L. In the left part of the channel, as $-\infty < x < x_0$, the wall is smooth.

At the initial instant of time, for $x < 0$, the gas has parameters P_3, $v_3 = 0$ and ρ_3 and the

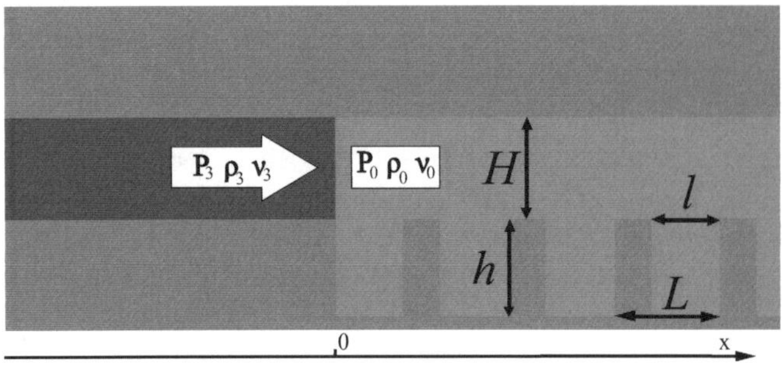

Figure 1. Geometry of the problem.

parameters P_0 , $v_0 = 0$, ρ_0 for $x \geq 0$. Here P is the pressure, v is the velocity, and ρ is the gas density. The normal component of the velocity of the gas is zero at the gas-channel-wall and gas-groove-wall boundaries . Adhesion on the walls is absent.

Similarly [20, 32], the effects of viscosity, the heat diffusion, and body force is neglected. We assume that the gas behaves like an perfect gas and $\gamma = 1.4$ is the gas specific heat ratio. So the basic equations governing the propagation of a shock wave along a pseudo-perforated tube are as follows

$$\frac{\partial}{\partial t}\begin{bmatrix}\rho \\ \rho v_i \\ \rho E\end{bmatrix} + \frac{\partial}{\partial x_j}\begin{bmatrix}\rho v_j \\ \rho v_i v_j \\ (\rho E + P) v_j\end{bmatrix} = 0, \qquad E = e + \frac{|v|^2}{2}, \qquad e = \frac{P}{(\gamma - 1)\rho}, \tag{1}$$

where E is the total specific energy of the gas and e is the specific internal energy.

The gas velocity has only two nonzero components (longitudinal v_x and vertical v_z), since the groove width is equal to the channel width.

3. Approximate analytical solution of shock tube problem

In our previous paper [32] we have demonstrated that the shock wave propagating in the right part of the channel represents a discontinuity with the structure.

According to [32], behind the leading gas-dynamic discontinuity there is a relaxation zone in which the flow parameters change rapidly and nonmonotonically. After some time after the passage of the leading wave, the gas velocity in the grooves is close to zero and the gas moves only in the region above the grooves. The resulting tangential velocity discontinuity is unstable, but the development of this instability occurs in a much longer time than that corresponding to the relaxation zone.

Behind the zone of relaxation is a zone in which flow parameters change slowly. A diagram of the simplifying separation of the flow into zones and regions is shown in Fig. 2. Averaging of the gas parameters over grooves in volumes equal to hl is performed, and the parameters of the gas in the space inside the grooves are over the corresponding volumes HL. For the average values of the gas parameters at the end of the relaxation zone, we use the index 1 if the gas is outside the groove and the index is 2 if the gas is inside the groove.

It is assumed that $P_1 = P_2$, $v_2 = 0$, $\rho_1 \neq \rho_2$.

If the wave has traveled a distance that is much greater than the width of the structure, then on the shock wave for the flow averaged over the cross-section the relations obtained in [32] are

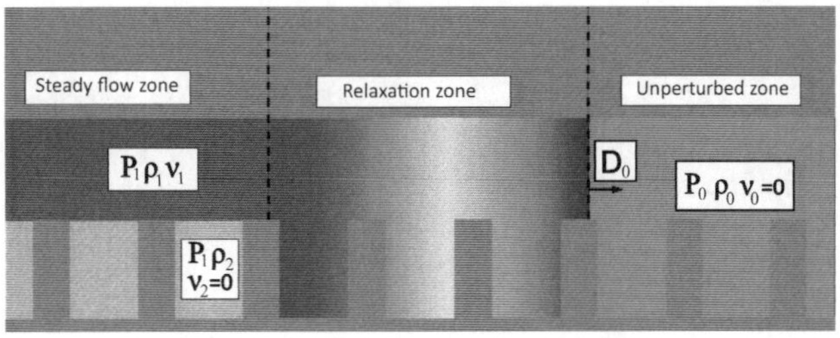

Figure 2. Division of the flow into zones used to obtain approximate formulas for the discontinuity structure.

satisfied proceeding from the conservation laws and physically grounded assumptions about the nature of the flow in the relaxation zone behind the shock discontinuity. These relations have the form

$$\begin{cases} \rho_0 D_0 + \alpha\rho_0 D_0 + v_1\rho_1 - \rho_1 D_0 - \alpha\rho_2 D_0 = 0 \\ \rho_1 v_1 D_0 + \alpha v_1(\rho_2 - \rho_0)D_0 + P_0 - \rho_1 v_1^2 - P_1 = 0 \\ \frac{(1+\alpha)(P_1-P_0)}{\gamma-1} D_0 - \frac{\rho_1 v_1^2}{2}(v_1 - D_0) - \frac{\gamma P_1 v_1}{\gamma-1} = 0 \\ \rho_2 = \rho_0 + (1 - \frac{\rho_0}{\rho_1})P_1\left(\frac{P_1}{\rho_1} + \frac{(\gamma-1)v_1^2}{2}\right)^{-1} \end{cases} \tag{2}$$

where $\alpha = hl/(HL)$.

The system (2) consists of four equations for five unknown parameters D, P_1, v_1, ρ_1 and ρ_2. Hence, the system of algebraic equations implicitly determines the dependence of P_1 on v_1. This dependence is an analogue of the effective shock adiabat for a shock wave with a structure propagating in a channel with pseudo-perforated walls.

A rarefaction wave propagates to the left along the smooth part of the channel. The pressure and velocity of the gas in this wave are related by the well-known relation

$$v_1 = \frac{2\sqrt{(\gamma P_3}}{(\gamma-1)\sqrt{\rho_3}}\left[1 - \left(\frac{P_1}{P_3}\right)^{\chi}\right], \ \chi = \frac{\gamma-1}{2\gamma} \tag{3}$$

Parameters of the flow P_1, ρ_1, ρ_2, v_1, D can be found by solving five algebraic equations (2) (3). Let $H = 10$ m, $L = 2$ m, $h = 10$ m and $l = 0.8$ m. For $P_0 = 100$ kPa, $\rho_0 = 1.17$ kg/m ^{3}, $P_3 = 1$ MPa and $\rho_3 = 11.7$ kg/m^3 the system of equations (2) and (3) has the solution $P_1 = 235$ kPa, $v_1 = 290$ m/s, $\rho_1 = 2.23$ kg/m^3, $\rho_2 = 2.10$ kg/m^3 and $D = 451$ m/s.

The pressure profiles at the time $t = 0.96$ s are shown in Figure 3 and Figure 4 by lines 4 and 5. If the right-hand side of the channel is smooth, then the shock wave has parameters $P_s = 285$ kPa, $v_s = 284$ m/c $\rho_s = 2.04$ kg/m^3 and $D_s = 556$ m/s. Consequently, in the case under consideration, the presence of the grooves leads to a decrease in the average velocity of the shock wave by 19% and the overpressure decreases by 17%.

4. Calculation results

The system of equations (1) was solved by the Godunov-Kolgan method [33]. A exact solution of the Riemann problem was used to calculate the fluxes across the boundaries of the cell. A structured grid was used. The cells of the difference grid are parallelepipeds of the same size. The calculation was carried out using the three-dimensional gas-dynamics code Avangard / SafetyFor [32] developed in NRNU MEPhI. The grid contains 12800x6x100 cells. The cell sizes

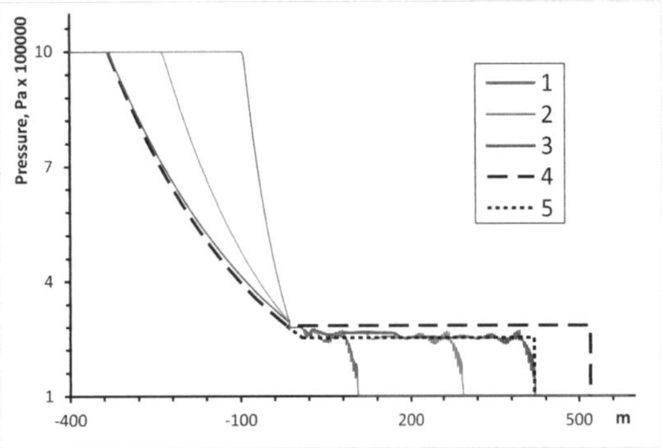

Figure 3. Numerical (1-3) and analytical (4,5) average pressure distributions at time t. 1 - t = 0.28 s, 2 - 0.69 s 3,4,5 - 0.96 s. Line 4 - analytical solution for a channel without perforation, line 5 - an approximate analytical solution for a channel with a pseudo-perforated wall.

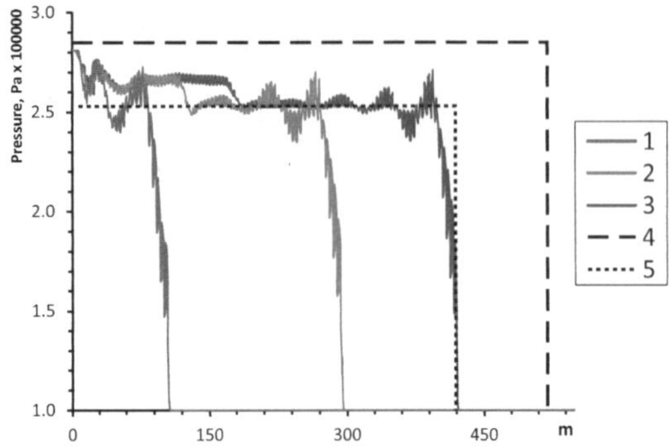

Figure 4. b Numerical (1-3) and analytical (4,5) average pressure distributions at time t. 1 - t = 0.28 s, 2 - 0.69 s 3,4,5 - 0.96 s. Line 4 - analytical solution for a channel without perforation, line 5 - an approximate analytical solution for a channel with a pseudo-perforated wall.

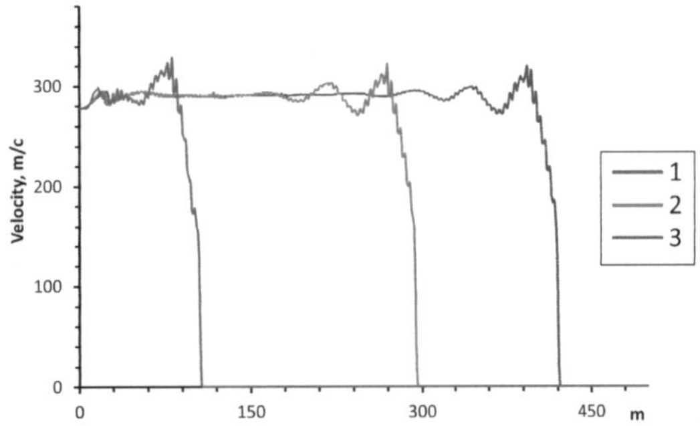

Figure 5. Numerical average velocity distributions at time t. 1 - $t = 0.28$ s, 2 - 0.69 s 3 - 0.96 s.

in dimensionless units $\Delta x/L$, $\Delta y/L$, $\Delta z/H$ are 0.05, 0.05 and 0.02, respectively. The time step was determined atomically for $CFL = 0.1$, which certainly provides the necessary accuracy and stability of the calculation results for the Godunov-Kolgan method.

Figure 3 and Figure 4 shows the distribution of the average pressure $\overline{P}$ as a function of the coordinate x, and the average velocity profile $\overline{v}$ is presented in Figure 5. The average pressure is calculated over the entire section of the channel, including the grooves, according to the formula

$$\overline{P}(x,t) = \int_{-h}^{H} P(x,z,t)dz/(H+h),$$

and the average velocity is calculated only in the part of the channel that is above the grooves:

$$\overline{v}_x(x,t) = \int_{0}^{H} v_x(x,z,t)dz/H.$$

The pressure and velocity profiles are shown for the times 0.28 s, 0.69 s and 0.96 s, when the shock wave passed 54, 147 and 270 grooves, respectively.

The rightmost characteristic of the centered rarefaction wave runs along the smooth part of the tube with a low velocity of 5 m/s, therefore for $x = x_0$ the pressure varies insignificantly.

A shock wave propagates to the right. This shock wave consists of a leading gas-dynamic discontinuity and a relaxation zone in which long-wavelength oscillations are damped (see Figure 4). The parameters on the leading gas-dynamic discontinuity and the structure of the flow behind it vary little over time. The pressure at the end of the relaxation zone remains constant. This assumption was used in [32] in deriving the equations (2). The results of gas dynamic calculations confirm this assumption.

The flow parameters averaged over the channel cross section have long-wave and short-wave oscillations. These oscillations are not introduced numerically, but are inherent in the solution of the system (1). The amplitude of the oscillations decays with distance from the leading gas-dynamic discontinuity. In the relaxation zone, long-wave pressure oscillations occur near the mean value, which agrees well with the value of P_1 (line 5, in Figure 4) obtained using the analytical approach described in Section 3.

The appearance of long-wavelength and short-wave oscillations of the flow parameters in the relaxation zone is associated with reflection of the waves from the lower and lateral walls of the groove. Figure 6 shows the pressure field at time $t = 0.96$ s. It can be seen that there are compression and rarefaction waves in the gas behind the leading discontinuity. Shortwave perturbations arise due to the interaction of the leading shock with the side surface of the groove, and long-wave oscillations are caused by the circulation of compression and rarefaction waves between the lower boundary of the groove and the upper channel wall. The first maximum of pressure is caused by the leading gas-dynamic discontinuity propagating along the upper part of the channel. The average speed of this discontinuity is 451 m/s. The pressure in a plane shock wave propagating at this velocity is 182 kPa and agrees well with the first maximum of pressure on Figure 4. Then the pressure increases as a result of the arrival of waves arising from the interaction of the leading discontinuity with the walls of the grooves.

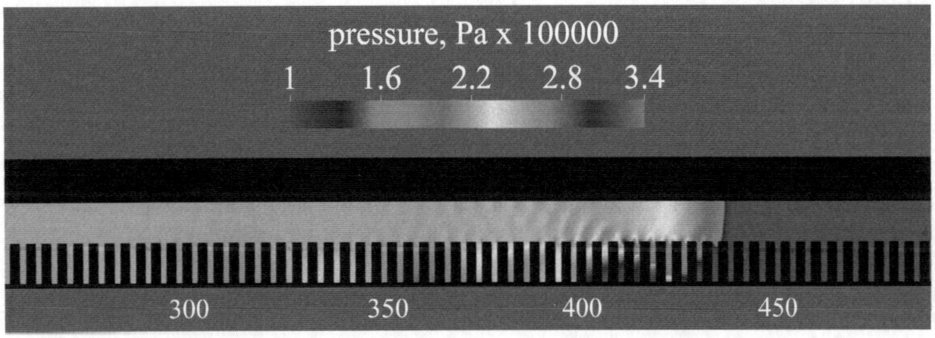

Figure 6. Pressure field at time $t = 0.96$ s.

5. Conclusions

An analytical solution is obtained for the shock tube problem for the decay of the initial pressure and density discontinuity in a rectangular channel with grooves in the wall. The analytical solution is based on the effective shock adiabat for a channel with periodically located barriers [32]. This problem is also solved by numerical modeling. It is shown that the approximate analytical solution correctly predicts the propagation velocity of the leading discontinuity and the flow parameters at this discontinuity. The leading gas-dynamic discontinuity is followed by a relaxation zone in which there are long-wave and short-wave flow oscillations. These oscillations are caused by waves arising from the interaction of the flow behind the leading shock wave with the side and bottom walls of the groove. An approximate analytical solution allows one to obtain the values of the flow parameters at the end of the relaxation zone. These values are in good agreement with the numerical calculation.

Acknowledgments

This work was supported by the Russian Science Foundation, project no. 16-19-00188.

References

[1] *Abe A., Takayama K., Itoh K.*. Experimental and numerical study of shock wave propagation over cylinders and spheres// WIT Transactions on Modelling and Simulation.2001 P. 10
[2] *Ben-Dor G., Igra O., Elperin T.*: Handbook of Shock Waves. Academic Press, 2000, P. 320-392
[3] *Igra, O., Falcovitz, J., Houas, L., Jourdan, J.*. Review of methods to attenuate shock/blast waves.//Progress in Aerospace Sciences. 2013. V.58, P. 135.
[4] *Szumowski A.P.* Attenuation of a shock wave along a perforated tube.// Shock Tube Research, Proc 8th Int Symp Shock Tubes./Ed. by J.L. Stollery , A.G. Gaydon, P.R. Wen Chapman & Hall. 1971. P.14.
[5] *Wu J.H.T., Ostrowski P.P* Shock attenuation in a perforated duct.// Shock Tube Research, Proc 8th Int Symp Shock Tubes./Ed. by J.L. Stollery , A.G. Gaydon, P.R. Wen Chapman & Hall. 1971. P.15
[6] *Sasoh A., Matsuoka K., Nakashio K., Timofeev E., Takayama K., Voinovich P., Saito T., Hirano S., Ono S., Makino Y.* Attenuation of weak shock waves along pseudo-perforated walls// Shock Waves. 1998. V.8. P.149159.
[7] *Skews B.* Shock Wave Interaction with Porous Plates// 2005. Exp. Fluids. 2005. V.39. N 5., P.875884.
[8] *Forolov S.M., Gelfand B.E., Medvedev S.P, Tsyganov S.A.* Quenching of shock Waves by barriers and screens// 17th hat Syrup Shock Waves Tubes/Ed. by Y.W. Kim. Bethlehem USA. 1989. P.314-320 (1989)
[9] *Frolov S.M., Gelfand B.E.* Shock wave attenuation in partially confined channels //Shock Waves. 1992. V.2, P.97101.
[10] *Frolov S.M.* Effectiveness of attenuating shock waves in channels by various methods// Journal of Applied Mechanics and Technical Physics. 1993. Vol. 34. N 1. P. 31-36.
[11] *Gelfand B.E., Frolov S.M., Medvedev S.P.* Measurement and computation of shock-wave attenuation in a rough pipe// Combustion Explosion and Shock Waves. 1990. Vol. 26. N 3. P. 335-338.
[12] *Il'ichev A.T.* Stability of solitary waves in membrane tubes: A weakly nonlinear analysis// Theor. Math. Phys. 2017. V.193. N 2. P. 1593-1601.
[13] *Sadot O., Ben-Dor G., Hadjadj A.* Experimental and Numerical Investigations of Shock Wave Attenuation over Obstacles// Book of Proceedings of ISSW-27. 2009. P.327.
[14] *Britan, A., Igra, O., Ben-Dor, G., Shapiro, H.* Shock wave attenuation by grid sand orifice plates//Shock Waves. 2006. V.16, P.115.
[15] *Berger, S., Sadot, O., Ben-Dor, G.* Experimental investigation on the shock-wave load attenuation by geometrical means// Shock Waves. 2010. V.20, P.2940.
[16] *Berger S., Sadot O., Ben-Dor G.* Numerical Investigation of Shock-Wave Load Attenuation by Barriers//28th International Symposium on Shock Waves/Ed. by K. Kontis. Springer, Berlin, Heidelberg. 2012. P.111-115.
[17] *Britan A.A., Karpov A.V., Vasilev E.I., Igra O.O., Ben-Dor G.G., Shapiro E.E.* Experimental and Numerical Study of Shock Wave Interaction with Perforated Plates// ASME. J. Fluids Eng. 2004. V.126. N 3. P.399-409.
[18] *Sha S., Chen Z., Jiang X.* Influences of obstacle geometries on shock wave attenuation// Shock Waves. 2014. V.24. P.573-582.
[19] *Chaudhuri A., Hadjadj A., Sadot O., Ben-Dor G.* Study of Shock-Wave Mitigation through Solid Obstacles. //28th International Symposium on Shock Waves/ K. Kontis. Springer. Berlin. Heidelberg. 2012. P.491-498.
[20] *Wang, C., Ding, J., Tan, S., & Han, W.* High Order Numerical Simulation of Detonation Wave Propagation Through Complex Obstacles with the Inverse Lax-Wendroff Treatment// Communications in Computational Physics. 2015. V.18. N 5, 1264-1281.

IOP Conf. Series: Journal of Physics: Conf. Series **1099** (2018) 012013 doi:10.1088/1742-6596/1099/1/012013

[21] *Igra D., Igra O.* (2015) Attenuating Shock Waves by Barrier Having Different Orientations: A Numerical Investigation.// 29th International Symposium on Shock Waves 2/Ed. by R. Bonazza, D. Ranjan. Springer. P.1273-1278.

[22] *Gongora-Orozco N., Zare-Behtash H., Kontis K.* Experimental Studies on ShockWave Propagating Through Junction with Grooves// Proceedings of the 47th AIAA Aerospace Sciences Meeting Including New Horizons Forum and Aerospace Exposition, Florida, USA: American Institute of Aeronautics and Astronautics. 2009. P.5-8.

[23] *El G. A., Hoefer M. A., Shearer M.* 2016 Dispersive and diffusive-dispersive shock waves for nonconvex conservation laws// SIAM Review. 2016. V.59. P.3-61.

[24] *Il'ichev A.T., Chugainova A.P., Shargatov V.A.* Spectral Stability of Special Discontinuities// Doklady Mathematics. 2015. V.91. P.347-351.

[25] *Chugainova A.P., Shargatov V.A.* Stability of discontinuity structures described by a generalized KdV-Burgers equation//Computational Mathematics and Mathematical Physics. 2016. V.56. P.263-277

[26] *Kulikovskii A.G., Chugainova A.P., Shargatov V. A.* Uniqueness of self-similar solutions to the Riemann problem for the Hopf equation with complex nonlinearity// Computational Mathematics and Mathematical Physics. 2016. V.56. P.1355-1362.

[27] LeFloch P G 2002 *Hyperbolic systems of conservation laws: The theory of classical and nonclassical shock waves, Lectures in Mathematics*, ETH Zurich, Birkhauser

[28] *Chugainova A.P., Il'ichev A.T., Kulikovskii A.G. and Shargatov V.A.* Problem of arbitrary discontinuity disintegration for the generalized Hopf equation: selection conditions for a unique solution// J. Appl. Math. 2017. V.82. P.496-525.

[29] *Chugainova A.P., Shargatov V.A.* Stability of nonstationary solutions of the generalized KdV-Burgers equation //Computational Mathematics and Mathematical Physics. 2015.V.55. P.251-263.

[30] *Chugainova A.P.* Special discontinuities in nonlinearly elastic media// Computational Mathematics and Mathematical Physics. 2017. V. 57 N 6. P.1013-1021.

[31] *Chugainova A.P., Shargatov V.A.* Flow Structure behind a shock wave in a channel with periodically arranged obstacles //Proceedings of the Steklov Institute of Mathematics. 2018.V.300. P.216-228.

[32] *Shargatov V.A., Chugainova A.P., Gorkunov S. V. and. Sumskoi S. I.* Flow Structure behind a shock wave in a channel with periodically arranged obstacles //Proceedings of the Steklov Institute of Mathematics. 2018.V.300. P.216-228.

[33] *Kolgan V.P.* Application of the principle of minimizing the derivative to the construction of finite-difference schemes for computing discontinuous solutions of gas dynamics //Journal of Computational Physics. 2011. V. 230, N 7. P. 2384-2390.

IOP Conf. Series: Journal of Physics: Conf. Series **1099** (2018) 012014 doi:10.1088/1742-6596/1099/1/012014

Modeling of shock wave propagation over the obstacles using supercomputers

V A Shargatov[1], A S Pecherkin[2], A S Sofin[2], A A Agapov[2], S V Gorkunov[1], S I Sumskoi[1], Yu A Bogdanova[1], A V Karabulin[1]

[1]National Research Nuclear University MEPhI (Moscow Engineering Physics Institute), Kashirskoye shosse 31, Moscow 115409, Russia
[2]CJSC Scientific technical center of industrial safety problems research, Block 14, Perevedenovsky Pereulok 13, Moscow 105082, Russia

E-mail: shargatov@mail.ru

Abstract. A computational technology has been developed for modeling the propagation of shock waves in volumes with obstacles. It was shown that the proposed approach allows to simulate real situations of shock wave propagation with an accuracy of several percent (for pressure wave amplitude). This accuracy is achieved through parallelization of the computational process and the use of restructured meshes. An example of calculation is given.

1. Introduction

The problem of the shock wave propagation in a cluttered space is of great interest from a practical point of view. In particular, to assess the consequences of the explosion of fuel-air clouds. Such clouds can form in the atmosphere during an emergency release of fire and explosive liquids and gases.

To solve this problem, it is necessary to describe a number of rather complex processes:

- release of gas, liquid or multiphase flow from equipment;
- propagation of the substance in the environment (dispersion in the air, spreading over the surface);
- combustion/explosion of clouds of the fuel-air mixture, generation and propagation of shock waves;
- action of shock waves and other damaging factors on various objects.

Each of these stages is described in terms of its own models, which in turn can introduce various errors in the solution.

It should be noted that potentially the largest errors are introduced at those stages of the calculation, where there may be abrupt changes in physical quantities that will determine the level of damage. In this case, for a correct assessment of the negative impact, it is necessary to resolve the space-time changes with the required accuracy.

To such abruptly changing parameters, first of all, one can attribute pressure Indeed, in the combustion/explosion of fuel-air clouds, the pressure sharply increases in the generated shock waves / compression waves, and then decreases in the rarefaction waves propagating after them. As a result, the pressure profile looks as shown in Figure 1 by the red line (a). In numerical simulations, if the spatial resolution is not sufficient the gradients in the pressure profiles are "smeared out", the pressure maxima are "cut off," and the picture of the "fine" wave propagation structure is lost. As a result, the wave profile (see Figure 1, the red line (a)), is reproduced approximately (see Figure 1, blue line (b)). The difference in the peak pressures between the "exact" shock wave and the "approximate" numerical solution (namely, this peak pressures determine the nature of the consequences) can reach several tens or even hundreds of percent.

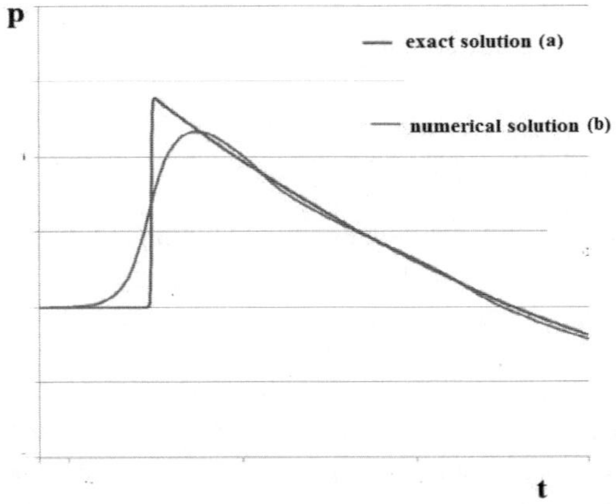

Figure 1. Shock wave pressure profiles: high-accuracy calculation (a) and calculation performed using a "coarse" mesh (b).

So it is clear that there is a need to develop methods and software for high-accuracy modeling of shock wave propagation processes in 3D volume with obstacles.

Until recently, the necessary accuracy of the calculation could be achieved only for one-dimensional [1] - [3] or two-dimensional [4] - [7] problems. But now calculations in a three-dimensional setting can be made with the appropriate accuracy. There is a relatively small number of software codes capable to adequately describe the processes of generation and propagation of shock waves; FLACS [8-10] and KFX [11, 12] are the most important codes for simulating industrial accidents. These codes are, in fact, "black boxes" and have a fairly high cost, so the development of their analogs, based on published models and methods, is an urgent task.

In this paper, we propose approaches to mathematical modeling of explosive processes in various emergency situations. In this case, the proposed approaches use such elements as:

- technology of parallel calculations;
- technology of restructured mesh.

Using the high-accuracy numerical methods, the proposed approach allows in many cases to achieve the reasonable accuracy necessary for predicting the parameters of shock waves with an error of several percent.

2. Basic equations and method of solution

To calculate the propagation of shock waves in a gas mixture, the following standard equations for the transfer of mass, momentum, energy, and individual components of the gas mixture are used (without taking into account chemical reactions, wall friction, turbulence and buoyancy):

The equation of mass conservation:

$$\frac{\partial \rho}{\partial t}+\frac{\partial(\rho u_i)}{\partial x_i}=0. \tag{1}$$

The equation of momentum conservation:

$$\frac{\partial(\rho u_i)}{\partial t}+\frac{\partial(\rho u_i u_j)}{\partial x_j}=-\frac{\partial p}{\partial x_i}. \tag{2}$$

The equation of energy conservation:

$$\frac{\partial(\rho E)}{\partial t}+\frac{\partial(\rho u_j E)}{\partial x_j}=-\frac{\partial(p u_j)}{\partial x_j}. \tag{3}$$

The equations of conservation of individual gas mixture components:

$$\frac{\partial(\rho Y_k)}{\partial t}+\frac{\partial(\rho Y_k u_j)}{\partial x_j}=0. \tag{4}$$

here x_i – spatial coordinates, t – time, ρ – density, u_i - velocity components, p – pressure, E – the specific total energy equal to the sum of the thermal component, chemical component and kinetic energy, Y_i – mass fraction of the i-th gas component of the mixture.

The system of equations (1) - (4) is closed by the equation of state for a mixture of ideal gases.

Equations (1) - (4) were integrated using the second-order Godunov-Kolgan method [13]. The Godunov-Kolgan method is a conservative monotone method. This method allows to resolve the discontinuous solutions (shock waves and contact discontinuities) that are smeared over the mesh. This smearing for the shock wave is usually 3-4 cells and does not depend on the wave intensity. This is acceptable with the available processing power.

3. Obstacle accounting

The correct formulation of the problem requires that equations (1) - (4) must be supplemented by boundary and initial conditions.

In the simplest case, the boundary conditions must be specified only at the boundaries of the computational domain, which for rectangular decomposition is a parallelepiped. Accordingly, the boundary conditions must be defined on each of the six faces of the original parallelepiped.

In the presence of obstacles, a part of the space within the calculation region is filled with objects that are bounded by impenetrable rigid surfaces (examples of such objects are, for example, buildings, equipment, fences, etc.). In this case, the computational domain can be treated as an initial parallelepiped, from which obstacles are removed. The surfaces on which it is necessary to set the boundary conditions are now not only the six surfaces of the original parallelepiped, but the entire set of surfaces that restricted obstacles. Accordingly, the definition of boundary conditions on these surfaces will allow modeling of the actual presence of obstacles. In practice, this is achieved by combining the boundaries of the obstacle with the boundaries between the difference cells (with the some error that naturally arises in this case) and setting conditions on the boundaries of the difference cells such as a "rigid wall" without sticking, which corresponds to setting zero normal velocity and tangential velocities from the center of the nearest cell from the calculation region.

4. Use of restructured meshes

As already noted above, the solution of most problems arisen in calculation of shock wave the propagation faces the problem of obtaining the required calculation accuracy.

The most obvious way to increase the accuracy of the calculation (for the given numerical method) is to reduce the size of the mesh cell. However, this method is associated with a significant increase in the cost of computational resources: since explicit schemes are usually used to calculate the propagation of shock waves, a double decrease in the space step size leads to 2^4 increase in the number of operations performed, and, consequently, to the same growth in the need for time of calculation.

So the question arises how to increase the accuracy of the calculation without increasing (or significantly not increasing) the costs (primarily computer memory and time of calculation).

The most effective way to improve accuracy is to use non-uniform meshes. Irregularity of the mesh suggests that the difference cells in different parts of space and at different times have different sizes.

In the case of 3D problems with parallelepiped computational region there are three main approaches for constructing an irregular mesh:

- concentration/rarefaction of spatial steps along one direction; such a construction of the mesh can lead to a decrease in the accuracy of the numerical method due to a decrease in the accuracy of the approximation of spatial derivatives;
- nesting of meshes ("patches"/mesh refinment) is organized by dividing the large cells into several smaller ones; such construction of a mesh requires the use of rather complex algorithms for hierarchical data organization; also, for such a mesh, complex algorithms of moving / moving in the flow of a region with a nested mesh may be required;
- the restructuring of the mesh is extremely effective when the process evolves over space over time, covering more and more new regions; in this case, there is no need to immediately perform a calculation on the entire calculation region, since the parameters may not change in some regions..

In this paper, the problem of propagation of shock waves is considered - a kind of a flow that progressively develops into newer and new regions, therefore, the reorganization of the mesh can be used: at the beginning of the calculation, a relatively small calculation region is chosen and calculation is carried out only in it with a small step size; when the flow develops to such an extent that it is about to go beyond the boundaries of the calculation region, the sizes of the calculation region and the sizes of the mesh are increased by a predetermined value to allow the further flow evolution. An example of this approach is shown in Figure 2.

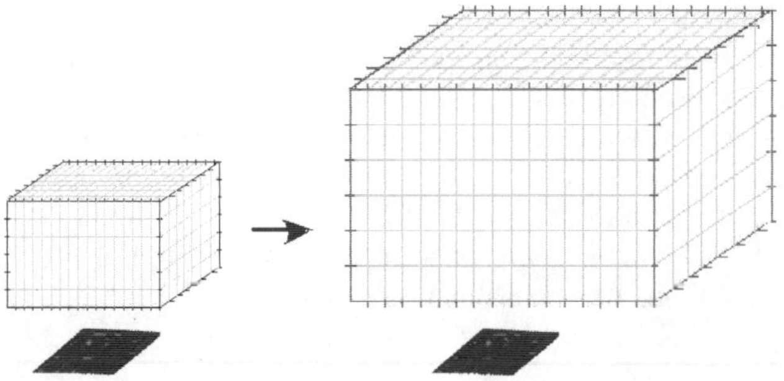

Figure 2. An example of a restructured mesh..

5. Parallel calculations

The numerical method chosen for solving the equations of gas dynamics uses a three-dimensional structured rectangular mesh. The complexity of the computations in each cell is similar, since the calculations are based on the same formulas of the numerical method. Thus, if each of the parallel processes performs calculations with the same number of cells, then the load on the processors will be the same, and they will not expect each other to exchange data. Therefore, in this case it is more efficient to parallelize the calculated algorithm not at the level of tasks that are the same for each cell of the difference mesh, but use the so-called "geometric parallelization".

With "geometric parallelization", each compute node will show approximately equal performance. In this case, the efficiency of the calculation depends on the time spent on exchanging information between the processors. The time spent on exchanging information is determined by the number of exchanges that each process must perform and the amount of information sent in one exchange.

Each process must, independently of other processes, perform calculations for one step in time in the cells that are part of its zone of responsibility. For this, the process must have information not only about cells in its parallelepiped, but also about cells that are in the zone of responsibility of other processes. When we use the Godunov-Kolgan method, these will be cells that are in two layers surrounding the parallelepiped (see Figure 3).

Formally, these two layers of cells provide the setting of "boundary conditions" for the area of responsibility of the processor. The process does not make calculations for these "boundary cells", they are only needed to perform calculations in the cells lying inside the parallelepiped (zone of responsibility).

Exchange of information on the parameters of "boundary cells" occurs between the processes of the area of responsibility, which have a common surface. For each process, whose parallelepiped does not have a common border with the boundary of the entire calculation area, it is necessary to perform up to 6 such exchanges, that is, 6 times to send data about its cells and 6 times to receive data from neighboring processes.

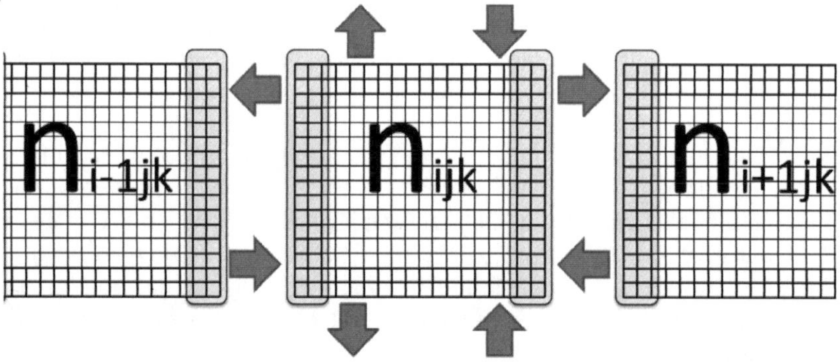

Figure 3. "Cross-linking" of computational domains, which are computed on different computational nodes/processors..

In this case, it is necessary to coordinate the sending and receiving of information between the processes. At the beginning, each process calculates the process numbers from which it should receive data, and to which it must send the data. After that, the exchange processes occur according to the following algorithm. First, the processes exchange information on the surfaces perpendicular to the X axis, then the Y axes and after that the Z axis. On the processed axis, the process sends information to the process that serves the "left" area from it, and receives data from this process After that, similar operations are carried out with the process, which is "on the right."

6. Calculation example

To test the proposed model and the developed program (including a parallel version), the following task was solved: on a flat surface there are several rectangular parallelepipeds, each of which is an object with absolutely rigid walls. The layout of the parallelepipeds is shown in Figures 4a (top view) and 4b (three-dimensional plan). From the scale segment in Figure 4a it can be seen that the length of the parallelepipeds is of the order of a several tens of meters. And their width is several meters. Such scales are typical for typical industrial buildings Usually on industrial sites there are buildings in 2-3 floors, therefore as height of considered parallelepipeds (buildings) was chosen to be 8 m.

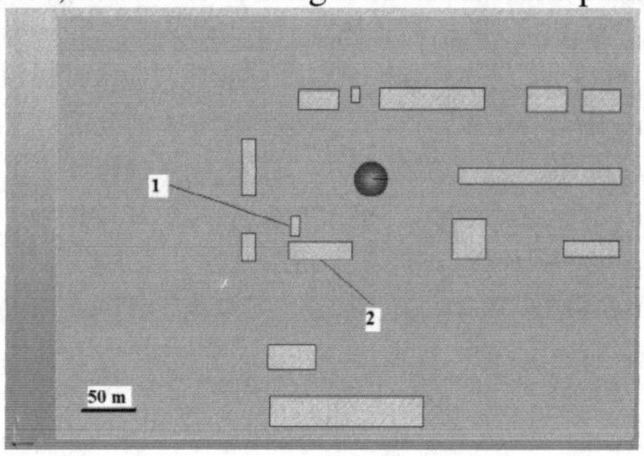

a)

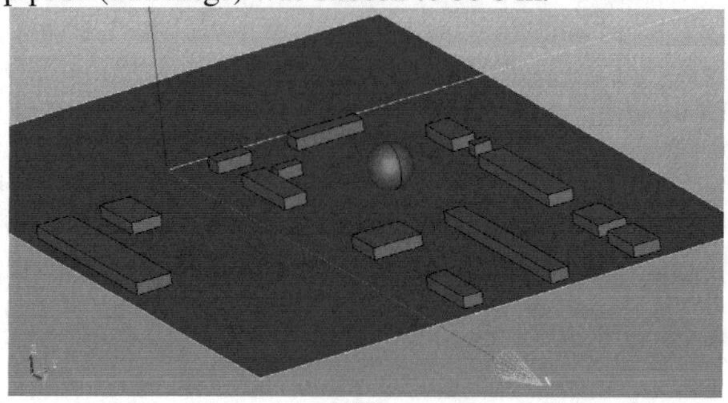

b)

Figure 4. 3D visualization of the location of buildings and cloud: a) top view, b) view from the southeast.

To generate the shock wave, a sphere containing compressed air was specified (in Figure 4, this sphere is marked in red). The expansion of such a sphere will allow to simulate realistic shock waves.

Next we will use the following agreement on directions in Figure 4a: the direction to the top will be considered as the north, the direction to the right to the east, and so on.

The lower point of the sphere with compressed gas touches the ground surface. The radius of the sphere is equal to 15 m. The pressure in the sphere is equal to 10 atm. This situation corresponds to the value of internal energy in the cloud in the amount of 456000 MJ, which corresponds to the combustion of approximately 9.9 tons of conventional fuel with a combustion heat of 46 MJ/kg. Combustion of this amount of fuel is possible with sufficiently large releases.

As already noted above, on the ground and on all surfaces of the "buildings" the "rigid wall" condition is specified without sticking.

Calculations using the approach outlined above were carried out in four versions:

- "coarse mesh" with a step of 0.96 m on a mesh of 520x520x400 cells; the calculation time for 256 processors was 4 hours 56 minutes; during this time the process was calculated for 1 s;
- "medium mesh" with a step of 0.4 m on the mesh of 744x744x500 cells; the calculation time for 320 processors was 11 hours 31 minutes; during this time, 0.33 s have been calculated; during this time the process was calculated for 0.33 s
- "fine mesh" with a step of 0.2 m on the mesh of 1000x1000x500 cells; the calculation time for 320 processors was 15 hours 43 minutes during this time the process was calculated for 0.19 s
- calculation on a mesh with a large step (2 m) on a single-processor computer; such a calculation can be considered typical from the point of view of obtaining information about the propagation of shock waves in a obstacle volume using a single-processor computer in an acceptable time frame (several hours).

Figure 5 shows the loading of the north side of building 2 (in the middle of the wall) at points located above the ground at altitudes of 0 m (Figure 5a), 4 m (Figure 5b) and 8m (Figure 5c), the last point is already on the edge of the roof. From this figure one can see that the loading along the lower part of the building (at heights of 0-4 m) is noticeably larger than on the upper part of the north wall. This is due to the fact that after a sudden increase in the pressure after the arrival of the shock wave, its decrease immediately begins, and this decrease takes place due to both the rarefaction wave propagating behind the incoming shock wave and the rarefaction waves generated at the boundaries of the buildings when they are flowing along the incident wave. And, if the decrease in pressure from a rarefaction wave "sitting" on an incident shock wave is the same for almost all points of the northern facade, the rarefaction formed at the edge of the "northern wall-roof" of building 2 begins to affect the upper part of the wall earlier and only after some time this rarefaction reaches the bottom. A similar loading takes place in the sections of the northern wall adjacent to the western and eastern walls of building 2.

The loading at the lower point of the northern wall of building 2 is 237 kPa, the loading at 4 m is 233 kPa, and the loading at the roof level (8 m) - 182 kPa (calculation data for a "fine mesh").

From the data presented, one can see that using the proposed approach, it is possible to significantly increase the number of cells (in tens or even hundreds of times), expand the calculation region and duration in comparison with the calculation on a single-processor system where it is possible to carry out calculations on meshes with only a few million cells..

As one can see from Figure 5 calculation on a single-processor computer in 2-2.5 times underestimates the peak amplitude of shock waves.

From Figure 5 it is clearly seen that a decrease in the size of the space step leads to a significant improvement in the accuracy of the solution. From Figure 5 it follows that at a step of 0.2 m accuracy is attained a value which can be considered acceptable: the difference between calculations on a "fine mesh" (0.2 m) and "medium mesh" (0.4 m) is several per cent - 5-10%; this value can be regarded as the accuracy of the calculation. This means that with the next step size decrease, the improvement in accuracy can reach a value of 3-5%. Accordingly, the size of the spatial step of not more than 0.2 m can be considered as the step, the calculations with which can be considered as reliable. So the value of the spatial step necessary to obtain a satisfactory accuracy in the calculation of shock waves can be taken equal to 0.05-0.1 m.

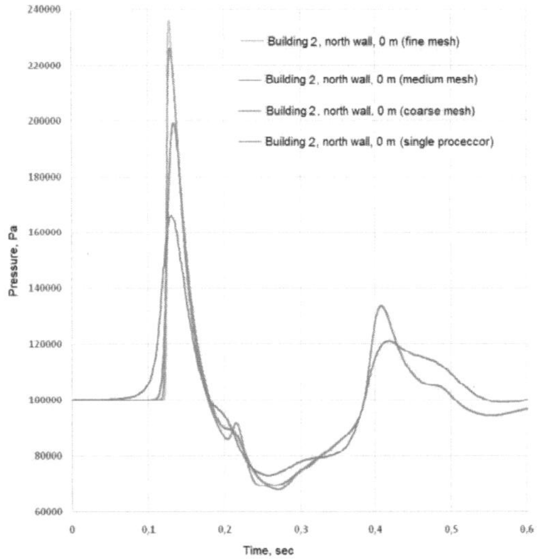

a)

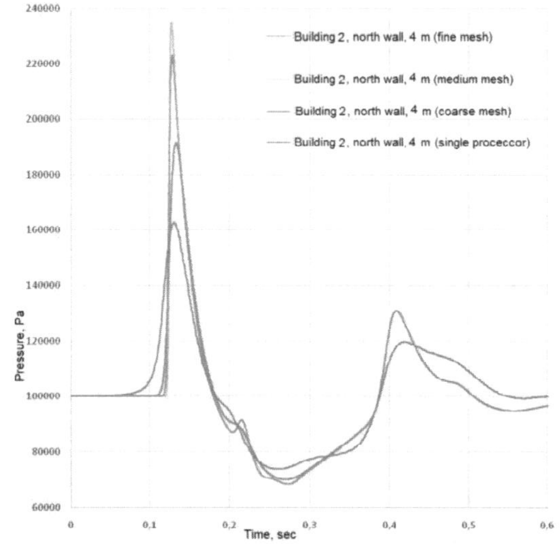

b)

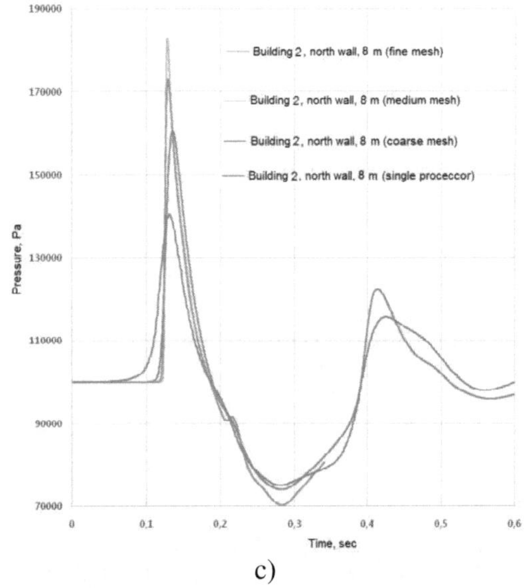

c)

Figure 5. Pressure versus time on the northern wall of building 2: at a height of 0 m - at the ground (a), at a height of 4 m (b) and at a height of 8 m - at the roof (c).

In Figure 5, in addition to the first peak of pressure (the effect of the primary wave formed in the atmosphere with the motion of the resting air), a second pressure peak is also observed. It is a secondary shock wave, arising upon repeated expansion of the cloud after it's collapse, which in turn follows the first expansion of the cloud (when the first shock wave was generated). This wave comes to the building 2 at 0.4-0.45 s and is calculated with some "smearing", which means a decrease in the gradient at the front and smoothing its maximum.

In addition to the primary and secondary waves, pressure dependencies on time can also be demonstrated waves reflected from neighboring buildings. This situation is observed for building 1, where the building itself is relatively small, and the nearby building 2, because of its larger size, is capable to reflect waves of greater intensity, since these waves will be weakly attenuated at distances comparable to the size of building 2. Figure 6 shows the pressure versus time dependences for a point located in the middle of the west wall of building 1 (at a height of 4 m from the ground). Due to the location of the buildings 1 and 2 and the small distance between them, a region of high pressure is created (see Figure 4), when shock wave arrives. Unloading of this corner space occurs through the gap between buildings 1 and 2. Accordingly, at the exit from the corner of buildings 1 and 2, a shock wave is formed, which creates additional loading on the western wall of building 1. This is clearly seen in Figure 6, where the primary shock wave hasn't a ordinary triangular profile, but has a trapezoidal profile with two local maxima. And the second pressure peak at this trapezium is higher than the first one. This second pressure peak was formed by the interaction of waves that rounded building 1, and above all by the wave coming from the northern wall of building 2.

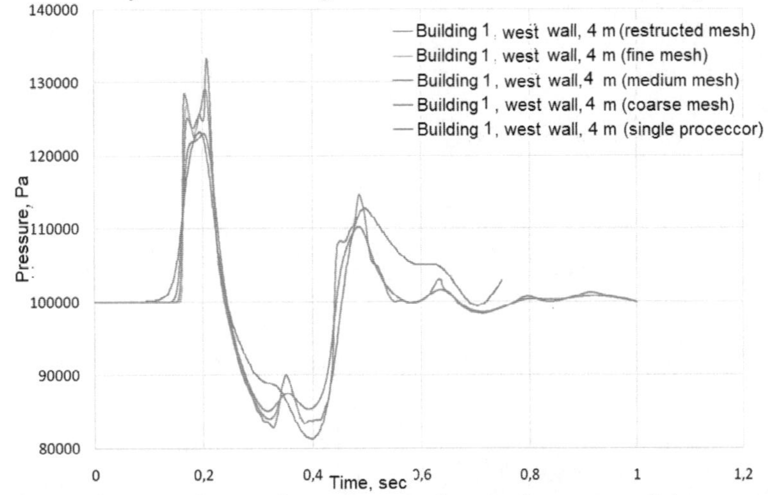

Figure 6. Dependence of pressure on time in the center of the western wall of building 1 (in the middle, at a height of 4 m from the ground)..

The results of the calculation using a restructured mesh are shown in Figure 6 also. Calculations on 256 processors took 8 hours, about the same as on a "coarse" mesh. However, in terms of accuracy, the solution on a restructured mesh significantly exceeds the solution on a "coarse" mesh in calculating the fine structure of the wave.

Actually, the restructured mesh initially had a size of 60x60x40 m with the initial spatial step of 0.111x0.111x0.1125 m (520x520x400 cells - the size same as for the "coarse" mesh). During the calculation of a process with a duration of 1 s, 5 rearrangements of the meshes were made. At the same time, each time the spatial step of the next mesh increased by 1.553092 times, reaching 1,039156497 m by the end of the calculation.

Similar loading patterns are observed for other buildings. In conclusion of this paper, it is interesting to consider not only the behavior of pressure at individual points, but also the spatio-temporal variation of pressure. The pressure fields at ground level are shown in Figure 7 at time moments of 0.11 sec, 0.20 sec, 0.30 sec and 0.45 sec. The moments of time illustrating the most characteristic features of the flow are chosen.

From Figure 7a for the time moment 0.113 c. one can see how the shock wave that came to building 2 (and also to the building located north-east of the site of the explosion) is reflected from the northern and eastern walls of the building. In this case, the pressure significantly increases at the point of reflection (this pressure corresponds to the first peak at the gage dependences). It is also clearly seen from Figure 7a that a region of low pressure forms at the point of the "explosion". It is also clear that the incident shock wave has an axisymmetric shape.

In Figure 7b for the moment of time 0.196 s one can see further reflection of the shock wave from two buildings in the north and south-east (areas of intense red color). These reflected waves, when approaching the edge of the building, exit into already unconfined volume, which somewhat amplifies the waves that go further through the air. This can also be seen in the case of a building located north-west of the "explosion" point (see Figure 7b and 7c)).

From Figure 7b one can see how the reflection occurs in the corner formed by the buildings 1 and 2. Due to the partial confinement of space in this corner, an area of high pressure exists for some time (see Figure 7b). And in this region, one can clearly see how the high-pressure region formed in the corner is unloaded through the gap between buildings 1 and 2. As a result of which there is a rather strong wave (area of high pressure) behind the building 1 on its western side, which increases the loading of building 1 on the western wall (this loading in Figure 6 can be related to the presence of a second peak at the pressure-time dependence)

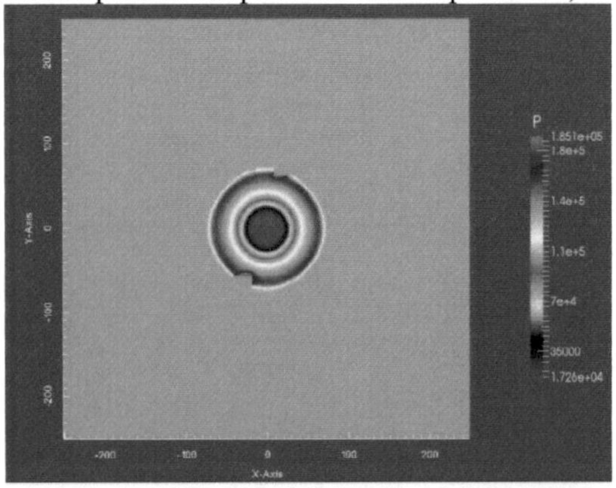

a)

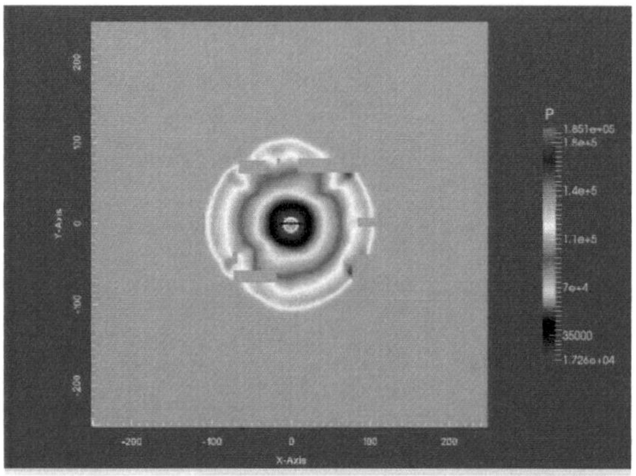

b)

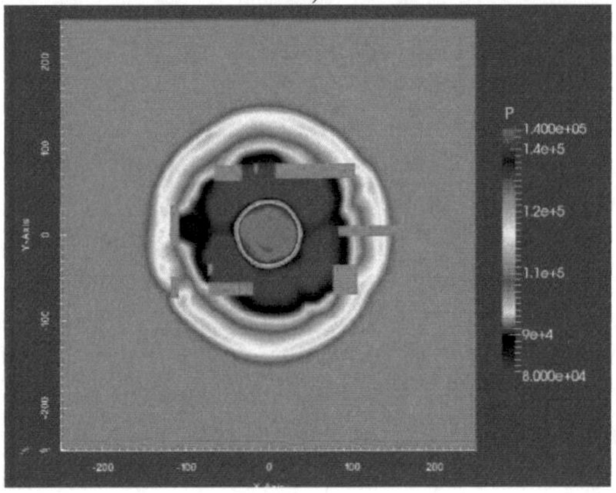

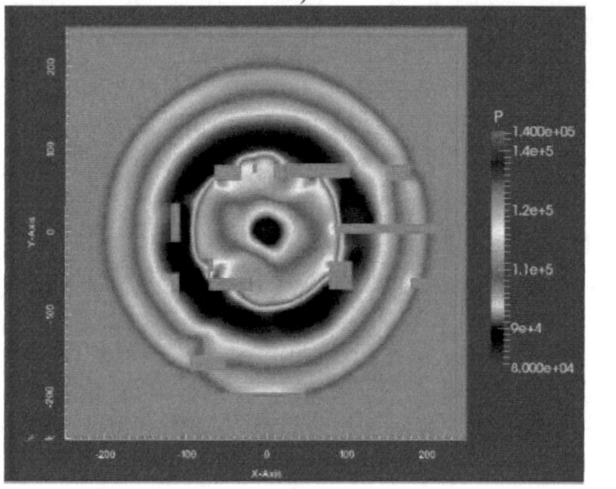

d).

c)

Figure 7. Pressure fields on the ground at time moments 0.11 s (a), 0.20 s (b), 0.30 s (c) ,. 0.45 s (d).

From Figure 7b it is also clear that loads on long buildings located so that the shock wave "slides" along the building (for example, a building in the east of the "explosion" point) are much less effective.

It should be noted one more feature, clearly visible in Figure 7b, by the time 0.196 s the expansion of the cloud in the place of the "explosion" is replaced by its collapse, as a result, in the epicenter of the "explosion" the pressure increase begins.. In Figure 7b, the zone of increasing pressure at the place of the "explosion" just begins to form, and in Figure 7c it has already formed in place of the initial cloud (red circle in Figure 7c. This area will subsequently expand and generate a secondary shock wave that is clearly visible on the dependences of most gauges (see Figure 5-6).

It should be noted one more feature of shock wave propagation in a space with obstacles: in the initially spherical wave, after its interaction with obstacles, there are sections of the front with different pressures on it. Thus, when the shock wave propagates in a volume with obstacles, the inhomogeneity of the parameters along the front of the shock wave remains after the release of the shock wave from the clutter. Thus, after the propagation of the shock wave in a volume with obstacles, even when the shock wave leaves this volume, the inhomogeneity of the parameters along the front of the shock wave is presented. The magnitude of pressure variation along the front can reach up to several tens of percent. Figure 7d shows the situation when the primary wave has interacted, practically with all the buildings. This figure also clearly shows the secondary shock wave propagating from the place of the "explosion" as a result of repeated (after the collapse) expansion of the cloud in the place of the "explosion". When the secondary wave propagates, the same effects are observed as in the propagation of the primary wave: reflection from the walls, the appearance of zones of increased pressure in partially confined spaces (for example, in the corner arrangement of buildings), generation of shock waves in the case of unloading of high pressure regions through the canyons between buildings, weaker loading of buildings when shock waves slide along them, etc.

7. Conclusions

A computational technology has been developed for calculating the propagation of shock waves in a space with obstacles that allows us to take into account not only the action of the primary wave with an acceptable accuracy, but also the effect of waves reflected from other obstacles, secondary shock waves, and also to see with good resolution the nature of the load of various parts and elements of buildings, which increases the reliability of assessing the impact of shock waves on them.

It was shown that

- for realistic estimation of shock wave parameters in modeling typical industrial accidents, a spatial resolution of the mesh of 5-10 cm is required (for explosions up to several fuel tons);
- the modeling of the shock wave propagation with a spatial step of several meters introduces an error in the amplitude of the shock wave of 200-300%; from this it follows that now simulation of shock wave propagation in a three-dimensional formulation is possible only on clusters with several hundreds of processors;
- the use of restructured (expanding) meshes is a promising way to optimize the solution of such problems and allows to resolve pressure peaks with an error of several percent.

Acknowledgments
This work is supported by the Russian Science Foundation under grant 16-19-00188

References

[1] Gelfand B E, Gubin S A, Mikhalkin V N, Shargatov V A 1985 *Fizika gorenia i vzryva* **21** 92.
[2] Gubin S A, Shargatov V A 1989 *Fizika gorenia i vzryva* **25** 111.
[3] Borisov A A, Gelfand B E, Gubin S A, Odintsov V V, Shargatov V A 1986 *Khimicheskaya fizika* **3** 435.
[4] Gostintsev Yu A, Gubin S A, Sumskoi S I, Shargatov V A, 1990 *Combustion, Explosion, and Shock Waves* **26** 473.
[5] Borisov A A, Gelfand B E, Gubin S A, Sumskoi S I, Shargatov V A,. 1988 *Fizika gorenia i vzryva* **24** 124.
[6] Sumskoi S I, Pchelnikov A V, Shamshin I O, Victorov S B, 2007 *Bezopasnost truda v promyshlennosti* **3** 46.
[7] Sumskoi S I, Pchelnikov A V, Shamshin I O, Victorov S B, 2007 *Bezopasnost truda v promyshlennosti* **4** 42.
[8] Agapova E A, Degtiarev D B, Lisanov M V, Kryukov A S, Kulberg S B, Sumskoi S I, 2015 *Bezopasnost truda v promyshlennosti* **9** 71.
[9] Dasgotra A, Teja G V V V, Sharma A, Mishra K B, 2018 *Journal of Loss Prevention in the Process Industries* In Press
[10] Hansen O R, Ichard M, Davis S G, 2010 Validation of FLACS for Vapor Dispersion from LNG Spills: Model Evaluation Protoco In: 12th Annual International Symposium of the Mary Kay O'Connor Process Safety Center
[11] Rian K E, 2016 *Chemical engineering transactions* **48** 175.
[12] Chang B J, Seungho J 2016 *Energy Science and Engineering* **4** 143.
[13] Kolgan V P, 1972 *Uchenye zapiski TSAGI* **3** 68.

IOP Conf. Series: Journal of Physics: Conf. Series **1099** (2018) 012015 doi:10.1088/1742-6596/1099/1/012015

Shock waves in porous media

S A Gubin[1, 2]

[1]National Research Nuclear University MEPhI (Moscow Engineering Physics Institute), Kashirskoye shosse 31, Moscow 115409, Russia

[2]State Center «Interphysica», 31, Kashirskoye shosse, Moscow, 115409 Russia

e-mail: gubin_sa@mail.ru

Abstract.The experimental data [1, 2] on the effect of the parameters of plane incident shock waves and the characteristics of a porous medium on the attenuation and reflection of waves from a rigid obstacle are examined. The propagation of waves with a pressure jumps of up to $\Delta P/P_0 = 20$ ($P_0 = 0.1$ MPa) in polyurethane foam blocks with a density of 20 kg/m3 and 35 kg/m3 and a length of up to 0.8 m, tightly adjacent to the walls and the end face of the driven section of the shock tube is studied. At lengths of the porous block greater than the extent of the shock wave, the porous block effectively attenuated the shock wave. For a wave of limited duration with a triangular pressure profile, a stronger attenuation of the wave in the polyurethane foam block was observed in comparison with that for an extended wave pulse. The reflection pressures of the waves in polyurethane foam with pressure jumps of $\Delta P/P_0 \geq 12$ exceed the reflection pressures of shock waves in air. The degree of attenuation or amplification of a wave depends on the material density and the ratio of the lengths of the foamed polyurethane block and the pressure pulse.

1. Introduction

Various applications of porous materials in industry motivate studies on the impact of shock waves on porous media. Porous materials differ from solid materials and gases in that their compressibility is higher than that of gases, but lower than that of solids or liquids. A high compressibility of porous materials and of gas−liquid bubbly, foamy, granular, and bulk-density systems makes the propagation of shock waves in these media essentially similar. Such media can be used to protect objects from the impact of explosions. To describe the characteristics of the interaction of a shock wave with a porous barrier and its propagation therein, it is necessary to know the velocities differences of the phases, viscoelastic behavior of the framework, and the dynamics of pore destruction at the stage of compaction of the porous material in the wave.

The characteristics of the propagation of shock waves in porous materials are of interest from a scientific point of view as well, not least because of the effect of interfacial interaction in velocities and stresses on the parameters and structure of shock waves in a porous material.

Porous media are used for attenuating and damping shock waves [1]. At the same time, high pressures behind reflected shock waves in porous materials have been measured [1, 2]. Thus, a porous layer can both weaken and amplify the impact of air shock waves on a rigid wall.

To determine the conditions of attenuation of a shock wave in a porous material, it is necessary to analyze the parameters of the incident and reflected shock waves at the given characteristics of the porous medium (gas volume fraction, pore size, density of foamed polyurethane).

The aim of the present work is to determine the conditions for changing the intensity of incident and reflected shock waves in a porous material. For this, the experimental data [1, 2] on the effect of the parameters of a plane incident shock wave and the characteristics of a porous medium on the reflection of the waves from a rigid wall are analyzed. The interaction of shock waves with porous materials [2–11] for a plane air shock wave incident on a polymer polyurethane block tightly adjacent to the walls and end of the pipe is studied experimentally and theoretically.

In [10], a method is proposed for calculating the characteristics of steady shock waves in a porous compressible medium the elasticity of which is determined by the elasticity of the gas in the pores. The numerical studies of the propagation and reflection of shock waves from a rigid wall covered with a porous material performed in [4–9] made it possible to determine the conditions for wave attenuation. In [10], the structure and behavior of reflected shock waves in porous media saturated with a liquid were studied experimentally and numerically. Recent experimental works [12, 13] have shown the effectiveness of nanoporous non-wetting media for the attenuation of compression waves.

2. Experimental setup

Experiments on the structure of pressure waves in a porous medium were carried out on a shock tube of rectangular cross section (45×30 mm, figure 1). Driven section 2, 1.5 m in length, was equipped with three piezoelectric pressure sensors, 3, 4, and 5. Sensors 4 and 5, located along the length of the driven section of the shock tube (figure 1) measured the parameters of the wave in foamed polyurethane. Pressure sensor 6 recorded the reflection pressure at the rigid end face of the tube. Driver section 1 (Figure 1) of length 0.5 m could be shortened to generate short shock waves with a triangular pressure pulse. In the experiments, the propagation of waves in foamed polyurethane blocks, up to 0.8 m in length, inserted tightly into the driven section of the shock tube, without a gap to the walls and end face. The initial pressure was $P_0 = 0.1$ MPa.

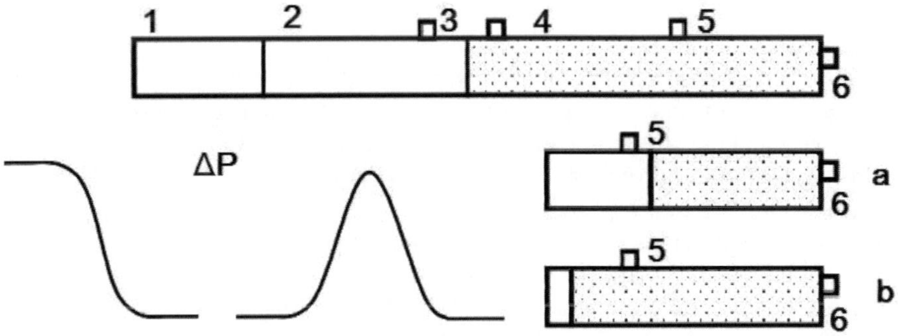

Figure 1. Schematic diagram of the shock tube and porous blocks.

The porous material samples were cellular-structure polyurethane foams with a low (~20 kg/m^3) and a higher density (35 kg/m^3), with up to 96.5% of the volume of which was occupied by the gas phase. The cell size in the low-density polyurethane foam was 2–6 mm. The denser polyurethane foam had 1-mm cells. For the given cell size, the number of longitudinal and transverse skeletal elements in the denser polyurethane foam exceeded three to four times their number in the low-density material of the same volume. Such a change in the internal structure increased the strength of the denser polyurethane foam.

Pressure waves in the polyurethane block were generated by reflection of an air shock wave at the polyurethane–air interface. We investigated waves with a pressure jump of $\Delta P_1/P_0$ = 1.5–20 under normal conditions. The measured parameters were the propagation velocity, pressure in the incident wave, the pressure of reflection of the wave at the end face of the driven section, and the degree of attenuation of the shock wave in the polyurethane foam.

3. Results

It turned out that the pressure in waves in porous media, in contrast to gaseous media, increases gradually over several hundred microseconds. There was no pressure jump at the front of the shock wave in the foamed polyurethane medium. The time of increase in pressure decreased with increasing wave intensity. The pressure in the reflected shock wave also increased gradually, without a sharp pressure jump.

The authors of [3, 8] explained the considerable length of the wave front in polyurethane foams is by filtering the gas through the permeable skeleton of the porous material into the unperturbed region ahead of the wave front. In addition, gas filtration helps reduce the pressure drop at the front of the wave and reduce its intensity and velocity.

The deceleration of extended waves was recorded in porous material blocks of lengths 0.6 and 0.8 m and density 35 kg/m^3. For example, over a distance of 0.6 m, at a front pressure maximum of 1.2 MPa, the wave velocity decreased from 320 to 200 m/s. The pressure ΔP_1 of a wave propagating through the porous material decreased at a rate of 0.6 MPa/m. The rate of attenuation of the wave depended on the wave intensity $\Delta P_1/P_0$. A wave with a pressure maximum of $\Delta P_1/P_0$ = 7.0 attenuated at a rate of 0.3 MPa/m. A porous inset with a length comparable to or longer than the length of the wave effectively attenuated the shock wave.

Weak waves with a pressure maximum of ~0.2 MPa were weakened more slowly. A decrease in the density of foamed polyurethane led to a decrease in the rate of damping of the wave. The attenuation of the wave with the distance traveled in the polyurethane foam sample of density 35 kg/m^3 turned out to be approximately twice as large as that for the low-density polyurethane foam, with a density of 20 kg/m^3.

That the pressure of a wave in the dense porous material was lower can be explained by increases in the energy spent on compression, pore destruction, deformation of skeletal elements, and acceleration of condensed particles to the gas velocity. A stronger weakening of shock waves in the dense polyurethane foam as compared to the less dense material can be used to effectively attenuate shock waves in dense porous barriers.

A wave of limited duration with a triangular pressure profile experienced a stronger attenuation in a block of polyurethane foam than a wave with an extended pressure profile did. For example, in a 0.5 m – long porous block, a $\Delta P_1/P_0 = 4$ wave with a triangular pressure profile of duration 1 ms and a length of 0.3 m was attenuated to $\Delta P_1/P_0 = 1.5$. Short pressure pulses were observed being effectively attenuated in long porous blocks. By contrast, extended-pulse waves, with a length exceeding the length of the porous inset remained practically unattenuated. To assess the effectiveness of wave attenuation in porous materials, the most important characteristics are the lengths of the porous material block and the pressure pulse length. The pressure pulse profile, as well as the density and cell size of the porous material influenced the attenuation of the wave to a lesser extent.

The results of measuring the parameters of reflected shock waves are shown in figure 2. The dependence of the dimensionless pressure of the reflected shock wave, $\Delta P_2/\Delta P_1$, on the pressure maximum ratio of the incident wave, $\Delta P_1/P_0$, is displayed in figure 2. As can be seen from figure 2, weak incident waves, with ΔP_1 ~ 0.2 MPa, are reflected with a reflection coefficient of $K = \Delta P_2/\Delta P_1 = 2$. The velocity of such waves is close to the speed of sound in foamed polyurethane, 200 m/s [1]. The intensity of the shock wave reflected from the rigid wall after passing through the polyurethane foam block ahead of it increases with the intensity of the incident wave. For example, for an incident shock wave in a gas with a pressure rise ratio of $\Delta P_1/P_0 = 5$, the dimensionless pressure reflection from a foamed polyurethane block at the end face of the shock tube is $\Delta P_2/P_0$ ~15. The reflection coefficients of increases with the intensity of the incident shock wave. For an incident wave with a pressure maximum of $\Delta P_1/P_0$ ~ 12, the reflection coefficient is comparable with that for an air shock wave. In the case of intense incident shock waves, the reflection coefficients $K = \Delta P_2/\Delta P_1$ for foamed polyurethane exceed those for air. The measured values of the reflection coefficient was $K = \Delta P_2/P_1 \geq 7$.

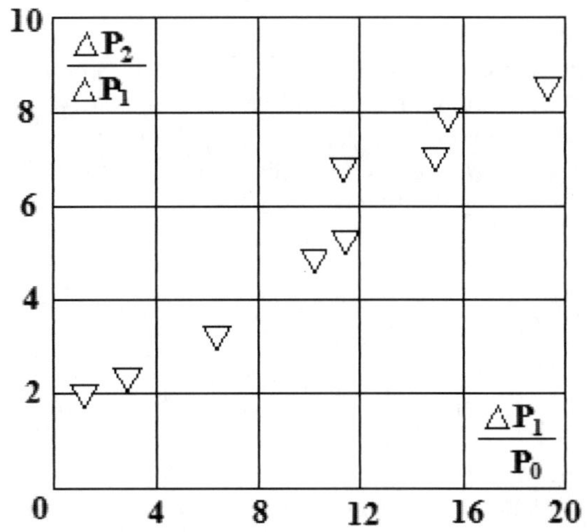

Figure 2. Dependence of the reflection coefficient for a shock wave in polyurethane foam, $K = \Delta P_2/P_1$ on the pressure maximum ratio of the incident shock wave, $\Delta P_1/P_0$. ΔP_2 is the pressure maximum of the reflected shock wave, ΔP_1 is the pressure maximum of the incident shock wave, and P_0 is the normal pressure.

4. Discussion

When the shock wave falls onto a boundary gas–polyurethane foam interface, the reflected wave travels back into the gas, whereas the refracted wave penetrates into the porous block, compressing it and transmitting an impulse to the condensed material. In the shock wave, the porous structure is destroyed and condensed particles (fragments) of the material are accelerated to the velocity of the gas flow behind the wave. The stoppage of the two-phase flow at the rigid end face of the shock tube produces pressure rise. The reflection of the two-phase flow pulse explains why a high pressure zone arises behind the reflected shock wave. This wave has an oscillatory structure, with the pressure amplitude exceeding that of the reflected wave at the gas–foamed polyurethane interface. The model [2, 3, 6] for a pseudo gas with an effective adiabatic exponent closely reproduces the high pressure recorded when a wave is reflected from the boundary between a porous block on a rigid wall.

Experiments with a porous material in which the cells were predestroyed showed that the reflection pressure of the wave in the skeletal material (without cells) was significantly lower than the reflection pressure in intact polyurethane foam. The skeletal structure of the material hinders the acceleration of the condensed material in the flow behind the wave because of a decreased resistance to the gas flow, thereby reducing the reflection pressure of the wave.

In the case of strong waves, the movement of the two-phase flow entrains the entire solid phase in which the bulk of the porous material is concentrated, a factor that contributes to increasing the pressure in the reflected wave as compared to that in air. In weak incident waves, solid material is entrained only partially. Therefore, the reflection coefficient for weak waves is $\Delta P_2/P_0$ ~ 2. The degree of attenuation or amplification of a wave depends on the length of the polyurethane foam, duration of the wave pressure pulse, cell size, and density (or gas volume fraction) in the porous material.

The reflection coefficients for short pressure pulses in a porous medium are lower than those for extended waves of the same intensity. This is also explained by an incomplete entrainment of solid material by the two-phase flow in a short pressure pulse. In this case, the impulse transferred from the two-phase flow to the rigid wall turns out to be smaller than that characteristic of an extended shock wave.

5. Conclusions

Thus, the presence of a foamed polyurethane layer on a rigid wall can lead both to an increase in the amplitude of the reflected extended pulse and to an attenuation of the incident shock wave. The porous layer enhances the dynamic effect on the barrier for extended waves and reduces it in the case of short pulsed impacts. The degree of attenuation or amplification of the wave depends on the length of the polyurethane foam block, pulse duration, and density (gas volume fraction) of polyurethane foam.
The reflection pressures for shock waves in polyurethane foam at pressure maximum ratios of $\Delta P_2/P_0 \geq 12$ exceed those for waves in the air.

Acknowledgments

This work was supported by the Russian Science Foundation, project no. 16-19-00188.

6. References

[1] Gelfand B E, Gubin S A, Kogarko S M, Popov O E 1975 *Zh. Prikl.Mekh.Tekn. Fiz.* **6** 74
[2] Gelfand B E, Gubanov A V, Timofeev E I 1983 *Izv. Akad. Nauk SSSR. Mekh.Zhidk. Gasa.* **4** 79
[3] Gvozdeva L G, Faresov Y M 1985 *Zh. Tekn. Fiz.* **55** 773
[4] Gubaidullin A A., Dudko D N 2000 *Combustion, Explosion, and Shock Waves* **36** 496
[5] Sorek S, Levy A, Ben-Dor G, Smeulders D 1999 *Transport in Porous Media* **34** 63
[6] Baer M R 1992 *Shock Waves* **2** 121
[7] Kitagawa K, Takayama K, Yasuhara M 2006 *Shock Waves* **15** 437
[8] Seitz M W and Skews B W 2006 *Shock Waves* **15** 177
[9] Skews B W 1995 *Shock Wave Impact on Porous Material* ed Brun R, Dumitrescu L Z (Berlin: Springer) p 11
[10] Dontsov V E, Nakoryakov V E, Pokusaev B G 1994 *Akustichyeskii zhurnal* **40** 683
[11] Kiselev S P 1998 *J. Prikl.Mekh.Tekn. Fiz.* **39** 27
[12] Belogorlov A A, Byrkin V A, Borman V D, Tronin V N 2016 *J. Phys. Conf. Ser.* **751** 012035
[13] Belogorlov A A, Borman V D, Byrkin V A, Paryohin D A, Tronin V N 2016 *J. Phys. Conf. Ser.* **751** 012031

IOP Conf. Series: Journal of Physics: Conf. Series **1099** (2018) 012016 doi:10.1088/1742-6596/1099/1/012016

Attenuation of Shock Waves in Bubbly Media

S A Gubin[1, 2]

[1]National Research Nuclear University MEPhI (Moscow Engineering Physics Institute), Kashirskoye shosse 31, Moscow 115409, Russia
[2]State Center «Interphysica», 31, Kashirskoye shosse, Moscow, 115409 Russia
e-mail: gubin_sa@mail.ru

Abstract. Experiments on the attenuation of spherical shock waves were carried out in a vessel and a shock tube filled with a two-phase gas babble–water mixture with a gas volume fraction of $\beta = 2.3$–26%. The attenuation of the spherical blast wave generated by a charge directly into a two-phase medium proved to be more effective than in the case of using bubble screens. Experiments have shown that, at $\beta = 2.3\%$, the intensity of spherical shock waves decreases 20-fold, whereas at $\beta = 26\%$, by a factor of 40–50. A bubble layer enhances the dynamic effect on the barrier for extended waves and weakens it in the case of short-pulse disturbances. The degree of attenuation or amplification of the blast wave depends on the thickness of the layer, pulse duration, and volume fraction of bubbles.

1. Introduction

Gas-liquid bubbly systems are used to attenuate blast waves. The development of technologies for cutting, welding, and processing of materials by explosion, as well as the need to carry out blasting operations during the liquidation of the consequences of accidents in oil and gas transport pipelines and in performing underwater explosions, requires the development of effective methods for attenuating the intensity of blast waves in the surrounding medium. Methods for localizing the effect of an explosion and diminishing explosive loads are mainly based on the use of compressible two-phase systems capable of efficiently absorbing the energy of blast waves. To do this, bubbly media are used, a high compressibility of which provides significant dissipative losses due to the interaction between phases during the propagation of shock waves in gas–liquid mixtures.

Two-phase bubbly mixtures arise because of cavitation in rarefaction waves arising during accidents on pipelines. Studying the dynamics of shock waves in such environments is important for predicting the consequences of accidents on pipelines.

Analysis of the parameters of shock waves in gas–liquid mixtures and investigation of the characteristics of reflection of wave pulses in bubbly liquids makes it possible to determine optimum modes of shock impact on objects. The results obtained can be used in preventing emergencies during blasting operations, creating effective protective barriers capable of localizing the energy of the explosion and reduce the consequences of accidents on pipeline transport.

The complexity of describing the propagation of shock waves in gas–liquid systems and the limited number of experimental studies on the attenuation of blast waves in bubbly media motivates studies on the dynamics of shock wave processes in liquid containing gas bubbles.

Shock waves in bubbly media have been studied experimentally and theoretically [1–14]. The results are overviewed in the monographs [2, 3]. It was experimentally shown [4–8] that bubbly liquids are an effective means of attenuating the intensity of shock waves. It was established that the characteristics of shock waves in a bubbly liquid are determined by the intensity of the wave and are dependent on the parameters of the bubbles, such as the gas volume fraction, bubble size, and viscosity of the liquid. When a shock wave propagates in a bubbly liquid, gas bubbles can break up [9].
The effect of the solubility of the gas in the bubbles on the structure and amplification of shock waves in two-phase bubbly mixtures was studied in [13]. An anomalous increase of the pressure in shock waves reflected from a rigid wall in bubbly media was observed [4, 11, 14]. Thus, the question arises, under what conditions blast waves are significantly attenuated and under what conditions a high pressure of reflection of shock waves reduces the efficiency of attenuation of incident shock waves in bubbly liquids.
The aim of the present work is to examine the experimental data from [4] on the dynamics of the variation of the shock wave parameters with the distance at various contents of gas babbles in water. We have studied the attenuation of incident and reflected blast waves generated in bubbly media.

2. Experimental setup

Experiments on the attenuation of spherical shock waves were carried out in three- and one-dimensional geometries in the apparatuses schematically displayed in figures 1a and 1b. In the 3D case, spherical shock wave 2 was generated by a charge at the center of cylindrical vessel 1 (figure 1a). In the one-dimensional case, shock wave 2 was generated in the upper part of a tube 1, 50 mm in diameter and 1.3 m in length (figure 1b); the pressure behind the wave was measured with piezoelectric sensors. The volume concentration of the gas, β = 2.3–26%, was determined from the height of the rise of the liquid after it was bubbled with nitrogen or carbon dioxide. The experiments were carried out under normal conditions. The experimental setups were described in detail in [4].

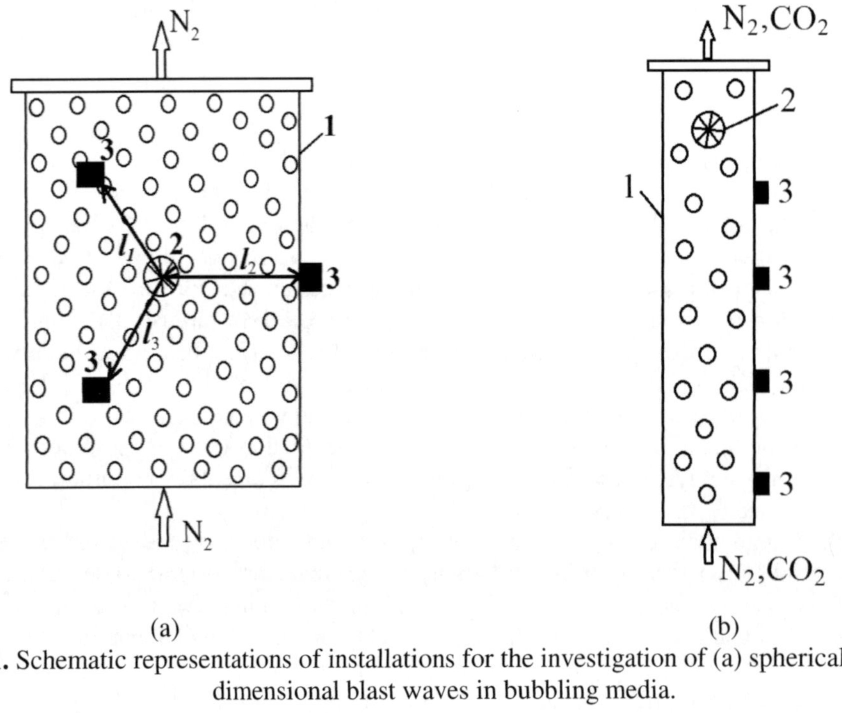

Figure 1. Schematic representations of installations for the investigation of (a) spherical and (b) one-dimensional blast waves in bubbling media.

The parameters of spherical shock waves in the vessel with the two-phase medium were measured by three knife pressure piezosensors 3 located at distances l_1 = 219 mm, l_2 = 134 mm, and l_3 = 180 mm from the charge (figure 1a). The sensor located at the distance of 180 mm measured the pressure in the shock wave reflected from the bottom of the vessel. The pressure behind the shock waves in the tube was recorded by four piezosensors 3 mounted in the walls of the tube 0.24 m apart (figure 1b).

3. Results

The structure of the spherical shock wave changed with the distance it traversed in the two-phase medium. The positive phase of compression of the triangular pulse of the spherical shock wave increased with the distance traveled by the wave in the two-phase medium. The region of constant average pressure in the shock wave widened. Oscillograms exhibited pressure fluctuations in the incident shock wave [4, 11]. The pressure in the incident shock wave was observed to increase with the volume fraction of bubbles (figure 2).

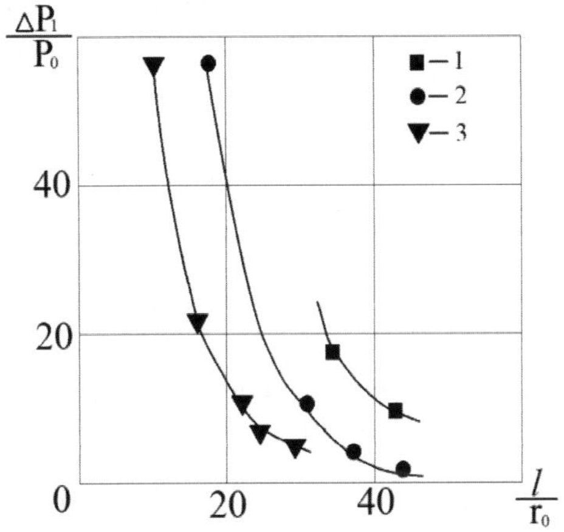

Figure 2. Dependences of the pressure decrease in spherical shock waves in bubbly media at various weights of the charge (1) 1, (2) 3, and (3) 6 g.

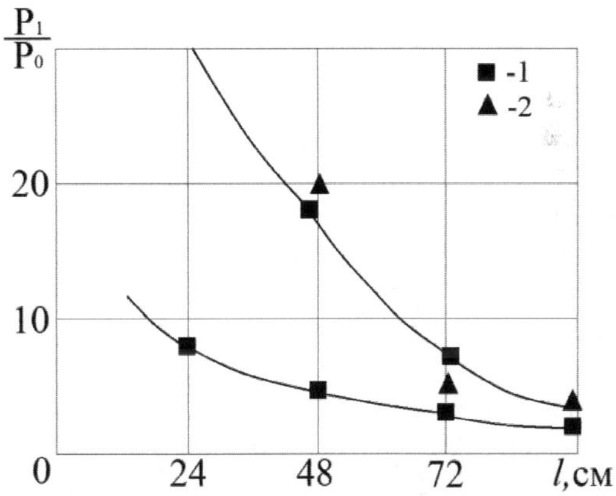

Figure 3. Variation of the wave front process with the distance along the tube length for (1) N_2 (β = 2.8%) and (2) CO_2 (β = 10.7%).

Figure 2 shows the change in the maximum pressure at the front of the incident blast wave ΔP as a function of the dimensionless distance l/r_0 (r_0 is the radius equivalent to the weight of the spherical HE charge). Points 1, 2, and 3 correspond to experiments with of 1-, 3-, and 6-g charges and gas volume fractions of 2.3, 16.6, and 26%. The accuracy of determination of the excess pressure of spherical shock waves did not make it possible to reliably distinguish between the experimental data obtained at β = 16.6 and 26%. At distances greater than (30–40) r_0 (points 3, 2 in figure 2), the pattern of change of the maximum pressure at the blast wave front with increasing distance in media with a volume fraction of bubbles of 16.6–26.0% is comparable with that for shock waves in air [15].
The degree of attenuation of the blast wave depended on the intensity, duration of the positive compression phase of the spherical wave, distance traveled by the wave in the bubble medium, and volume fraction of bubbles. The attenuation of a blast wave generated directly inside a two-phase medium turned out to be more significant than in the case of using bubbly screens [1]. Experiments have shown that, at β = 2.3%, the intensity of spherical blast waves decreases approximately 20-fold, whereas at β = 26%, by a factor of ~50.
A strong attenuation of blast waves in a liquid with gas bubbles occurs due to the specifics of the expansion and interaction of the gaseous explosion products with the two-phase medium. The rate of formation and expansion of explosion products in a two-phase bubbly medium differs from the conditions of formation of the domain of explosion products in a homogeneous gas or liquid. The

products of an explosion expand non-isentropically in a two-phase medium. The detachment of the shock wave from the explosion products occurs at a shorter distance than in a gas medium.

The pressure pulse of a blast wave generated in the tube (figure 1b) acquired the form of a step with a constant average pressure at a distance of ~0.3 m. The propagation velocity of incident shock waves decreased with increasing gas volume fraction in the bubbly medium and increased with the shock wave intensity. Figure 3 shows how the pressure at the wave front changes along the tube length (curve 1, $\beta = 2.8\%$; curve 2, $\beta = 10.7\%$). Just as in the spherical case, the pressure at the wave front decrease with increasing volume fraction of gas bubbles (figure 3). Points 1 and 2 were obtained for nitrogen- and carbon dioxide-filled bubbles. No influence of the thermophysical properties of the gas on the attenuation of strong shock waves in bubbly media has been revealed.

4. Discussion

As the intensity of the blast wave decreases, so does the extension of its positive compression phase. For an initial blast wave pulse shorter than the thickness of the bubble layer, it weakened, producing a lesser effect on the obstacle. With increasing shock wave intensity, the extent of the positive phase of compression of the blast wave increases. When the length of the initial blast wave pulse is greater than the thickness of the bubble layer, the intensity of the wave pulse decreases little. In all the experiments performed in [9], the mass velocity and dynamic thrust of parameters of the blast waves were high enough to break up the bubbles.

The average pressure in the blast wave reflected from the rigid wall increased by a factor of 3–4. The pressure of the reflected wave increased noticeably with the intensity of the incident wave, so that the impact on the barrier was greater.

The reflected shock waves propagated through a finely dispersed medium of small bubbles, already fragmented in the incident wave (with an average size of $d_{10} = 0.05d_0$, where d_0 is the diameter of the initial bubbles.) The amplification of the reflected shock wave occurs because the secondary bubbles are rapidly compressed during the compression phase of the shock wave, with the kinetic energy of the radial motion of the liquid transforming into the potential energy of the gas pressure in the compressed bubbles. The compression of gas bubbles behind the shock wave front is accompanied by the generation of numerous pressure pulses, which produce an increase in the amplitude of the reflected shock wave.

Note that the fragmentation of bubbles sharply increases their surface area, thereby enhancing the intensity of heat transfer between the phases. However, a relatively weak compression of the small secondary bubbles in the shock wave (due to a sharp decrease in the size of the initial bubbles during breakup) does not allow the gas to become very hot during the compression of the secondary bubbles behind the shock wave. The fragmentation of bubbles in the shock wave prevents a significant heating of the gas in them. This makes the process of compression of the secondary bubbles in the shock wave isothermal. Experimentally, no intensification of heat transfer between the phases on the parameters of the shock wave during the fragmentation of bubbles was detected. Under conditions that impede the fragmentation of bubbles (powerful short pulses, a high viscosity of the liquid, and a high initial pressure), the intensity of the heat exchange between the phases decreases, and the broken-up bubbles contract adiabatically. The weak influence of the thermophysical properties of different gases in bubbles on the parameters of the shock wave, found in [4], is indicative of the isothermal compression of bubbles in shock waves. This conclusion was confirmed in [11, 12].

The solubility of the gas further enhances the intensity of weak reflected shock waves [3, 13]. The solubility of gases is accelerated by the fragmentation of bubbles, a process that intensifies the mass transfer between the phases (large interphase gas–liquid surface area) and convective diffusion due to the relative turbulent movement of the gas bubbles and the liquid behind the wave. The amplification of the reflected shock wave can be explained by sharp pulsations of pressure in the liquid generated by gas bubbles collapsing because of their complete dissolution.

It was shown [14] that the energy of the pressure pulse accumulates in the bubbly medium zone if its duration exceeds the time of propagation of the wave through the bubbly liquid area. A similar effect of energy accumulation is used in nanoporous medium–incompressible fluid systems for attenuating pressure pulses [16, 17].

5. Conclusions

In view of the foregoing, it can be expected that the intensity of incident and reflected shock waves in media with cavitation-prone bubbles will be higher than that in media containing air bubbles. The amplification of shock waves in two-phase media containing cavitation-prone bubbles occurs because the bubbles completely dissolve during the compression phase, thereby producing the high-velocity radial motion of the liquid and, consequently, a sharp increase in pressure during bubble collapse, which explains the significant increase in the amplitude of the incident and reflected shock waves. This effect of amplification of incident and reflected shock waves must be taken into account when analyzing the risk of the accidental destruction of pipelines accompanied by cavitation of liquid products (hydrocarbons).

Thus, for the protection of objects from blast waves by means of bubbly media, it is necessary to take into account the transformation of shock waves in the protective medium and to correctly estimate the real parameters of the incident and reflected shock waves at various distances and volume fractions of the gas by using an isothermal or adiabatic model.

Acknowledgments

This work was supported by the Russian Science Foundation, project no. 16-19-00188.

6. References

[1] Kudinov V M, Palamarchuk B I, Vakhnenko V A 1983 *DAN SSSR.* **273** 1080
[2] Kedrinsky V K 2000 *Hydrodynamics of explosion. The experiment and model* (Novosibirsk: SB RAS) p 434
[3] Nakoryakov V E, Pokusaev B G, Schreiber I R 1990 *Wave dynamics of gas - and vapor-liquid media* (Moscow: Energoatomizdat) p 238
[4] Gelfand B E, Gubanov A V, Gubin, S A, etc. 1977 *Izv. AN SSSR Mech. Zhidk. i Gaza* **1** 173
[5] Gubaidullin A A, Ivandacv A I, Nigmatulin R I 1976 *DAN SSSR* **226** 1299
[6] Ohtani K, Sugiyama H, Mizobata K, Sasaki Y 2017 *Proc. Conf. Hokkaido Branch* doi.org/10.1299/jsmehokkaido.2001.41.188
[7] Frolov S M, Avdeev K A, Aksenov V S etc. 2017 *Int. J. Multiphase Flow* **92** 20
[8] Avdeev K.A., Aksenov V.S., Borisov A.A. etc. 2015 *J. Phys. Chem. B* **9** 895
[9] Gelfand B E, Gubin S A, Nigmatulin R I etc. 1977 *DAN SSSR* **235** 292
[10] Sugiyama H, Ohtani K, Mizobata K, Ogasawara H 2005 *Shock Waves* 1085
[11] Sichev A I 2010 *Zh. Tekh. Fiz.* **80** 31
[12] Surov V S 1998 *Zh. Tekh. Fiz.* **68** 12
[13] Dontsov V E 1998 *Zh. Prikl. Mech. Tekh. Fiz.* **39** 19
[14] Agisheva U O, Bolotnova R Kh, Buzina V A, Galimzianov M N 2013 *Fluid Dynamics* **48** 151
[15] Adushkin V V 1963 *Zh. Prikl. Mech. Tekh. Fiz.* **5** 107
[16] Belogorlov A A, Borman V D, Byrkin V A, Paryohin D A, Tronin V N 2016 *J. Phys. Conf. Ser.* **751** 012031
[17] Borman V D, Belogorlov A A,Tronin V N 2018 *Colloids and Surfaces A: Physicochemical and Engineering Aspects* **537** 540

IOP Conf. Series: Journal of Physics: Conf. Series **1099** (2018) 012017 doi:10.1088/1742-6596/1099/1/012017

Analytical equation for the damping coefficient of the acoustic wave in the strong centrifugal field

S.V. Bogovalov, V.A. Kislov, I.V. Tronin

National Research Nuclear University "MEPhI" Kashirskoye shosse, 31, Moscow, 115409, Russia

E-mail: IVTronin@mephi.ru

Abstract. In the fast rotating gas (for example, in the gas centrifuge) three families of linear waves exist with different polarizations and law of dispersion. Analytical equation for the damping coefficient of the wave with the weakest damping is determined. This wave propagates strictly along the axis of rotation and the energy density of this wave concentrates at the wall of the rotor.

1. Introduction

Wave dynamics in rotating liquids and gases attracts attention for a long time. Although the waves in the rotating liquids were studied in many works [1, 2, 3], the properties of the waves in the gases were studied only in few works and the investigations have been limited by rather moderate rotational velocities [4]. Properties of these waves are interesting not only from the theoretical point of view rather than for practical use, in particular in physics of gas centrifuges (GC).

GC are explored for uranium isotope production starting with 50th years of the last century. In spite of this, the physics of the gas flow in the GC is still not fully understood. In our previous paper [5] we argued that the scoops used for the expulsion of the gas from the GC produce waves which propagate along the rotor. The properties of the waves in ideal gas compressed in the centrifugal field of the order 10^6 g (g – acceleration of gravity at the Earth surface) are rather specific.

The velocities of the conventional GC is about $600 - 700$ m/s at the rotor radius about $6 - 9$ cm [6]. The centrifugal acceleration is about $10^5 - 10^6$ g at these parameters. Therefore, pressure of the working gas changes on 3-4 orders of magnitude at 1 cm radius variation. In these conditions conventional acoustic waves split into three families having different dispersion and polarization [5].

The waves in the gas are able to produce so-called "acoustic flow" which has been discussed starting with the work by Lord Rayleigh [7, 8]. The waves produced by the scoops are certainly able to produce an additional axial circulation in the gas centrifuge. Recently this was shown in direct numerical experiments [9]. Moreover, the numerical experiments [10] have shown that the waves change the gas flux from the waste chamber of GC provided that the waves can reach the opposite end of the rotor and product baffle. For the gas centrifuges this may result into reduction of the gas content in the gas centrifuges.

The waves with the weakest damping are especially interesting for us because they can strongly modify the impact of the scoops on the axial circulation and gas content in the GC. In this paper we focus our attention on these waves. They have dispersion law similar to the dispersion law of the conventional acoustic waves and are polarized along the rotational axis.

2. Analytical estimation of the damping

According to [11], the damping coefficient γ_d of a linear wave can be obtained as

$$\gamma_d = \frac{|\dot{E}_{mec}|}{2Ec}, \tag{1}$$

where $\dot{E}_{mec}$ – rate of the mechanical energy dissipation, E – average energy of the wave. The rate of the mechanical energy dissipation equals to [11]

$$\dot{E}_{mec} = -\frac{\lambda}{T}\int \overline{(\nabla T)^2} dV - \frac{\eta}{2}\int \overline{\left(\frac{\partial v_i}{\partial x_k} + \frac{\partial v_k}{\partial x_i} - \frac{2}{3}\delta_{ik}\frac{\partial v_l}{\partial x_l}\right)^2} dV, \tag{2}$$

and

$$E = \int \left(\frac{\rho_0 \overline{v^2}}{2} + \frac{c^2 \overline{\rho'^2}}{2\rho_0}\right) dV. \tag{3}$$

Energy dissipation (2) is determined by velocity and temperature gradients. Average energy in the wave can be calculated by substitution of the axial velocity profile for dissipationless rotating gas (equation (54) from the work [5]) into integral (3)

$$v_z = v_0 \exp\left(\frac{(1-\gamma)\omega^2(r^2-a^2)}{2c^2}\right) e^{i\Omega t}, \tag{4}$$

where $v_0 = \frac{c}{\gamma}\frac{\bar{p}_w}{p_w}$ – amplitude of velocity in the wave, and γ is the adiabatic index of the gas. Initial density profile has a form

$$\rho_0 = \rho_w \exp\left(\frac{\gamma\omega^2(r^2-a^2)}{2c^2}\right). \tag{5}$$

Then we obtain

$$E = 2\pi L \int_0^a \frac{\rho_w e^{A\left(\frac{r^2}{a^2}-1\right)} v^2}{2} r dr = \frac{E_0}{A}\frac{\gamma}{2-\gamma}\left(1 - \exp\left(-A\frac{(2-\gamma)}{\gamma}\right)\right), \tag{6}$$

where $A = \frac{M\omega^2 a^2}{2RT_0}$, E_0 – energy of the wave with amplitude v_0 in uniform gas with density ρ_w.

Calculation of integral (2) can be divided on two parts corresponding to the volume and surface contribution in to the damping as follows

$$\dot{E}_{mec} = \int_{r<a-h} (...)dV + \int_{a-h<r<a} (...)dV, \tag{7}$$

where h is the thickness of layer where the amplitude of the plane wave decays at approaching to the wall due to friction and heat exchange with the wall. The characteristic thickness of this layer is [11]:

$$h \approx \sqrt{\frac{2\nu}{\Omega}}, \tag{8}$$

where ν – kinematic viscosity, Ω – wave frequency.

2.1. Surface damping

In the region $a - h < r < a$ the perturbation of velocity and temperature go to zero due to friction and heat exchange with the wall. At the condition

$$h = \sqrt{\frac{2\nu}{\Omega}} \ll \frac{a}{A}. \tag{9}$$

the thickness of the surface layer is much less the characteristic scale of variation of density and pressure in the gas. In this case we can neglect variation of these values at the calculation of the surface integral in eq. (7). Then, this integral $\dot{E}_{mec}^{surf}$ exactly coincides with the similar integral for an ordinary plane wave interacting with the wall and the damping coefficient equals

$$\gamma_d^{surf} = \frac{|\dot{E}_{mech}^{surf}|}{2Ec} = A\frac{(2-\gamma)}{\gamma}\frac{|\dot{E}_{mec}^{surf}|}{2E_0 c}\left(1 - \exp\left(-A\frac{(2-\gamma)}{\gamma}\right)\right). \tag{10}$$

But

$$\frac{|\dot{E}_{mec}^{surf}|}{2E_0 c} = \gamma_0^{surf}, \tag{11}$$

where γ_0^{surf} is the surface damping coefficient for the wave in the quiescent rotor (tube) given in [11]. Therefore, the surface damping in the rotating gas equals

$$\gamma_d^{surf} = \gamma_0^{surf} A\frac{(2-\gamma)}{\gamma}\left(1 - \exp\left(-A\frac{(2-\gamma)}{\gamma}\right)\right). \tag{12}$$

2.2. Volume damping

For the calculation of $\dot{E}_{mec}$ in the volume it is necessary to have correct solution for the velocity and temperature variation in the wave. Exploration of the solution obtained in the dissipationless approximation gives strongly overestimated damping. Unfortunately, we have no analytical solution of the problem. Therefore, we give approximate estimate of the volume damping. Integral in (2) is accumulated in all the volume of the rotor. However, viscosity and thermal conductivity strongly suppress the wave reducing the its amplitude to zero below radius a_{min} where the viscous terms in the Navier-Stocks equations dominate over the inertial terms. Equation for the axial velocity v_z of the rotating gas in the cylindrical coordinate system in the axisymmetric approximation has the form [11]:

$$\rho\frac{\partial v_z}{\partial t} + \rho\left(v_r\frac{\partial v_z}{\partial r} + v_z\frac{\partial v_z}{\partial z}\right) = -\frac{\partial p}{\partial z} + \eta\left(\Delta v_z + \frac{1}{3}\frac{\partial\,\mathrm{div}\,\vec{v}}{\partial z}\right). \tag{13}$$

The viscous terms in eq. (13) dominate over the inertial terms when

$$\rho c < \eta k, \tag{14}$$

where c is the conventional sound velocity, η dynamical viscosity of the gas and k is the wave vector of the wave. Near the wall $\rho c > \eta k$. The density of the gas reduces with the radius. At $r < a_{min}$ this inequality fails. This gives equation for a_{min} in the form

$$c\rho_w \exp\left(\frac{M\omega^2 a^2}{2RT_0}\left(\frac{a_{min}^2}{a^2} - 1\right)\right) = k\eta. \tag{15}$$

In the region $0 < r < a_{min}$ the amplitude of the wave is close to zero due to strong viscosity. In the interval between a_{min} and a the wave velocity is described by eq. (4). The temperature perturbation is also described by solution from [5]. Therefore it is naturally use the solution (4) for calculation of the volume damping but cutting the integral below a_{min}. Equation (4) does not take into account the interaction with the wall. Therefore the integration on r can be extended from $a-h$ to a because $h \ll a$. Then, integrating (2) in the interval from a_{min} to a we obtain the following equation for $\dot{E}_{mec}$

$$\dot{E}^{vol}_{mec} = -\frac{\pi L}{2}\frac{\bar{p}^2_w}{\rho^2_w}\left(\left(\frac{\lambda}{c^2_p T_0}+\frac{\eta}{c^2}\right)\left(2A\Gamma+1+\frac{1}{A}\left(\frac{k\eta}{c\rho_w}\right)^{-2\Gamma}\ln\frac{k\eta}{c\rho_w}+\right.\right.$$
$$\left.\left. + k^2\frac{a^2}{2A\Gamma}\left(\frac{\lambda}{c^2_p T_0}+\frac{4}{3}\frac{\eta}{c^2}\right)\left(1-\left(\frac{k\eta}{c\rho_w}\right)^{-2\Gamma}\right)\right)\right), \tag{16}$$

where $\Gamma = \frac{\gamma-1}{\gamma}$. Energy of the wave E has been calculated according to eq.(6). Then, volume damping coefficient equals

$$\gamma^{vol}_d = \frac{|\dot{E}^{vol}_{mech}|}{2Ec} = \frac{A}{a^2}\frac{2-\gamma}{\gamma}\left(1-\exp\left(-A\frac{(2-\gamma)}{\gamma}\right)\right)^{-1}\cdot$$
$$\cdot\left(\left(\frac{\lambda}{c^2_p T_0}+\frac{\eta}{c^2}\right)\left(2A\Gamma+1+\frac{1}{A}\left(\frac{k\eta}{c\rho_w}\right)^{-2\Gamma}\ln\frac{k\eta}{c\rho_w}+\right.\right.$$
$$\left.\left. + k^2\frac{a^2}{2A\Gamma}\left(\frac{\lambda}{c^2_p T_0}+\frac{4}{3}\frac{\eta}{c^2}\right)\left(1-\left(\frac{k\eta}{c\rho_w}\right)^{-2\Gamma}\right)\right)\right). \tag{17}$$

3. Conclusions

The analytical equation for the damping coefficient of the acoustic wave in the gas placed in to the strong centrifugal field is determined. The total damping consists of two part: the volume and surface damping. The volume damping is defined by the gradients of the velocity and temperature in the volume of the gas. The surface damping is defined by interaction of the wave with the wall of the rotor. Surface dumping strongly dominates the volume damping at the conventional parameters of the GC. Comparison with our numerical experiments [12] shows that our analytical equation agrees with numerical results in the limits of a few percents in the frequency range conventional for GC. The accuracy of estimation of the volume damping is essentially worse. But since the surface damping strongly dominates the volume damping, the total damping is estimated practically with the same accuracy in a few percents appropriate for the practical exploration.

Acknowledgments

The present work was supported by Russian science foundation, project N 18-19-00447.

References

[1] Baines P G 1967 *Journal of Fluid Mechanics* **30** 533–546
[2] Greenspan H P 1968 *The theory of rotating fluids* (Cambridge University Press, New York)
[3] Kobine J J 1995 *Journal of Fluid Mechanics* **303** 233–252
[4] Duguet Y, Scott J F and Le Penven L 2005 *Physics of Fluids* **17** 114103 pp. 1-13
[5] Bogovalov S, Kislov V and Tronin I 2015 *Theoretical and Computational Fluid Dynamics* **29** 111–125
[6] Glaser A 2008 *Sci. Glob. Secur.* **16** 1–26
[7] Rayleigh L 1884 *Philos. Trans. Roy. Soc. London Ser. A* **175** 1–21
[8] Rayleigh L 1896 *The Theory of Sound* (MacMillan, London)
[9] Bogovalov S, Kislov V and Tronin I 2016 *Applied Mathematics and Computation* **272** 670–675
[10] Bogovalov S, Kislov V and Tronin I 2016 *AIP Conference Proceedings* **1738**
[11] Landau L D and Lifshitz E M 1987 *Fluid Mechanics* (Butterworth-Heinemann, Oxford)
[12] Bogovalov S, Kislov V and Tronin I 2015 *Physics Procedia* **72** 310 – 317

IOP Conf. Series: Journal of Physics: Conf. Series **1099** (2018) 012018 doi:10.1088/1742-6596/1099/1/012018

Dynamical stability of running solitary waves in fluid-filled elastic membrane tubes

A T Il'ichev[1,2]

[1]Steklov Mathematical Institute, Gubkina str. 8, 119991 Moscow, Russia
[2] National Research Nuclear University "MePHI",Kashirskoye sh. 31,115409 Moscow, Russia

E-mail: ilichev@mi.ras.ru

Abstract. We examine the problem of stability of solitary waves, propagating in a fluid-filled membrane tube. We consider waves with speeds starting from those given by the linear dispersion relation (it is known that there may exist four families of solitary waves having such speeds), i. e. the waves of a finite amplitude bifurcating from the quiescent state of the system. It is shown that if a solitary wave speed is bounded away from zero the solitary wave itself is orbitally stable, when the fluid initially stationary (the mean flow is absent).

1. Introduction

The governing equations for quasi-one dimensional motion of the perfect fluid in the axisymmetric membrane tube were obtained in [1] by means of straightforward derivation. Study of spectral stability of a branch of steady solitary-wave solutions (so-called aneurysm solutions) in the absence of the fluid inside the tube (pressure-controlled case) is given in [2]. A bifurcation parameter was the inflation pressure, and the authors found that all family of solitary waves is always spectrally unstable (i. e. a perturbation of a wave form exponentially grows with time). In [3] stability of the whole branch aneurysm solutions is studied when the fluid inside the tube is present, but a mean flow (a constant speed of the fluid at infinity) is zero. It was found there that the aneurysm is still unstable, but the presence of fluid has a strong stabilizing effect. The authors of [4] undertook a stability analysis of aneurysm solution in the presence of the mean flow and found that if a speed of the fluid at infinity is bounded away from zero, then the aneurysm is spectrally stable.

In the present paper we examine the problem of stability of solitary waves, propagating with a non-zero speed in a fluid-filled axisymmetric membrane tube. In this case a bifurcation parameter is not the inflation pressure any more (it may take arbitrary values), but a solitary wave speed. We consider only waves with speeds close to those given by the linear dispersion relation, i. e. the waves of a small (but finite) amplitude bifurcating from the quiescent state of the system. In other words we adopt weakly nonlinear description of solitary waves.

2. Formulation of the problem

We model the tube as incompressible, isotropic, hyperelastic, cylindrical membrane. The tube has a constant undeformed radius R and a constant undeformed thickness H. The tube is assumed to be infinitely long, and end conditions are imposed at infinity. We use cylindrical coordinates, and undeformed configuration is given by coordinates R, Θ, Z.

We assume that the axisymmetry remains throughout the entire deformation; the deformed configuration is expressed using cylindrical polar coordinates r, θ, z, where $r = r(Z,t)$, $\theta = \theta(Z,t)$, $z = z(Z,t)$, and t denotes time.

The principal directions of the deformation correspond to the lines of latitude, the meridian and the normal to the deformed surface, and the principal stretches are given by

$$\lambda_1 = \frac{r}{R}, \quad \lambda_2 = (r'^2 + z'^2)^{\frac{1}{2}}, \quad \lambda_3 = \frac{h}{H}, \tag{1}$$

where the indices 1, 2, 3 are used for the circumferential, axial and radial directions respectively, a prime represents differentiation with respect to Z, and h denotes the deformed thickness.

The principal Cauchy stresses σ_1, σ_2, σ_3 in the deformed configuration for an incompressible material are given by

$$\sigma_i = \lambda_i W_i - p, \qquad i = 1, 2, 3 \quad \text{(no summation)},$$

where $W = W(\lambda_1, \lambda_2, \lambda_3)$ is the strain-energy function, $W_i = \partial W/\partial \lambda_i$, and p is a Lagrange multiplier, associated with the constraint of incompressibility. Utilizing the incompressibility constraint $\lambda_1\lambda_2\lambda_3 = 1$ and the membrane assumption of no stress through the thickness direction $\sigma_3 = 0$, we find

$$\sigma_i = \lambda_i \hat{W}_i, \qquad i = 1, 2$$

where $\hat{W}(\lambda_1, \lambda_2) = W(\lambda_1, \lambda_2, \lambda_1^{-1}\lambda_2^{-1})$ and $\hat{W}_1 = \partial \hat{W}/\partial \lambda_1$ etc. The pipe material has a density ρ.

As an example we give here frequently used three strain-energy functions, the Varga, Ogden and Gent materials, given respectively by,

$$W = 2\mu(\lambda_1 + \lambda_2 + \lambda_3 - 3), \tag{2}$$

$$W = \mu \sum_{r=1}^{3} \mu_r(\lambda_1^{\alpha_r} + \lambda_2^{\alpha_r} + \lambda_3^{\alpha_r} - 3)/\alpha_r, \tag{3}$$

$$W = -\frac{1}{2}\mu J_m \ln\left(1 - \frac{\lambda_1^2 + \lambda_2^2 + \lambda_3^2 - 3}{J_m}\right), \tag{4}$$

where μ is the shear modulus for infinitesimal deformations, $J_m > 0$ is a material constant representing the maximum stretch of the material and $\alpha_1 = 1.3$, $\alpha_2 = 5.0$, $\alpha_3 = -2.0$, $\mu_1 = 1.491$, $\mu_2 = 0.003$, $\mu_3 = -0.023$. The Ogden and Gent materials were proposed in [6] and [7] respectively, and are popularly used to model rubber.

The pressure in the ideal fluid filling tube is denoted by P, the fluid itself has a velocity v_f equal to $v_{f\infty}$ at infinity, the density of the fluid is denoted by ρ_f.

The equations of motion for the tube can be derived using the linear momentum balance applied to an infinitesimal material form of tube's wall [1] . We shall non-dimensionalize the governing equations using the following scales: R for Z, z and r, μ for the Cauchy stresses, $\mu/(RH)$ for P, $\sqrt{\mu/\rho}$ for v_f, and $R\sqrt{\rho/\mu}$ for the time. Using the same notation for the scaled variables, we have

$$\left[\sigma_2 \frac{z'}{\lambda_2^2}\right]' - Prr' = \ddot{z}, \quad \left[\sigma_2 \frac{r'}{\lambda_2^2}\right]' - \frac{\sigma_1}{\lambda_1} + Prz' = \ddot{r}, \tag{5}$$

where

$$z = z_\infty Z + u(Z,t), \quad r = r_\infty + w(Z,t), \quad u, w \to 0, \text{ as } Z \to \infty,$$

and dot denotes the differentiation with respect to dimensionless time.

Expressed in terms of the Lagrangian coordinate Z and the time t the dimensionless equations of motion for the fluid inside the tube read [1], (for derivation, see [3, 4])

$$\dot{r}z' - r'\dot{z} + v_f r' + \frac{1}{2} r v_f' = 0, \quad b_f \left[\dot{v}_f z' - v_f' \dot{z} + v_f v_f'\right] + P' = 0, \tag{6}$$

where b_f is defined by

$$b_f = \frac{\rho_f R}{\rho H}.$$

The governing equations (5) and (6) admit the uniform solution

$$r = r_\infty, \quad z' = z_\infty = \lambda_{2\infty}, v_f = v_{f\infty}, P = P_\infty \equiv \frac{W_1(c, r_\infty, \lambda_{2\infty})}{r_\infty \lambda_{2\infty}}. \tag{7}$$

We look for a general localized traveling wave solution for which the dependence on Z and t is through $Z - \hat{c}t$, where $\hat{c}$ denotes the wave speed of the wave. Localization means that as $Z - \hat{c}t \to \pm\infty$, the fluid-filled tube is in a uniform state given by (7).

We shall also need to refer to the dispersion relation for small-amplitude traveling waves superimposed on the uniform state 7. Assuming that the small-amplitude perturbations are proportional to then the scaled wavenumber k and wave speed $c = \hat{c}\lambda_{2\infty}$ satisfy the dispersion relation [4]

$$\begin{aligned}&\left(k^2 m + 2\right) c^4 - 4 v_{f\infty} c^3 - \left(m\alpha_0 k^2 + m\gamma_1 k^2 - 2v_{f\infty}^2 - m\beta_0 + m\beta_1 + 2\gamma_1\right) c^2 \\ &+4 v_{f\infty}\gamma_1 c - 2v_{f\infty}^2 \gamma_1 + m\gamma_1(k^2\alpha_0 + \beta_1 - \beta_0) - m(\alpha_1 - \beta_0)^2 = 0,\end{aligned} \tag{8}$$

where $m = 1/(b_f r_\infty^2 \lambda_{2\infty})$, and the expressions for the quantities α_0, α_1, γ_1, β_0, β_1 are given in [5] (with remarks in [4]). When $v_{f\infty} = 0$, the above equation becomes a bi-quadratic on c at the limit $k \to 0$, and the roots can be written as

$$4c^2 = 2\gamma_1 + m(\beta_1 - \beta_0) \pm \sqrt{[2\gamma_1 - m(\beta_1 - \beta_0)]^2 + 8m(\alpha_1 - \beta_0)^2}, \tag{9}$$

so that the four roots are all real. These characteristic speeds correspond to four branches of long waves: two of them propagate to the right, the other two propagate symmetrically to the left.

It can be easily shown [1] that the fluid equations (6) in this case can be integrated to yield

$$P = P_\infty + v_{f0}\left(1 - \frac{r_\infty^4}{r^4}\right), \quad v_f = \frac{v_{f\infty} r_\infty^2}{r^2}, \tag{10}$$

where the constant v_{f0} is defined by

$$v_{f0} = \frac{1}{2} b_f (v_{f\infty} - c)^2, \quad c = \hat{c}\lambda_{2\infty}.$$

It is also known [1, 4] that the equations in (5) together with (10) have two integrals. They are given by

$$W - \lambda_2 W_2 + \frac{1}{2} c^2 \lambda_2^2 = C_1, \quad \frac{W_2 z'}{\lambda_2} - \frac{1}{2} P^* r^2 - c^2 z' = C_2, \tag{11}$$

where a prime again denotes differentiation with respect to $Z - \hat{c}t$,

$$P^* = P_\infty + v_{f0}\left(1 + \frac{\lambda_{1\infty}^4}{r^4}\right),$$

and the constants C_1 and C_2 can be determined by evaluating the corresponding left hands at the uniform state (7).

3. Running solitary waves

The two equations (11) can be written as a system of first-order ordinary differential equations [8, 9]:

$$\begin{aligned}\lambda_1' &= \lambda_2 \sin\phi,\\ \lambda_2' &= \frac{W_1 - \lambda_2 W_{12}}{(W_{22} - \hat{c}^2)} \sin\phi,\\ \phi' &= \frac{W_1}{W_2 - \hat{c}^2\lambda_2}\cos\phi - \frac{P\lambda_1\lambda_2}{W_2 - \hat{c}^2\lambda_2},\end{aligned} \tag{12}$$

where the prime denotes differentiation with respect to $\xi = Z - \hat{c}t$, ϕ is the angle between the meridian and the z-axis (so that $\sin\phi = r'/\lambda_2$, $\cos\phi = z'/\lambda_2$). Without loss of generality, we may assume that the center of the symmetric localized travelling wave is located at $\xi = 0$ so that $\phi(0) = 0$. Then if $\lambda_1(0)$ and $\lambda_2(0)$ are also known, the solitary wave solution can be determined by integrating the above system as an initial value problem.

According to (9) at fixed values of r_∞, z_∞ we determine the bifurcation value of the speed c and compute numerically a solitary wave solution for c varying. Hence, we obtain the solitary wave family, parameterized by c. There are two such families for solitary waves running in each direction, and these families begin with corresponding bifurcation values of c and end with some finite values of c.

4. Spectral stability

We denote such fully nonlinear bulging solutions of (3) by $r = \bar{r}(\xi)$, $z = \bar{z}(\xi)$, $P = \bar{P}(Z)$, $v_f = \bar{v}_f(\xi)$. To study their stability, we consider axisymmetric perturbations, and write

$$r(\xi,t) = \bar{r}(\xi) + \Psi(\xi)e^{\eta t}, \quad z(\xi,t) = \bar{z}(\xi) + \Phi(\xi)e^{\eta t},$$

$$P(\xi,t) = \bar{P}(\xi) + \Pi(\xi)e^{\eta t}, \quad v_f(\xi,t) = \bar{v}_f(\xi) + V(\xi)e^{\eta t},$$

where the mode functions $\Psi(\xi), \Phi(\xi), \Pi(\xi), V(\xi)$ and the growth rate η are to be determined. The equations for perturbations can be written in the form

$$\mathbf{y}' = \mathcal{M}\mathbf{y}, \tag{13}$$

where dot denotes differentiation with respect to ξ, $\mathbf{y} = (\Phi, \Phi', \Psi, \Psi', \Pi, V)^T$ and $\mathcal{M}$ is a 6×6 matrix whose components are (numerically) known functions of ξ and η. This system of equations are to be solved subject to the decay conditions $\mathbf{y} \to \mathbf{0}$ as $Z \to \pm\infty$. Denoting by $\mathcal{M}_\infty$ the limit of $\mathcal{M}$ as $\xi \to \pm\infty$, and substituting a trial solution of the form $\mathbf{y} = e^{\hat{k}\xi}\mathbf{r}$ into $\mathbf{y}' = \mathcal{M}_\infty\mathbf{y}$, we obtain the eigenvalue problem

$$(\mathcal{M}_\infty - \hat{k}I)\mathbf{r} = \mathbf{0}, \tag{14}$$

where I is the 6×6 identity matrix. Denote by $\hat{k}_1$, $\hat{k}_2$, $\hat{k}_3$ the three eigenvalues of (2) with negative real parts, and by $\mathbf{r}_1$, $\mathbf{r}_2$, $\mathbf{r}_3$ the associated right eigenvectors. There then exist solutions $\mathbf{y}_i(\xi)$ and of (1) such that

$$\lim_{\xi\to\infty} e^{-k_i\xi}\mathbf{y}_i(\xi) = \mathbf{r}_i, \quad i = 1, 2, 3.$$

Alternatively, we may consider the exterior system [10]

$$\mathbf{y}^{\wedge\prime} = \mathcal{M}^{\wedge}\mathbf{y}^{\wedge}, \tag{15}$$

and its adjoint system

$$\mathbf{x}^{\wedge\prime} = -\mathbf{x}^{\wedge}(\mathcal{M}^{\wedge}), \tag{16}$$

To construct the Evans function we need to solve (15) and (16) subjected to the conditions at both infinities

$$\lim_{\xi\to\infty} \mathrm{e}^{-k^{\wedge}\xi}\mathbf{y}^{\wedge}(\xi) = \mathbf{r}^{\wedge},$$
$$\lim_{\xi\to infty} \mathrm{e}^{k^{\wedge}\xi}\mathbf{x}^{\wedge}(\xi) = \mathbf{l}^{\wedge}, \tag{17}$$

where $k^{\wedge} = \hat{k}_1 + \hat{k}_2 + \hat{k}_3$, $\mathbf{r}^{\wedge}$ and $\mathbf{l}^{\wedge}$ are right and left eigenvectors of $\mathcal{M}^{\wedge}_{\infty}$ associated with the eigenvalue $k^{\wedge}$, correspondingly. The Evans function is defined by

$$D(\eta) = \mathbf{x}^{\wedge}(\eta, \xi) \cdot \mathbf{y}^{\wedge}(\eta, \xi), \tag{18}$$

and it can be shown that the condition of existence of an unstable eigenvalue η_0 is equivalent to $D(\eta_0) = 0$.

The problem of finding the unstable discrete spectrum η (eigenvalues located in the right half complex η-plane Ω^+) is analogous to determining the zeroes of the Evans function $D(\eta)$ lying in Ω^+. The number of zeroes of $D(\eta)$ can be computed with the help of the argument principle. The number of zeroes of $D(\eta)$ in Ω^+ is determined by the number of rotations of the image of the imaginary axis $\{\eta : \eta = \mathrm{i}\tau,\ \tau \in \mathbb{R}\}$ under the mapping $D(\cdot)$. Therefore, we need to construct the function $D(\eta)$ on the imaginary axis, i. e., numerically solve the ordinary linear equations (3), (4) under the conditions (5) with η running in a considerably large segment of the imaginary axis [11].

Our numerical computations showed that for bifurcation values of c increasing from zero two existing unstable real eigenvalues in Ω^+ converge, at some value $c = c_0$ coincide, and for $c > c_0$ cease to exist, so all solitary wave family, bifurcating from c is spectrally stable. This corresponds to the nonlinear orbital stability result of solitary waves of small amplitude [12], i. e. “beginners” of the respective family of solitary waves.

Acknowledgments

This work was supported by the Russian Science Foundation, project no. 16-19-00188.

References

[1] Epstein M and Johnston C On the exact speed and amplitude of solitary waves in fluid-filled elastic tubes 2001 *Proc. Roy. Soc. Lond.* A **457** 1195–213
[2] Pearse S P and Fu Y B Characterization and stability of localized bulging/necking in inflated membrane tubes 2010 *IMA J. Appl. Math.* **75** 581–602
[3] Il'ichev A T and Fu Y B Stability of aneurysm solutions in a fluid-filled elastic membrane tube 2012 *Acta Mech. Sin.* **28** 1209-18
[4] Fu Y B and Il'ichev A T Localized standing waves in a hyperelastic membrane tube and their stabilization by a mean flow 2015 *Math. Mech. Solids* **20** 1198-214
[5] Fu Y B and Il'ichev A T Solitary waves in fluid-filled elastic tubes: existence, persistence, and the role of axial displacement 2010 *IMA J. Appl. Math.* **75** 257–68
[6] Ogden R W Large deformation isotropic elasticity-on the correlation of theory and experiment for incompressible rubber-like solids 1972 *Proc. Roy. Soc. Lond.* A **326** 565–84
[7] Gent A N A new constitutive relation for rubber 1996 *Rubber Chem. Nechnol.* **69** 59–61
[8] Fu Y B, Pearce S P and Liu K K Post-bifurcation analysis of a thin-walled hyperelastic tube under inflation 2008 *Int. J. Non-linear Mech.* **43** 697–706
[9] Fu Y B and Xie Y X Effects of imperfections on localized bulging in inflated membrane tubes 2012 *Phil. Trans. R. Soc.* A **370** 1896–911
[10] Alexander J C and Sachs R Linear instability of solitary waves of a Boussinesq-type equation: A computer assisted computation 1995 *Nonlinear World* **2** 471–507

IOP Conf. Series: Journal of Physics: Conf. Series **1099** (2018) 012018 doi:10.1088/1742-6596/1099/1/012018

[11] Pego R L, Smereka P and Weinstein M I Oscillatory instability of traveling waves for a KdV-Burgers equation 1993 *Physica* D **67** 45–65
[12] Il'ichev A T Stability of solitary waves in membrane tubes: A weakly nonlinear analysis 2017 *Theoret. and Math. Phys.* **193** 1593–601

The modeling of anomalous transport in the diffusion-ballistic regime

P A Likhomanova[1,2], I V Tronin[1], A M Grekhov[1], V N Tronin[1] and V D Borman[1]

[1] National Research Nuclear University «MEPhI», Kashirskoe highway 31, Moscow, Russia
[2] National Research Center «Kurchatov Institute», pl. Akademika Kurchatova 1, Moscow, Russia

likhomanovapa@gmail.com

Abstract. The anomalous transport of a point-like particle within the disordered heterogeneous medium with accelerating traps was investigated using Monte Carlo simulation within the CTRW model. Outside the traps a particle moves according to the diffusion regime of motion. Inside the traps the ballistic regime takes place. Results of the Monte Carlo simulation shows that the final distribution of particle jumps belongs to the class of the stable distributions which corresponds to the anomalous diffusion regime. It is shown that the characteristic exponent of the stable distribution depends on the trap density and particle velocity inside the traps.

1. Introduction

In recent years the anomalous diffusion attracts the increasing interest in the research of diverse physical phenomena: delayed neutrons [1], cardiovascular systems [2], proteins transport in living cells [3] etc. In the first instance this interest is caused by the circumstance that structural features of the environment may be the reason of the anomalous transport [4]. Therefore a lot of works dedicated to the anomalous diffusion are focused on the diffusion in disordered media [5-10]. Such anomalous processes frequently appear within heterogeneous media because of the difference in velocity of particle motion in each phase. They are connected with acceleration (or deceleration) of a particle in media. For today a set of works are known in which the anomalous diffusion is studied in different heterogeneous systems: hindered transport of proteins in cell structures [3], cooling of silicon semiconductors, fast transport through CNT membranes [5] etc. Consequently the structural features of the medium and velocity of a particle inside the medium may determine the transport regime of particles.

Numerical simulations are frequently used for determination of the transport regime [6-10]. Various diffusion regimes (from sub-diffusion to ballistic) were obtained using numerical simulation for media with obstacles of different geometric shapes. Various objects, considered as obstacles and placed in a medium, were studied: spheres [6], rods, solid sphere and a permeable shell - a sphere of a larger radius (within so-called cherry-pit model [7]), a model of a porous medium formed with Voronoi tessellation [6, 8, 9], various lattice configurations with obstacles [10] etc. Lattice models are relatively simple to calculate and very well studied. Continuum models, on the contrary, are more complex in terms of modeling, however, they better describe real media - for example, polymers. Therefore, recently, continuous medium models are more often used. In these models transport processes can be described using the CTRW model [11]. CTRW makes it possible to describe jump stochastic processes in terms of probability density functions (PDFs) of length of particle jumps and waiting time between them. A lot of processes with different probability density functions of times and jumps were considered and described within the CTRW model [12-18]. At the same time, these PDFs might not be always known initially. The question is how to describe the anomalous diffusion in systems where the PDF of jumps and/or times isn't known a-priori. The answer can be obtained from a numerical modelling.

In this work we present the results of modeling of the two dimensional heterogeneous medium with rectangle shaped traps with a solid core and a permeable shell (as a cherry-pit model). Outside the trap particle motion is the ordinary diffusion with jumps and jump times distributed by the Gauss and the Poisson PDFs, respectively. Particle motion inside a trap represents the ballistic regime -- after getting inside a trap a particle moves to the opposite side of the trap. Time of flight inside the trap was calculated from the fixed particle velocity inside the trap and the distance of flight. The results of modeling shows that the PDF of particle jumps in such heterogeneous medium corresponds to the stable distributions. A characteristic exponent of the stable PDF is determinated by the density of traps and decreases characterized by the increase in trap density.

2. Numerical modeling

Calculations were performed on a square system with the linear dimension $L=15$ within periodic boundary conditions. The system is randomly filled with rectangles. All rectangles have an impermeable solid core with dimensions $l=1$, $d=0.05$, and a permeable shell with dimensions $l=1.05$, $d=0.1$. The density of the rectangles is defined as $\eta=\frac{lmN}{L^2}$, where l, m is the length and the width of the rectangle, respectively, and N is the number of rectangles. Density was varied in the range of $0\le\eta\le0.66$. For each density calculation was performed using the Monte Carlo method.

For each step of simulation there are two possible situations: the particle is located in the medium outside the trap or captured into the trap. Outside the trap particle motion is the ordinary diffusion and the space and the time steps are chosen from the Gaussian and the Poisson PDFs, respectively. When a particle attempts to jump into the area of impermeable rectangles, it is reflected from a solid core into the region of a permeable shell. If a particle belongs to one of the permeable rectangles (a particle is trapped), it moves to the opposite side of the rectangle. The flight time is calculated from the specified dimensionless velocity V ($V=100$, $V=0.1$ and instantaneous displacements are investigated) and the distance corresponding to the flight length. For each simulation we calculated the final position of the particle at time $t=10000$. Totally we performed 10^6 simulations for each value of the trap density.

3. Analysis and results

We used the following parametrization of the stable distribution [19]:

$$\varphi(t,\alpha,\beta,c,\mu)=\exp\left[it\mu-|ct|^{\alpha}\cdot(1-i\beta\cdot\operatorname{sgn}(t)\cdot\tan\left(\frac{\pi\alpha}{2}\right)\right],\ \alpha\neq1$$

Where $-\infty<\mu<\infty$, $c\ge0$, $0<\alpha\le2$, $-1\le\beta\le1$ are the distribution parameters.

The characteristic exponent α determines the form of the distribution function, location parameter μ - position of the maximum of the distribution, scale parameter c - width of the distribution, and skewness parameter β determines distribution asymmetry.

For $\alpha=2$ the stable distribution changes into the Gaussian distribution. Since the main difference between the Gaussian and stable distributions corresponds to the characteristic exponent α, we focused on its calculation. α was calculated using the maximum likelihood method. In our case, $1<\alpha<2$.

Figure 1 shows the dependence of the characteristic exponent α on the trap density η for the case of finite velocity and for instantaneous particle displacements in traps. In the absence of the traps ($\eta = 0$) characteristic exponent $\alpha = 2$ corresponds to the Gauss distribution, whereas for every $\eta > 0$ we obtain stable distribution with $\alpha < 2$. This result corresponds to the effective acceleration of the particle caused by traps and transition from the diffusion to the superdiffusion regime. Effect of the finite particle velocity in the traps slightly slows down the acceleration process in comparison with the case of instantaneous displacements, as shown in [20]. This slowdown is illustrated in Figure 1. It is seen that decreasing of the velocity corresponds to increasing of the characteristic exponent $\alpha(\eta)$. This increasing corresponds to the decreasing of the mean square displacement of the particle during the walk in the heterogeneous medium. However, in spite of the fact that the finite velocity of motion slows down the particle in comparison with the case of instantaneous displacements, the form of the distribution of jumps in traps remains stable [20].

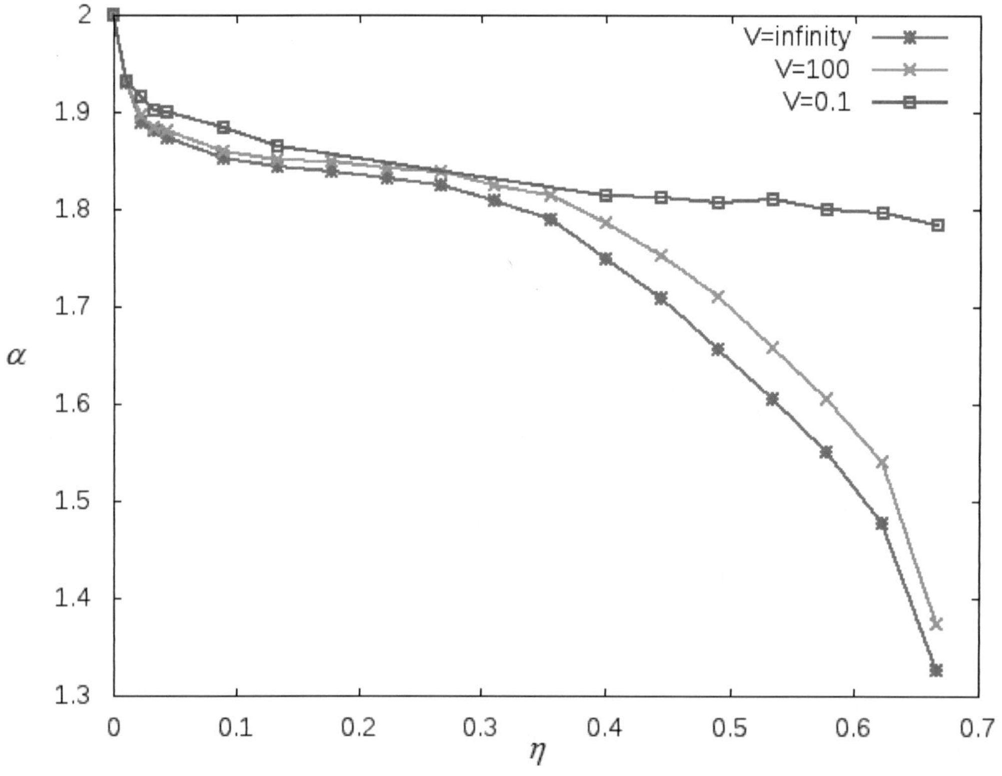

Figure 1. The characteristic exponent of stable distribution α as function of the density of rectangular traps for the case of finite velocities *V=100, V=0.1* and instantaneous displacements.

From Figure 1 it can be seen that the velocity impacts weakly on the parameter α for $\eta < 0.3$. However, when $\eta > 0.3$, change of speed has a significant impact on the shape of stable distribution showed on Figure 2. This effect may be associated with formation of the clusters of connected traps with increasing traps density: clusters create a path along which the particle will fly through the cluster and that increase the effective speed of movement. The formation of the connected traps through which ballistic transport takes place allows the particle to fly away on long distances, unlike the case of unconnected traps. Thus, two factors apparently influence on the mean square displacement of a particle: the formation of clusters of connected traps and the ballistic regime of motion. However, stable distributions are caused by the ballistic regime of motion inside the traps.

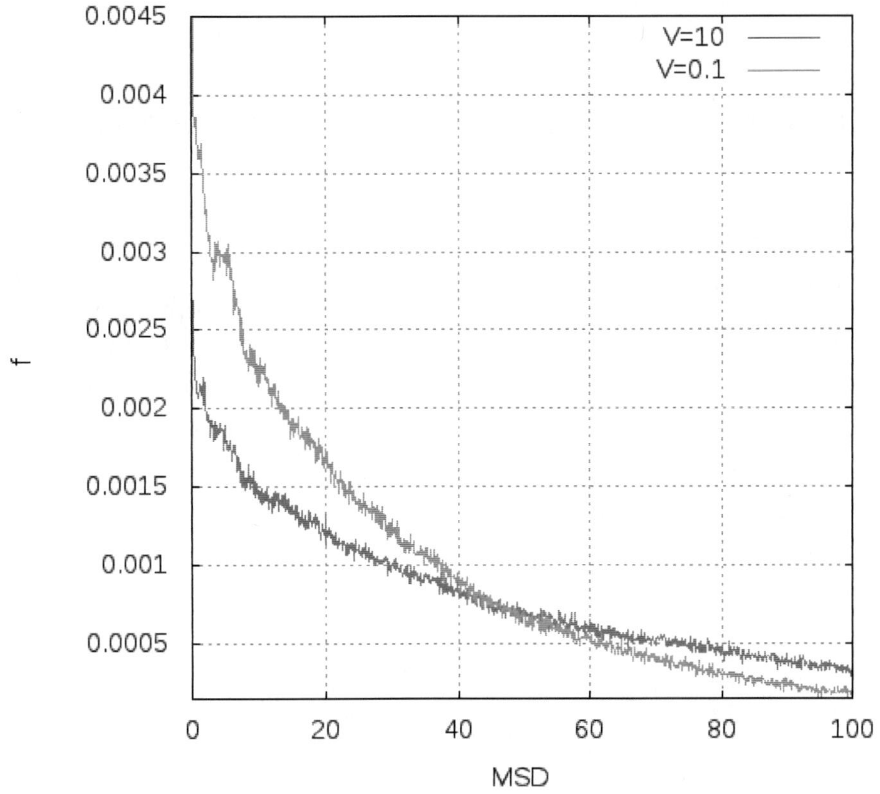

Figure 2. The distribution function of the mean square displacement of a particle for the velocities *V=0.1, V=10* for a fixed density $\eta = 0.578$.

4. Conclusions

In this paper the diffusion of a particle inside the heterogeneous media with accelerating traps has been investigated using the Monte Carlo method. A particle inside accelerating traps moves in the ballistic regime. The results of simulation show that the distributions of particle jumps belong to the class of stable distributions. Therefore we present one of the possible mechanisms how stable distributions in the heterogeneous media are formed. Also, the influence of the main parameters – trap density and particle velocity in traps, on the form of obtained stable distribution has been investigated.

Acknowledgements

Calculations were partially performed using the high-performance computing center of the NRNU MEPhI (Basov farm).

References

[1] Yarovikova I V and Uchaikin V V 2003 *Journal of computional mathematics and mathematical physics* **43** 1478-1490
[2] Hughes A D 2015 *Artery Research* **10** 1–10
[3] Wachsmuth M, Rippe K, Baum M and Erdel F 2014 *Nature Communications* **5** 1–12
[4] Delhommelle J, Hicks E and Desgranges C 2014 *Journal Molecular Simulation* **40** 656–663
[5] Li J, Verweij H and Schillo M C 2007 *Small* **3**(12) 1996–2004
[6] Frey E, H¨ofling F and Munk T 2008 *The Journal of Chemical Physics* **128**(16) 164517-164517
[7] Hofling F, Franosch Th, Horbach J, Schnyder S K and Spanner M 2015 *Soft Matter* **11** 701–711
[8] Saxton M J 2010 *Biophys J.* **99**(5) 1490–1499
[9] Wright J L and Hastedt J E 2006 *Pharmaceutical Research* **23** 2427–2440

IOP Conf. Series: Journal of Physics: Conf. Series **1099** (2018) 012019 doi:10.1088/1742-6596/1099/1/012019

[10] Saxton M J 1994 *Biophys J.* **66** 394–401
[11] Gorenflob R and Abdel-Rehima E A 2008 *Journal of Computational and Applied Mathematics* **222**(2) 274–283
[12] Klafter J and Metzler R 2000 *Physics Reports* **339** 1–77
[13] Metzler R and Tejedor V 2010 *Journal of physics A: Mathematical and theoretical* **43** 082002
[14] Metzler R, Schulz J H P and Chechkin A V 2013 *Journal of physics A: Mathematical and theoretical* **46** 475001
[15] Metzler R, Jeon J H and Chechkin A V 2014 *Phys.Chem.Chem.Phys.* **16**(30) 15811– 15817
[16] Metzler R, Barkai E, Burov S and Jeon J H 2011 *Phys.Chem.Chem.Phys.* **13**(5) 1800–1812
[17] Barkai E and Froemberg D 2013 *The European Physical Journal B* **86**(331) 1–13
[18] Metzler R, Cherstvy A G and Chechkin A V 2014 *Soft Matter* **10**(10) 1591–1601
[19] Nolan J P 2016 *Stable Distributions Models for Heavy Tailed Data* (Birkhäuser, Boston)
[20] Saenko V V, Zolotarev V M and Uchaikin V V 1999 *JETP* **115** 1411–1425

IOP Conf. Series: Journal of Physics: Conf. Series **1099** (2018) 012020 doi:10.1088/1742-6596/1099/1/012020

Non-classical transport in highly heterogeneous porous media

P S Kondratenko[1,2] **and L V Matveev**[1,2]*

[1] Nuclear Safety Institute, Russian Academy of Sciences, Bolshaya Tul'skaya St. 52, Moscow, Russia, 115191
[2] Moscow Institute of Physics and Technology (State University), Institutskii per. 9, Dolgoprudny, Moscow Region, Russia, 141700
*E-mail: matweev@ibrae.ac.ru

Abstract. The review is presented on non-classical impurity transport in highly heterogeneous media with various correlations of structure characteristics. Regularly heterogeneous, statistically homogeneous double-porous and fractal media are considered. Concentration distributions in the main body and at asymptotically large distances are described. The effects of sorption onto colloids are analyzed.

1. Introduction

In this paper the physical models developed by the authors for non-classical impurity transport are reviewed. By non-classical (anomalous) regimes are implied the regimes for which the size of main body localization depends on time according to

$$R \sim t^{\gamma}, \qquad \gamma \neq 1/2. \tag{1}$$

We have super-diffusion when $\gamma > 1/2$ and sub-diffusion at $\gamma < 1/2$. Anomalous transport processes occur in a variety of natural Sciences, such as plasma physics [1, 2], semiconductor physics [3, 4], astrophysics [5], biophysics [6, 7], hydrology [8, 9], and many others [10]. A special place in this series is occupied by geological media. An extensive amount of field observations accumulated in the last decades evidences that in many cases classical regularities based on Darcy and Fick laws can not describe contaminant transport processes in geological media so that discrepancies may be of several orders [11]. It is to this range of tasks that our studies were oriented.

Historically, the first models describing non-classical transport date back to the 30s of the last century. Apparently, the first was the work of Khinchin and Levy [12], in which the migration of a particle was considered as a result of its successive jumps of varying length and duration. The probabilities of the jumps were determined by the distribution function, and if the given function decreased sufficiently slowly with increasing these parameters, then the mean square displacement of the particle grew with time according to a nonclassical law. Later, this approach was actively used in other models for non-classical transport, such as CTRW [13], equations with fractional derivatives [14], models with non-Markov processes [15], nonlinear Fokker-Planck equation [16].

In our studies, a different approach was adopted. It is based on the averaging of equations in which impurity transport is caused by classical physical mechanisms, taking into account the distribution of inhomogeneities characteristic of a particular type of medium. The specificity of transport in geological environments is that the transport of impurities occurs as a solution in groundwater through channels formed by voids in the rock (pores, fractures). The mechanisms of transport are advection, that is, transport with the speed of local flow, and molecular diffusion in solution. As a result, the equation for the concentration in the channels has the form of a law of conservation of mass, with a flow of the classical form, including advection and diffusion.

To further construct the theory, it is necessary to average this equation over space, or over an ensemble of realizations. At the same time, the following properties of the medium are the key: first, the geometry of the channels, and secondly, a strong contrast of the properties characterizing the individual regions.

As for the first factor, conditionally all media from the point of view of correlation in the distribution of inhomogeneities can be divided into three classes: 1) the simplest type is regularly heterogeneous sharply contrasting media, such as a fracture in a porous medium; 2) further, statistically homogeneous media - when , starting from a certain averaging scale, the result does not depend on the scale, so that we can introduce average permeability, average velocity, etc., and 3) these are percolation media, or media that, within a certain large scale have fractal properties.

The second factor, fundamental for the emergence of nonclassical regimes, is a sharp contrast between the properties of neighboring regions. Indeed, if we have a fractured-porous medium and the fractures in it are of sufficiently large apertures, then the velocity of percolation through the fractures will be much greater than over the surrounding porous matrix. The difference, as a rule, is so great that in the main approximation we can neglect the leakage in the porous matrix and consider here only the diffusion. As a result, we can assume that the medium consists of two subsystems - strongly and slightly permeable. This approximation is called a double-porous medium [17]. Examples of such a medium are isolated cracks in the porous matrix, fractured-porous media with a continuous distribution of fractures, sandy-clay formations. Next, non-classical transport will be considered in regularly inhomogeneous (section 2), percolation (section 3) and statistically homogeneous sharply contrasting (section 4) media. Section 5 is devoted to an analysis of the effect of sorption effects on transport. We summarize our results in section 6.

2. Regularly heterogeneous media

2.1. Statement of the problem

Transport in regularly inhomogeneous media was considered in the following formulation. The medium consists of two regions - a geometrically regular, strongly permeable region I, which we call a crack, and the rest of the space (region II) filled with a porous but weakly permeable matrix. Two types of cracks were considered: a) a plane-parallel layer of thickness, and b) an infinite cylinder.

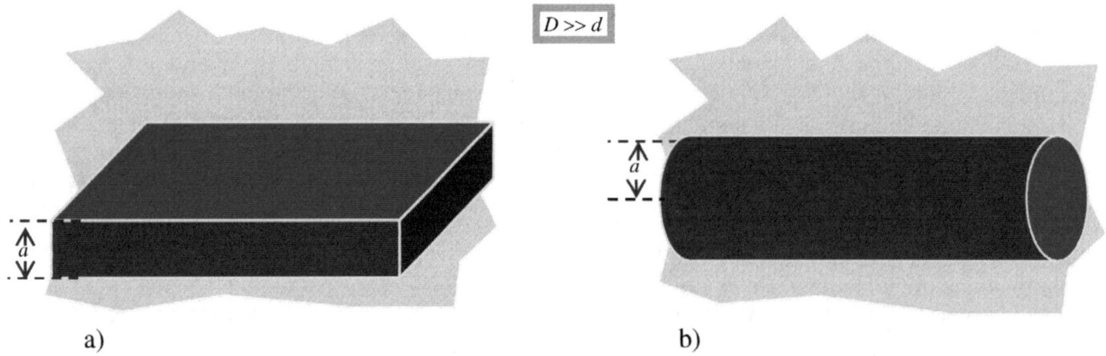

Figure 1. Schematic representation of a regularly heterogeneous medium.

Let us consider case a). In both regions, the transfer is determined by the diffusion mechanism, so that for the concentration $n(\vec{r},t)$ in the region *I* the equation is valid

$$\frac{\partial n}{\partial t} = D\Delta n\,, \tag{2}$$

and for the concentration $c(\vec{r},t)$ in the region *II* we have

$$\frac{\partial c}{\partial t} = d\Delta c . \quad (3)$$

There is a strong inequality $D >> d$. On the boundary of two regions, the conditions for equality of concentrations and normal flows are valid. At the initial time, the entire impurity is concentrated in the crack at the origin. This model was first proposed by A.M. Dykhne, therefore we call it the Dykhne model [18].

2.2. Behavior of the main body of impurity localization

Qualitative analysis has shown that the nature of the transfer in such a system depends on the time interval. At short times, $t << t_1$, where

$$t_1 = a^2/4d , \quad (4)$$

the impurity particles do not yet have time to go beyond the crack, and the transport along the crack occurs according to the classical diffusion law with a coefficient D. However, with time, particles will spend in the fracture less and less time, and basically they will be in the matrix. The fraction of time that the particles spend in the fracture and diffuse in the fracture plane with a coefficient D will be approximately equal to $\tau(t) \approx \frac{a}{\sqrt{dt}}$. In accordance with this, the size of the impurity cloud in the crack plane will grow according to

$$R^2 \approx \int_{t_1}^{t} D\tau(t')dt' \approx D\sqrt{t_1 t} . \quad (5)$$

The transfer along the matrix at this stage gives a very small contribution, and thus the matrix acts as a trap. This leads to the appearance of a nonclassical, sub-diffusion regime: $R \sim t^{1/4}$. Similarly, for a fast medium in the form of an infinite cylinder, the size of the impurity cloud along the axis of the cylinder increases according to the law

$$R \approx \sqrt{Dt_1 \ln(t/t_1)} . \quad (6)$$

At long times, the fraction of time spent by the particles in the crack is negligible, so that the particle practically always spends in a slow medium. As a result, the transfer is determined by the properties of the slow medium, that is, classical diffusion, but with a small coefficient d.

Similarly, a model can be considered in which, along with diffusion, advection with constant velocity u is also taken into account in the crack [19]. In this case, the behavior of the system depends on the Peclet number $Pe = \frac{ua}{d}$, which determines the relative speed of the diffusion and advection processes. When $Pe << 1$, the impurity behavior is described by the same laws as in the simple Dykhne model. When $Pe >> 1$ new non-classical regimes arise which are quasi-diffusive ones. They are characterized by the fact that the average displacement of particles increases with time according to the same laws as the size of the cloud. The velocity of the front motion depends on the shape of the crack. For a plane crack, the displacement of the front is proportional to the root of time, and for an infinite cylinder, to the logarithm of time.

2.3. Concentration asymptotics at large distances

All the above regimes can be obtained from the exact solution of the problem, which confirms the results obtained above on the basis of semi-qualitative arguments [20, 21]. But an exact solution also makes it possible to calculate the behavior of the concentration at asymptotically large distances from the source, $r >> R(t)$. The calculation shows that the asymptotical behavior at large distances, again, depends on the time interval. At short times, the concentration decreases due to Gauss, and is characterized by a high diffusion coefficient, D. When in the main cloud the transport regime changes to sub-diffusion, the asymptotics becomes two-stage. The near segment is described by a stretched exponential and is determined by sub-diffusion, and the far segment is described by a Gaussian exponential with a large diffusion coefficient. Namely, it is determined by the mode from the previous time interval. At the latest times, the asymptotics become three-stage: the two remote parts are the same as in previous time interval, and the nearest part is determined by the current regime of slow classical diffusion. The structure of the asymptotics at large times is shown schematically in Fig. 2.

Thus, from the Dykhne model it is possible to draw the following conclusions: in sharply contrasting media, the weakly permeable regions serve as traps that greatly retard the rate of impurity transport. This slowing leads to the appearance of sub-diffusion regimes, which essentially depend on the geometry of the medium and the mechanisms of transport. The change in the transport regimes that occurs in time in the main cloud leads to the formation of a multi-stage distribution of the concentration at asymptotically large distances. The following regularity was established for the concentration distribution in asymptotics: the more distant parts of the asymptotics are determined by the earlier modes of transport in the main cloud.

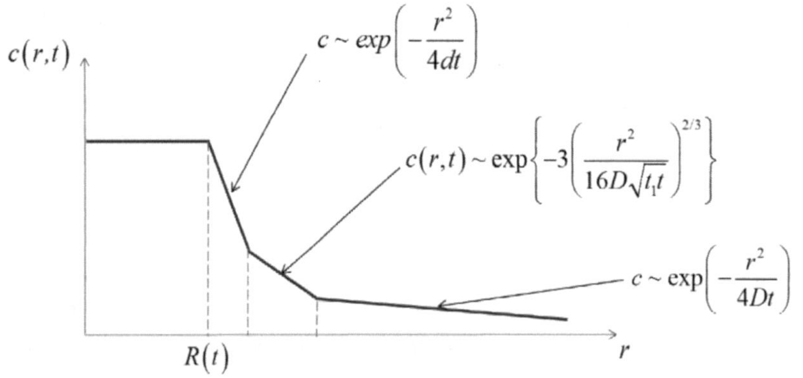

Figure 2. Multi-stage structure of concentration at asymptotically far distances from the source.

3. Percolation media

Now we consider the media in which the system of channels with high permeability has the form of a percolation cluster. In nature, this situation is often found in fractured formations, as well as in porous media, if there is a strong scatter of pores in size. A peculiarity of percolation media is that in the range of large spatial scales up to the correlation length ξ they are fractal and have the property of scale invariance. At scales greater than correlation length ξ they are statistically homogeneous.

Note that the percolation clusters themselves have an internal structure - they consist of a backbone that connects remote parts of the clusters providing long-range transport, and dead ends that are finite and join the core at a single point.

The equation for the microscopic concentration $\hat{c}$ during the transport through the percolation cluster is the standard conservation law, where the flux is determined by advection in a random velocity field:

$$\frac{\partial \hat{c}}{\partial t} + div\left(\vec{V}\hat{c}\right) = 0 \quad (7)$$

Our goal is to describe the impurity transport in term of the concentration averaged over the ensemble of realizations, $c = \langle \hat{c} \rangle$. The velocity field was considered to be incompressible, which is valid in describing the flow of groundwater. As the first step, an isotropic medium with a correlation length tending to infinity was considered. In this case, the velocity averaged over the ensemble of realizations is zero, so that the main quantity determining the transport is the velocity correlator.

Due to the property of scale invariance, under the transformation of similarity $\vec{r} \to \lambda \vec{r}$, an arbitrary quantity A transforms according with the relation $A \to \lambda^{\Delta_A} A$. The value Δ_A is called as the scale dimension of the quantity A. Applying this property to the pair velocity correlation function $\left\langle \vec{V}(\vec{r}_1)\vec{V}(\vec{r}_2) \right\rangle$, we conclude that in an isotropic homogeneous space this function has the form

$$\left\langle \vec{V}(\vec{r}_1)\vec{V}(\vec{r}_2) \right\rangle \propto V^2 \left(\frac{a}{|\vec{r}_1 - \vec{r}_2|} \right)^{2h}, \qquad (8)$$

where h is the velocity scale dimension.

Turning to the description of the concentration averaged over the ensemble of realizations, we arrive at an integro-differential equation, the study of which was carried out in different ways: on the basis of an analysis of scale dimensions [22], and using the Feynman diagram technique [23]. Both methods lead to similar results, which consist in the following.

The impurity transport mode depends on the rate of decrease of the correlation function, i.e. on the value of parameter h. When $h > 1$, the transport corresponds to the classical diffusion regime, so that the concentration is described by a Gaussian profile, including the asymptotically large distances. If $h < 1$, the super-diffusion regime is realized, so that the size of the impurity cloud increases with time according to the law

$$R \approx \left(a^h V t \right)^{\frac{1}{1+h}}. \qquad (9)$$

In this case, the decrease of concentration asymptotics is described by a compressed exponential, that is, it turns out to be faster than in the classics [22].

For the applicability of the random advection model to real media, the influence of the following factors was considered: the finite value of the correlation length, the anisotropy of percolation media, and the effect of traps (dead ends and a weak permeable porous matrix surrounding percolation cluster).

If the correlation length is finite, $\xi < \infty$, the behavior of the impurity is described as follows [24]. At times $t < t_\xi$, when the size of the cloud of impurity is smaller than ξ, the transport mode is the same as in an infinite fractal medium. Therefore, the transport under the condition $h < 1$ occurs in the super diffusion regime, according to (10). At times $t > t_\xi$ when the size of the cloud exceeds ξ, the medium becomes statistically homogeneous, and, consequently, the transport becomes a classical diffusion. The effective diffusivity D_ξ and the characteristic time t_ξ are given by the equalities

$$D_\xi \approx \frac{\xi^2}{t_\xi}, \qquad t_\xi \approx \frac{\xi}{V a^h}. \qquad (10)$$

Asymptotics of the concentration at large distances were also obtained. At times less than t_ξ, the decrease with distance was described by a compressed exponent, as before. At long times, $t >> t_\xi$, the asymptotics are two-stage. The near part corresponds to the current regime of classical diffusion, and the distant part corresponds to the previous regime of super-diffusion. In this case, the behavior in the asymptotics corresponds to the general laws obtained for regularly inhomogeneous media.

Consider now anisotropic fractal media. In a general case, the anisotropy of geological media is due to the gravity or the pressure gradient. In the presence of anisotropy, a fractal medium behaves differently under the scale transformations along and across the anisotropy axis. Therefore, it is necessary to introduce an additional index β for the coordinates in the basal plane, so that instead of $r \to \lambda r$, we consider the transformation $\{ z \to \lambda z, \ r_\perp \to \lambda^\beta r_\perp \}$. The study [25] showed that, depending on the parameter β, two cases are possible, namely, strong and weak anisotropy. When the anisotropy is weak ($h<1, \frac{1}{1+h}<\beta<\frac{2}{1+h}$), the transport goes in the super-diffusion regime in all directions, albeit at different rate:

$$R_{\parallel}(t)\sim\left(a^{h}Vt\right)^{\frac{1}{1+h}},\quad R_{\perp}(t)\sim\left(b^{\beta(1+h)-1}Vt\right)^{\frac{1}{\beta(1+h)}} \tag{11}$$

In the case of strong anisotropy ($h<1, \beta>\frac{2}{1+h}$), the transport in the longitudinal direction is still super-diffusion, and in the basal plane it is the classical diffusion:

$$R_{\parallel}(t)\sim\left(a^{h}Vt\right)^{\frac{1}{1+h}},\quad R_{\perp}(t)\sim\sqrt{D_{\perp}t} \tag{12}$$

It is also important to note that in the presence of anisotropy, an anomalous drift takes place, when the average particle displacement increases in accordance with the superdiffusion law, so as is the size of the impurity cloud along the anisotropy axis $\langle z\rangle\sim\sqrt{\langle z^{2}\rangle}=R_{\parallel}(t)$.

In real percolation environments, traps are always contained. One can suggest at least two sources of traps. First, as already indicated, this is due to the very structure of the percolation cluster, which, along with the backbone, has also dead ends. Secondly, the percolation cluster itself is often immersed in a porous, although weakly permeable medium. In this case, the term describing the exchange of impurities between the backbone and the traps enters the transport equation for the concentration averaged over the ensemble of realizations, in a manner analogous to that in the Dykhne model.

This term has the shape

$$Q(r,t)=\frac{\partial}{\partial t}\int_{0}^{t}\varphi(t-t')c(r,t')dt' , \tag{13}$$

where the properties of the integral kernel are determined on the basis of the properties of scale invariance of the medium [26, 27]. It follows that there is a large time interval, when the kernel has a power-law form $\varphi(t)\sim\frac{1}{t_{1}}\left(\frac{t_{1}}{t}\right)^{\alpha}$, where α characterizes the decay of the kernel in the self-similar interval. At large times, $t>t^{*}$, when the traps are saturated, the kernel decreases exponentially with time. As a result, the index determining the rate the impurity cloud growth is determined both by the properties of the velocity field and by the properties of the traps [28]:

$$R(t)\approx\left(a^{h}Vt_{1}^{\alpha}t\right)^{\frac{1-\alpha}{1+h}}\propto t^{\frac{1-\alpha}{1+h}}. \tag{14}$$

Depending on the relationship between the parameters (but under the condition $h<1$), the transport can occur in both super- and sub-diffusion modes.

To check the theory's conclusions, a comparison was made between the results of the theory and experiment [29]. The experiment [29] was carried out in fractured crystalline rocks. Injection and pumping out of the tracer were produced in a converging weakly dipole configuration. As tracers were chosen heavy water, bromides, pentafluorobenzoic acid, the diffusion properties of which are very different. Concentration behavior at large times is the same for all tracers. Consequently, the diffusion exchange of particles between a weakly permeable matrix and a strongly permeable medium does not affect the impurity transport processes. Hence, to compare the conclusions of the theory with experiment, an interpolation formula was proposed that describes the behavior of the concentration at some distance from the source as a function of time

$$c(r,t)=\left(\frac{t_{0}}{t}\right)^{\frac{3}{1+h}}\exp\left(-B\left(\frac{t_{0}}{t}\right)^{\frac{1}{h}}\right). \tag{15}$$

On the basis of this formula, the results of the theory were compared with experiment. This comparison is shown in Fig. 3.

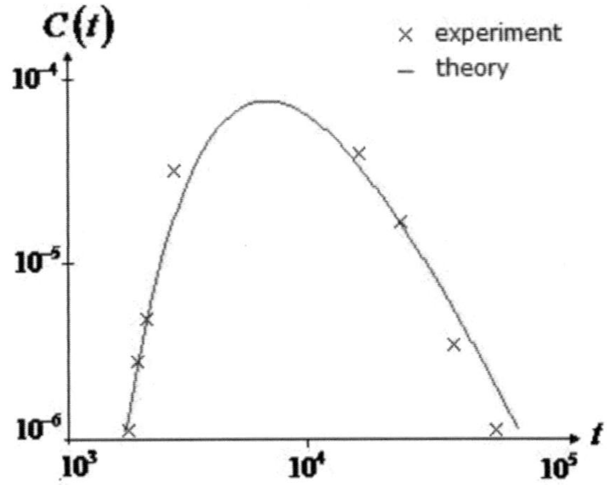

Figure 3. Comparison of the theory (formula (15)) with the results of tracer experiments [29].

4. Statistically homogeneous sharply contrasting media

An important (and perhaps most common) class of highly heterogeneous media are sharply contrasting, statistically homogeneous media. An example is fractured porous media when the fractures are uniformly distributed in a porous matrix. To describe transport in such media, there is a Gerke-Genuhent model [30], in which the impurity exchange between the fast and slow subsystems is considered in the "mean-field approximation". This implies that the exchange fluxes are determined by the difference in local average concentrations in blocks and fractures. However, it is not difficult to estimate that even for small blocks of the order of 10 cm, the time for concentration equalizing on the block sizes is quite large (tens and hundreds of years). And for an adequate description of transport during this time, it is necessary to take into account the remaining concentration gradients on the scales of individual blocks. For this, we developed a nonequilibrium model of double porosity [31, 32], in which the following results were obtained.

There is a large time interval during which the transport is described by anomalous regimes: along the average velocity, the transport occurs in the quasi-diffusion regime, when the average displacement $\langle r_{\parallel} \rangle$ and the size of the impurity cloud $R_{\parallel}$ grow in proportion to the root of time

$$\langle r_{\parallel} \rangle \approx R_{\parallel} \approx \sqrt{D_u t}, \quad D_u = u^2 t_1, \tag{16}$$

and in the transverse direction there is power sub-diffusion, described by the formula (5). In these formulas t_1 is the introduced above characteristic time of the beginning of the action of the traps, and u is the average velocity of groundwater filtration. At small and long times, the transport occurs in the classical advection-diffusion regime, but with different mean velocities and an effective dispersion coefficient.

5. Sorption effects

The advantage of our approach is that when describing non-classical regimes, it is possible to take into account in a natural way additional processes that affect the transfer, for example, sorption processes. It is known that sorption of impurities in the matrix during transport in porous media leads to a significant slowdown of the transport. For example, for long times, the average transport velocity and the effective dispersion coefficient decrease *K* times, where $K = 1 + k_D$ is a so-called delay factor, and the distribution coefficient k_D relates the impurity concentration in the solution with the impurity concentration adsorbed in the matrix [33]. On the other hand, if colloidal particles are present in the medium (microscopic particles ranging in size from tens of nanometers to tens of microns, which may be present in groundwater, adsorb and carry impurities), this can lead to a significant acceleration of transport in double porosity media. The reason for this is that, because of the large dimensions of these particles compared to the pore sizes in a weakly permeable matrix, the process of their escape into the matrix is suppressed. As a result, they are freely transported with groundwater velocity through the fractures and carry the impurity adsorbed on them.

The account of a colloidal subsystem in describing impurity transport was considered for different types of double porous media [34-36]. An important parameter in the model is the distribution coefficient σ, which connects the impurity concentration in the solution with the concentration adsorbed on colloids under equilibrium conditions

$$c_{eq} = \sigma m_{eq}. \tag{17}$$

The calculation shows that the set of possible transport modes is preserved, but the effective transport parameters and the time interval boundaries change. The colloids exert the greatest influence at $\sigma << 1$. In Fig. 4 the results of calculations of the average displacement of impurity particles for systems with colloids (upper curve) and without them (the lower curve) for statistically homogeneous double-porous media are shown. It is evident that the presence of colloids greatly accelerates the transport process. Accounting for this process is especially important when assessing the rates of environmental pollution.

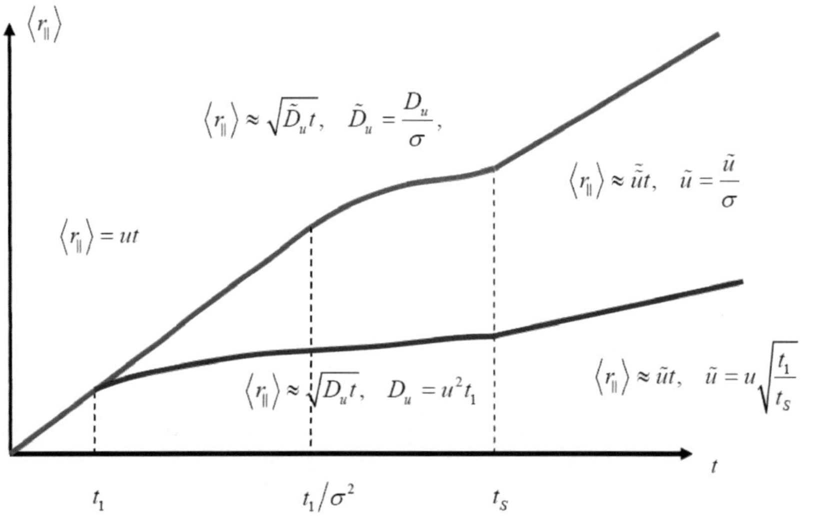

Figure 4. Average displacement of impurity for its migration in a statistically homogeneous double-porous medium in the presence of colloids. t_S corresponds to the characteristic time of equalization of the impurity concentration at the scales of the porous blocks.

6. Conclusions

In this paper, we consider models of nonclassical transport in strongly heterogeneous media with a sharp contrast of properties and various types of correlations in the distribution of structural media characteristics. Three possible media types are analyzed: regularly heterogeneous, statistically homogeneous double-porous and percolation media. It is shown that in a general case the transport regime in the main body region of impurity localization depends on the time interval. This is due to the fact that the fraction of time the impurity particles spend in the channels with rapid transport and in the traps depends on time interval. The exponential decrease in concentration at asymptotically large distances takes place for all types of anomalous transport. The change in the transport regimes in the main body region leads to a multi-step asymptotics, the more distant parts being determined by earlier transfer regimes.

For regularly heterogeneous media consisting of a well conducting region bounded in one or two directions and surrounded by a weakly permeable matrix, quasi-diffusion, power-law, and logarithmic subdiffusion modes can be realized.

For percolation media, with a sufficiently slow decrease in the correlator of flow velocity fluctuations, the transport occurs first in the super-diffusion regime, and then, in later times, the super-diffusion regime is changed by the classical diffusion regime. The effective diffusivity is of the order of the product of the average advection velocity and the correlation length of the medium.

The transport mode in anisotropic media with fractal properties depends on the degree of anisotropy. For weak anisotropy, super-diffusion takes place in all directions. For strong anisotropy, the classical diffusion regime is realized in the transverse direction. In addition, an anomalous drift takes place in anisotropic media, so that the average displacement of particles increases in accordance with the super-diffusion law.

For statistically homogeneous sharply contrasting media, there are long time intervals when the transport along the flow velocity occurs in the quasi-diffusion mode (the average particle displacement and the size of the impurity cloud grow with time according to the root law), and sub-diffusion takes place in a plane perpendicular to the mean velocity.

Sorption on colloids in sharply contrasting media can lead to a significant acceleration of the transfer process.

Acknowledgment

The authors acknowledge the support from the Russian Foundation for Basic Research (RFBR) under the Project 18-08-01229-a.

References

[1] Dreizin Yu A and Dykhne A M 1972 *JETP* **36** 127
[2] Bakunin O G 2015 *Phys. Usp.* **58** 252
[3] Qing Gu, Schiff E A, Grebner S, Wang F and Schwarz R 1996 *Phys. Rev. Lett.* **76** 3196
[4] Sher H and Lax M 1973 *Phys. Rev.* B **7** 4491
[5] Zelenyi L M and Milovanov A V 2004 *Phys. Usp.* **47** 749
[6] Weiss M, Hashimoto H and Nilsson T 2003 *Biophys. J.* **84** 4043
[7] Banks D S and Fradin C 2005 *Biophys. J.* **89** 2960
[8] Neuman S P 1990 *Water Resour. Res.* **26** 1749
[9] Sahimi M 1998 *Phys. Rep.* **306** 213
[10] Bouchaud J P and Georges A 1990 *Phys. Rep.* **195** 127
[11] Bolshov L, Kondratenko P, Pruess K and Semenov V 2008 *Vadose Zone J.* **7** 1135
[12] Khintchine A and Levy P 1936 *C. R. Acad. Sci. Paris* **202** 374
[13] Montroll E W and Weiss G 1965 *J. Math. Phys.* **6** 167
[14] Chukbar K V 1995 *JETP* **81** 1025
[15] Calvo I and Sanchez R 2008 *J. Phys. A: Math. Theor.* **41**, 282002
[16] Lutsko J F and Boon J P 2013 *Phys. Rev.* E **88** 022108
[17] Barenblatt G I, Zheltov Iu P and Kochina I N 1960 *J. Appl. Math. Mekh.* **24** 1225
[18] Dykhne A M, Dranikov I L and Kondratenko P S 2005 *J. Hydraul. Res.* **43** 2
[19] Dvoretskaya O A, Kondratenko P S and Matveev L V 2010 *JETP* **110** 58
[20] Kondratenko P S and Matveev L V 2007 *JETP* **104** 445
[21] Dykhne A, Dranikov I, Kondratenko P and Matveev L 2008 *Vadose Zone J.* **7** 1145
[22] Dykhne A M, Dranikov I L, Kondratenko P S and Matveev L V 2005 *Phys. Rev.* E **72** 061104

IOP Conf. Series: Journal of Physics: Conf. Series **1099** (2018) 012020 doi:10.1088/1742-6596/1099/1/012020

[23] Dranikov I L, Kondratenko P S and Matveev L V 2004 *JETP* 98 945
[24] Kondratenko P S and Matveev L V 2007 *Phys. Rev.* E **75** 051102
[25] Kondratenko P S and Matveev L V 2011 *Phys. Rev.* E **83**, 021106
[26] Dykhne A M, Kondratenko P S and Matveev L V 2004 *JETP Letters* **80** 410
[27] Matveev L V 2014 *JETP* **118** 662
[28] Bolshov L, Kondratenko P, Matveev L and Pruess K 2008 *Vadose Zone J.* **7** 1152–1160
[29] Becker M W and Shapiro A M 2000 *Water Resour. Res.* **36** 1677
[30] Gerke H H and van Genuchten M T 1993 *Water Resour. Res.* **29** 305
[31] Matveev L V 2012 *JETP* **115** 829
[32] Matveev L V 2014 *Physica* A **406** 119
[33] Rumynin V G 2011 *Subsurface Solute Transport Models with Application to GroundwaterHhydrology* (St. Petersburg : NAUKA) (in Russian)
[34] Matveev L V 2009 *JETP* **108** 1044
[35] Bolshov L A, Kondratenko P S and Matveev L V 2011 *Phys Rev* E **84**, 041140
[36] Kutsepalov V A and Matveev L V 2015 *Chaos Soliton Fract* **81** 480

Molecular-dynamic simulation of isotherm of cyclotetramethylenetranetramine with the nanopore type defect

S A Kozlova[1,*], S A Gubin[1], I V Maklashova[1], and A A Selezenev[1,2]

[1]National Research Nuclear University MEPhI (Moscow Engineering Physics Institute), Kashirskoye shosse 31, Moscow 115409, Russia

[2]The Russian Federal Nuclear Center All-Russian Research Institute of Experimental Physics (VNIIEF), Sarov, Russia

*e-mail: Sandra969@yandex.ru

Abstract.Molecular dynamic simulation of isothermal compression parameters are performed for a cyclotetramethylenetetranitramine single crystal ($C_4H_8N_8O_8$, HMX) using a modified ReaxFF-lg reactive force field. The unit cell consisted of eight $C_4H_8N_8O_8$ molecules (144 atoms). The calculated supercell consisted of an optimized unit cell and was created with a periodicity of 6×6×6 (48384 atoms). Our calculations were performed for the pressure range of 1–40 GPa, and the state of the single crystal $C_4H_8N_8O_8$ under the conditions of P = 0.05 GPa and T = 300 K was chosen as the initial state. The calculated isotherm of the material is consistent with experimental data. Static compression calculations of molecular crystal $C_4H_8N_8O_8$ with the nanopore type defect were carried out to study the effect of defects on the compressibility of crystals. Nanopores two sizes were considered. By results of calculations it was performed that the material with nanopore type defect is deformed more strongly, and the greater the defect, the stronger the deformation. At the same pressure value, the nanopore crystal has a higher compression ratio than an ideal single crystal.

1.Introduction

Molecular dynamics (MD) is based on direct calculation of the coordinates and velocities of a fairly large ensemble of particles (atoms or molecules) as they evolve over time. MD is widely used to simulate different physical processes in molecular systems, enabling us to obtain reliable information on the properties of investigated substances. The simulations in this work were performed using the LAMMPS MD software package that allows us to study processes of chemical interaction and predict the behavior of different materials under the action of extreme pressures and temperatures. These transformations depend strongly on the loading of a material and include deformation, rising temperature, and the emergence of defects in a crystal. The ReaxFF-lg [1] modified reactive force field allowed us to describe the thermophysical properties of the isothermal compression of organic single crystals over a wide range of pressures and temperatures, including molecular crystals with nanopore type defects. The software implementation of this interatomic force field is available in LAMMPS.

The goal is study the effect of the nanopore type defects on the compressibility of $C_4H_8N_8O_8$ crystal via MD simulations.

The results of the response of three nanofluid systems to an impact loading are presented in [2]. The authors conducted a series of experiments where they studied the dynamics of filling of nanoporous granular media Libersorb 23 (L23) and Polysorb-1 with water and silochrome (SCh-1.5) with Wood's alloy. The pore volumes were from 60 to 120 cm^3 with the filling start pressures from 0.6 to 15 MPa. In work [2] the compressibility of the liquids was determined from experiments with the chamber filled only with the liquid without any porous medium. The compressibility of the liquids coincides within the measurement accuracy with the tabulated values. As a result, the authors obtained the experimental time dependences of the pressure the internal volume in the chamber for different values of the impact energy (E = 5-70 J). Thus, the filling pressure is constant during the entire time of loading absorption at filling, it means the force acting on the protected object is constant in the presence of a nanoporous medium and is weaker by a factor of 1.3–8.5 than the force created by external loading in the absence of the nanoporous medium. In [2] was also found that the possibility of choosing a (nanoporous medium–nonwetting liquid) system and the conditions of their application that ensure the highly efficient operation of devices at various energies and momentum of external loading.

The paper [3] presents the results of molecular-dynamic modeling of the formation of a hot spot and the initiation of a chemical reaction in a PETN crystal under shock loading. Conditions under which a shock wave with a velocity of 2 km/s passes through a PETN crystal containing a spherical void of 20 nm in diameter were simulated. The results of MD modeling showed that a hot spot forms on the boundary of the void in the crystal, after the formation, chemical reactions are initiated, during which final decomposition products such as N_2 and H_2O are formed. In other words, the chemical reactivity of the PETN crystal significantly increases in the presence of there are voids in it.

2. Reactive force field ReaxFF-lg

In this paper, the reaction force field ReaxFF (Reactive Force Field) was used for the MD of modeling the properties of a HMX. The ReaxFF force field was developed by Andri van Duin [4], as a realization of reactive molecular dynamics, to describe the chemical interaction, dissociation and formation of chemical bonds, and to describe the properties of defects and imperfect crystal surfaces. This approach allows modeling modern systems with a number of atoms of the order of 10^6-10^8 on modern high-performance parallel computing clusters. The essence of the reaction force field ReaxFF is reduced to the parameterization of interatomic interactions in the molecules present in the ensemble under consideration. For each of the components of the atomic ensemble, a certain set of parameters is set, including parameters determined experimentally, parameters calculated using *ab-initio* methods or calculated by processing experimental data on the properties of the materials in question. The MD simulations in this work were performed using a modified version of the ReaxFF reactive force field [4] that considered the Van der Waals interactions between molecules, allowing us to reproduce molecular crystals thermophysical properties more accurately.ReaxFF-lg, corrects London dispersion interaction (van der Waals attraction) by adding an extra term using low-gradient model, E_{lg}, as follows:

$$E_{ReaxFF-lg} = E_{ReaxFF} + E_{lg}; \tag{1}$$

$$E_{ReaxFF} = E_{bond} + E_{lp} + E_{over} + E_{under} + E_{val} + E_{pen} + E_{coa} + E_{C2} + E_{tors} + \\ + E_{conj} + E_{H-bond} + E_{vdWaals} + E_{Coulomb}; \tag{2}$$

$$E_{lg} = -\sum_{ij,i<j}^{N} \frac{C_{lg,ij}}{r_{ij}^6 + dR_{eij}^6}, \tag{3}$$

IOP Conf. Series: Journal of Physics: Conf. Series **1099** (2018) 012021 doi:10.1088/1742-6596/1099/1/012021

where $E_{vdWaals}$ and $E_{Coulomb}$ are the energy of non-bonded interactions, namely the van der Waals and Coulomb interactions. In addition to E_{H-bond}, the energy of hydrogen bond, the rest of the energy terms are the bond order dependent and account for the bonded interactions that consists of E_{bond}, the energy of bond, E_{lp} for energy of lone-pairs, E_{over} of over-coordination, E_{under} of under-coordination, E_{val} of valence angle, E_{pen} of energy penalty for handling atoms with two double bonds, E_{coa} of coalition (three-body conjugation), E_{conj} of conjugated bonds (four-body conjugation), E_{tors} of torsion angles. E_{lg} to account for long range London Dispersion and r_{ij} is the distance between atom i and atom j, R_{eij} is the equilibrium vdW distance, and $C_{lg,ij}$ is dispersion energy coefficients, d is a scaling factor.

3. Calculation parameters

Firstly, we calculate isothermal parameters for a HMX single crystal via MD simulations. The unit cell consisted of 8 HMX molecules, i.e. 224 atoms. Further from the optimized unit cell, a calculated super-cell with a period of 6x6x6: 1728 molecules (48384 atoms) was made. Figure 1 shows the type of the calculated super-cell. The simulation was carried out in the pressure range from 0.05 to 40 GPa at a constant temperature of T = 300 K.

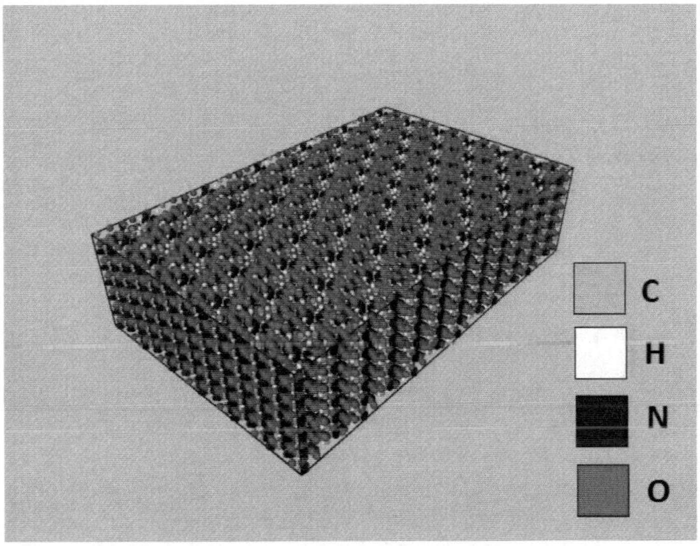

Figure 1. The calculated super-cell of HMX.

Static compression calculations of molecular crystal $C_4H_8N_8O_8$ with the nanopore type defect were carried out to study the effect of defects on the compressibility of crystals. Two sizes nanopores were considered. In the first case, to create a nanopore with the volume 0.209 nm^3, 163 atoms were removed from the calculation supercell (6048 atoms) and the porosity of the crystal was estimated at 2.7% of the total volume of the calculated supercell. The same way, it was created by removing 66 atoms of identical estimated supercell to obtain crystal $C_4H_8N_8O_8$ with a porosity of about 1%, in this case, the pore volume was 0.085 nm^3. Figure 2 shows the calculated super-cell of HMX crystal with two sizes nanopores.

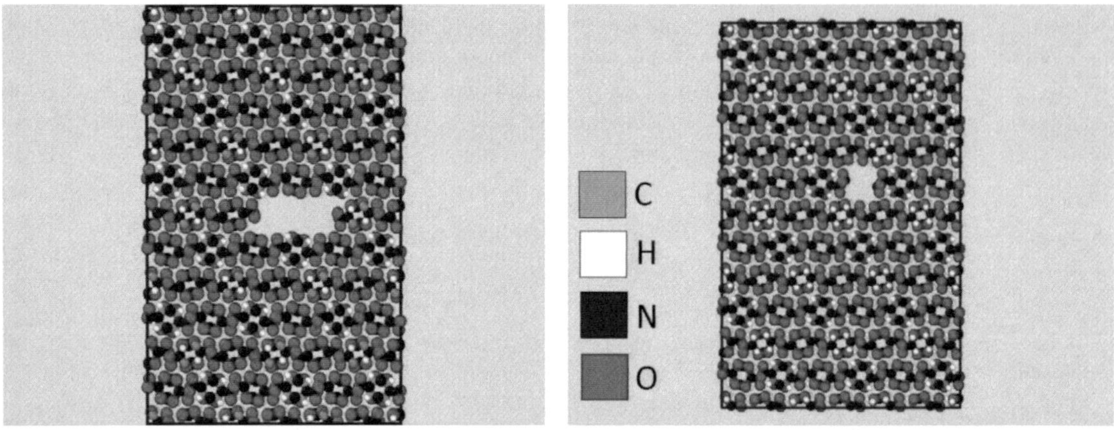

Figure 2. The calculated super-cell of HMX with nanopore: a porosity of 2.7% and 1%, respectively.

4. Results and discussion

Obtained in this work HMX isotherm is shown in figure 3, it is also in comparison with the known experimental and calculated results. At a pressure of 40 GPa, the compression ratio of the HMX reached a value of $V/V_0 = 0.7$.

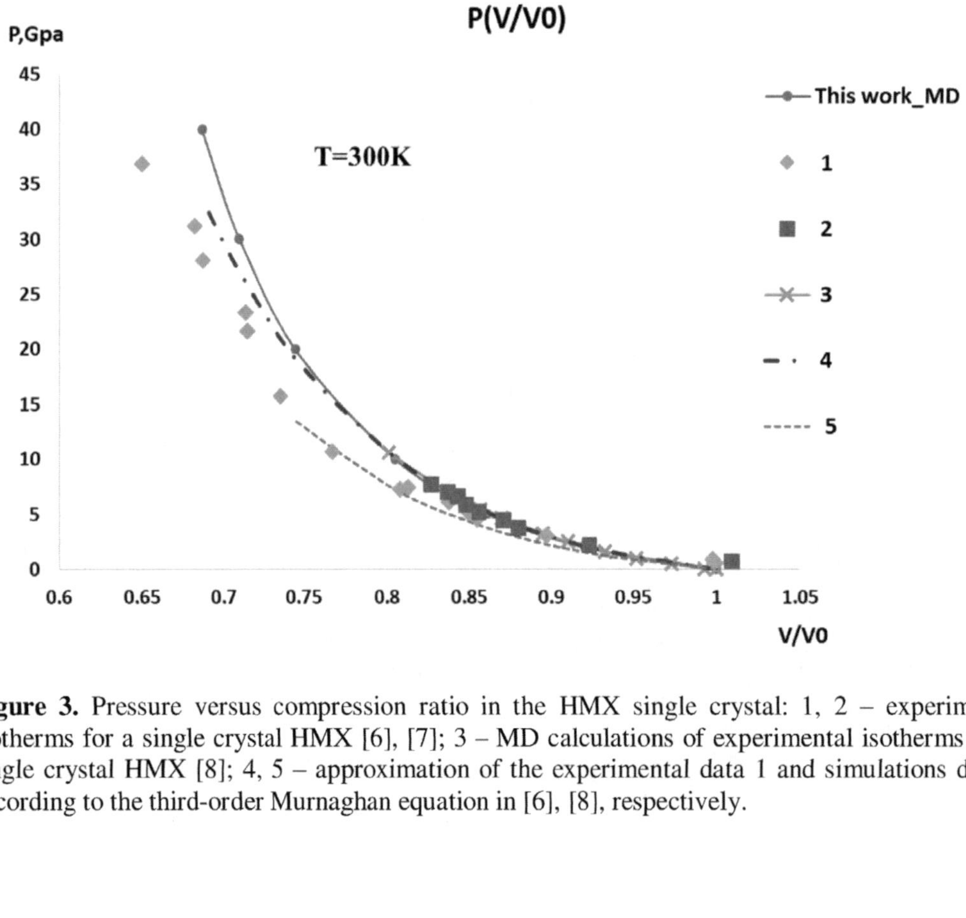

Figure 3. Pressure versus compression ratio in the HMX single crystal: 1, 2 – experimental isotherms for a single crystal HMX [6], [7]; 3 – MD calculations of experimental isotherms for a single crystal HMX [8]; 4, 5 – approximation of the experimental data 1 and simulations data 3 according to the third-order Murnaghan equation in [6], [8], respectively.

Figure 3 shows that the calculated pressure values in the isotherm of HMX single crystal obtained in this paper agree well with experimental data over a wide pressure range (0.05 -40 GPa). Also we can note an excellent agreement with the results of [8], where MD simulation of the isothermal compression of HMX single crystal was carried out using the quantum-chemical approach. The approximation of the results of the MD simulation of work [6] to 30 GPa also agrees well with the results of the MD simulation of this work.

MD simulations results of isotherms of crystal $C_4H_8N_8O_8$ with nanopore type defects (with different porosity values) are shown in figure 4.

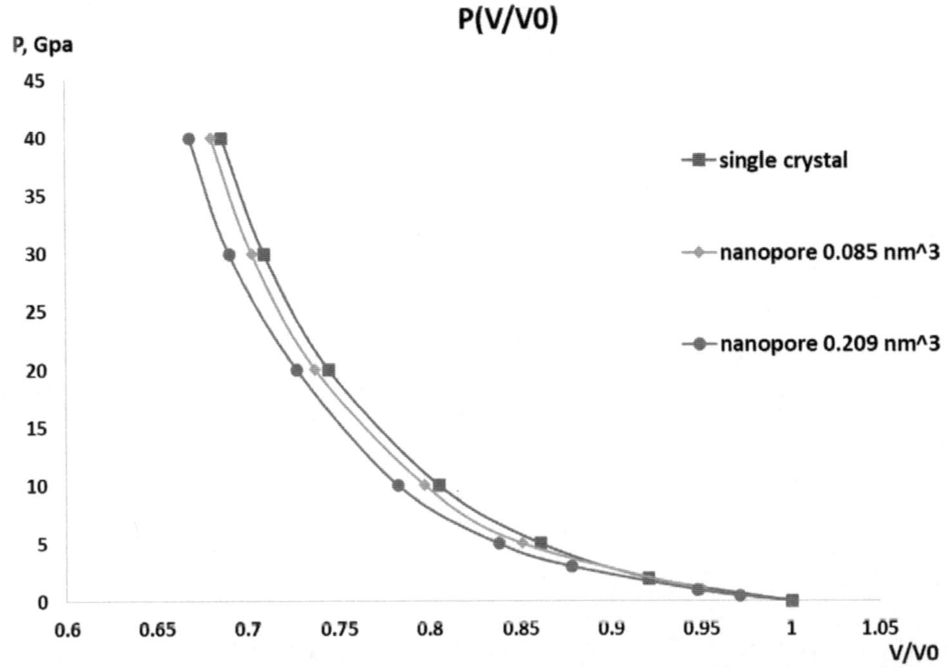

Figure 4. Isotherms of HMX crystals with nanopore type defects in comparison with the isotherm for a single crystal HMX.

5. Conclusions

By results of calculations it was performed that the material with nanopore type defect is deformed more strongly, and the greater the defect, the stronger the deformation. At the same pressure value, the nanopore crystal has a higher compression ratio than an ideal single crystal. The ReaxFF-lg modified reactive force field allowed us to describe the thermophysical properties of the isothermal compression of organic single crystals over a wide range of pressures and temperatures, including molecular crystals with nanopore type defects.

Acknowledgments

This work was supported by the Russian Science Foundation, project no. 16-19-00188.

6. References

[1] Liu L, Liu Y, Zybin S V, Sun H and Goddard W A 2011 *J. Phys. Chem. A* 115 11016
[2] Borman V D, Belogorlov A A, Tronin V N 2018 *Colloids and Surfaces A* 537 540
[3] Shan T R, Thompson A P 2014 *J. Phys. Conf. Ser.* 500 172009
[4] Dr. Adri van Duin, Department of Mechanical & Nuclear Engineering: URL: http:// www.engr.psu.edu/adri/Home.aspx

IOP Conf. Series: Journal of Physics: Conf. Series **1099** (2018) 012021 doi:10.1088/1742-6596/1099/1/012021

[5] Zhou T, Zybin S V, Liu Y, Huang F and Goddard III W A 2012 *J. Appl. Phys.* 111 124904
[6] Yoo C S, Cynn H., Howard W M, Holmes N 1998 *Proc. 11th Int. Det. Symp. (Colorado)* p 951
[7] Olinger B, Roof B and Cady H 1978 *Proc. Int. Symp. on High Dynamic Pressures (Paris)* p 3
[8] Sewell T D, Menikoff R, Bedrov D, and Smith G D 2003 *J. Chem. Phys.* 7417

The law of a stretched exponential and the crossover of the behavior of a disordered nanoporous medium with a non-wetting liquid with its anomalously slow relaxation

V D Borman, A A Belogorlov, V N Tronin and R R Khabibullin

Molecular Physics Department, National Research Nuclear University MEPhI, Kashirskot sh. 31, Moscow, 115409, Russian Federation

VNTronin@mephi.ru

Abstract. Relaxation of a disordered nanoporous medium filled with a nonwetting liquid is studied. An anomalously slow relaxation of the volume of the trapped liquid in such a medium is established, which is characterized by the law of a stretched exponential $\sim\exp\left[-(t/\tau_0)^{\beta}\right]$. At a temperature of $T_c \approx 280$ K, a dispersion transition occurs, namely, the index β changes sharply from $\beta > 0.4$ at $T < T_c$ to $\beta < 0.3$ at $T > T_c$. With such a transition, the non-wetting liquid becomes effectively wetting. An explanation for this crossover is proposed as the result of additional jamming of the liquid outflow from the porous medium due to the rupture of the percolation cluster of filled pores, over which liquid escapes.

1. Introduction

One of the general characteristics of disordered systems is their anomalously slow relaxation. These are such systems as granular matter, polymers, metallic glasses, molecular glasses, colloidal media, nanocomposites, etc. The relaxation in such systems is described by Kohlrausch's law (law of a stretched exponential). Namely, the basic property changes with time in accordance with the dependence $\sim\exp\left[-(t/\tau_0)^{\beta}\right]$, where the constant τ_0 denotes the relaxation time and parameter $\beta < 1$, non – exponentionality index or stretching exponent [1,2]. This behavior (jamming) is typical for such media as micellar polycrystals, colloidal fractal gels, surfactant phases, lamellar gels, concentrated emulsions polymeric nanocomposites [3-10], granular material [11], for which the motion from the initial state turns out to be difficult for various reasons in comparison with the usual motion of equilibrium state relaxing systems.

Such jamming is the apparent result of "blockage" or a significant slowdown of the possible motion of the system along part of the path in the configuration space of it is states, which leads to the nonergodic or quasi – nonergodic behavior of such systems [12]. It is thought that, anomalously slow relaxation belongs to the class of cooperative phenomena [3,13-15]. At the present time, several phenomenological models have been proposed for describing anomalously slow relaxation. Their analysis can be found in several recently published reviews [3,13-15]. These models are based on the assumption of the existence of local correlated particle configurations. However, in experiments to measure the correlator of statistical density, they are not recorded.

The disordered nanoporous medium investigated in the present work refers to silica gels that are formed upon a sol – gel transition in a colloidal medium. The random structure of the pores and the frame of the medium "freezes" upon gel formation as a result of a chemical reaction between the colloidal particles. If the pores of the random structure are interconnected, they can be completely filled with liquid. This takes place at porosity values exceeding the percolation threshold, at which an infinite percolation cluster is created in the medium from the associated pores with other pores and thus the pore fluid is filled with the medium. Parameters of the porous medium: the pore size distribution, the specific volume of the pore, and the pore surface, were measured by adsorption-desorption methods of gas [17,18] and liquid porosimetry [19,20,21]. In liquid porosimetry, non – wetting liquid (water) is used, which fills the porous medium at elevated pressure. In this case, the liquid is dispersed and transformed into an ensemble of interacting clusters. The fact of capture of a non – wetting liquid by a random pore structure can serve as a source of properties about a disordered structure. [22-24]. In accordance with [23] and [24] this fact is explained by the effective attraction of clusters. The mutual arrangement of clusters is ultimately determined by correlations in the location of the pores. An abnormally slow relaxation of a liquid captured by a porous medium at times $10^2 < t < 10^5$ was established in [25]. Such a relaxation is observed as a slow flow of liquid from the porous material. The dependence of the degree of filling on time can be approximated by an inverse power law with exponent less than 0.1. With different characteristics of a porous medium non-wetting liquid may not outflow from the porous medium for a several days or even months [26-29]. Small deviations from the critical values of temperature (T_c) and degree of filling (θ_c) lead to a dispersion transition (rapid outflow of liquid) [22]. In the present paper it is shown that, in the time range $10^2 < t < 10^5$ s, anomalously slow relaxation of trapped liquid state in a medium with pores of nanometer size is described by the law of a stretched exponential. On this occasion, the exponent (β) in this law and the relaxation time (τ_0) change discontinuously in the dispersion transition from values $\beta > 0.4$ at $T < T_c$, to values $\beta < 0.3$ at $T > T_c$.

It is presumed that the outflow of liquid can occur only in the presence of bonds between clusters of filled pores. Then, when the percolation cluster breaks, the liquid outflow is jammed.

2. Experiment

2.1. Porous medium

In the experiments, a nanoporous medium, Libersorb 23 (L23), was used. It is a silica gel KSK – G (Large Grain Silica Gel), the disordered pore structure of which is formed as a result of the sol-gel process. The surface of the pores of KSK – G was modified with alkyl silanes to impart hydrophobic properties. Parameters of samples of the porous materials (Libersorb 23) were studied using low-temperature sorption at the Autosorb IQ (Quantachrome, USA) and the helium pycnometer Micro – Ultrapyc 1200e (Quantachrome, USA). The density of the porous medium L23 was $\rho = 1.7798 \pm 0.0016$ g cm^{-3}, the specific pore volume V_p- $= 0.62 \pm 0.02$ cm^3 g^{-1}, the porosity of the material was $\varphi = 0.52$, the specific surface area $S_p = 199 \pm 7 \frac{m^2}{g}$ $S_p = 199 \pm 7$ m^2 g^{-1}, the average granule size of the L23 powder was ~10 μm. The function of the distribution of pores by volume according to the Barrett – Joyner – Halenda (BJH) Analysis is shown in (Figure 1). This function gives only a qualitative representation of the distribution of the pore volume along the radius. The average pore radius is $\langle R \rangle = 5.0 \pm 0.2$, nm in the vicinity of the maximum half – width of the distribution at half – height $\Delta R = 0.4 \pm 0.1$ nm, and the ratio $\Delta R / \langle R \rangle \leq 0.1$. Distribution tails in the region of large and small pore sizes are also observed. Therefore, due to the smallness of the ratio $\Delta R / \langle R \rangle \ll 1$, the porous medium L23 was used to test the power relaxation model [25]. In such a system (trapped liquid – a fractal infinite percolation cluster of pores), the filled pores form a potential barrier field of local metastable states. It is expected that the relaxation of such states will occur by inverse to the power law.

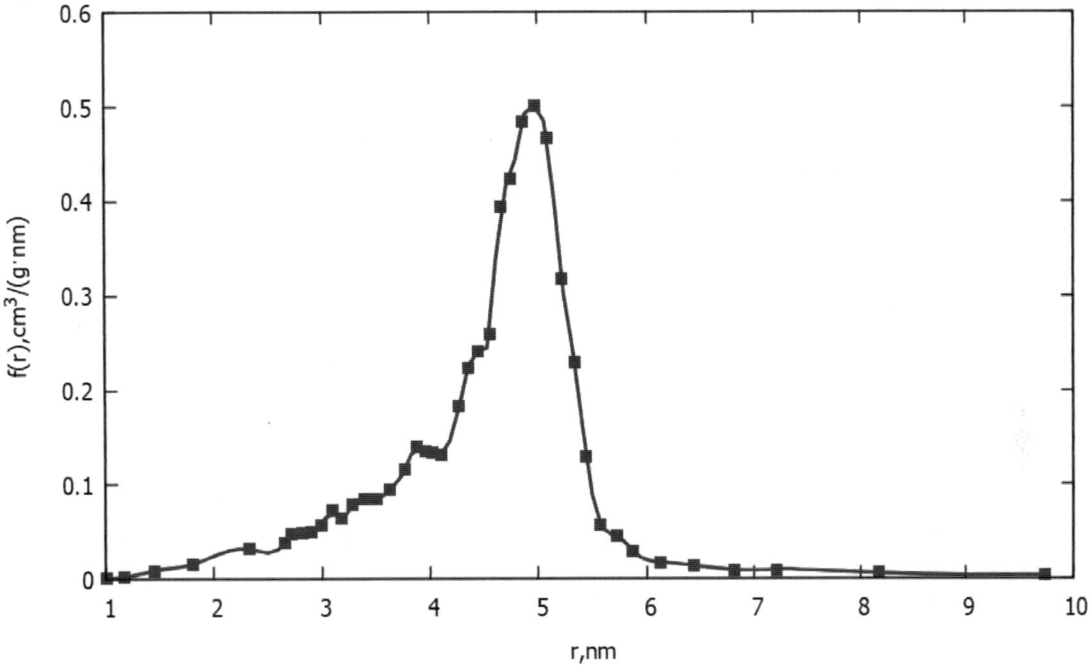

Figure 1. Pore size distribution function for the porous medium Libersorb 23 by the BJH method.

2.2. Measurement procedure

In the experiment, the degree of filling was measured as a function of time. The confined liquid is a liquid that remains in the pores (at the initial state $\theta = 1$) when the total pressure is removed. In this case, the other part of the liquid in the pores rapidly, during the time of pressure decrease, outflow from the pores. For the system L23 – water, the outflow time of this part of the liquid can be less than 1 s [25]. In this paper we have studied the relaxation of a state (slow outflow) of a liquid confined in a porous medium. The scale of the characteristic outflow of the liquid time may differ depending on the mechanism of transport of the liquid through the system of filled pores, or by capillary evaporation and subsequent capillary condensation at the boundary of the bulk liquid and granule of the powder of the porous medium immersed in this liquid. This latter mechanism can work efficiently, obviously, provided that an "infinite" percolation cluster is formed from empty pores, i.e. at the degree of filling of the confined liquid by $\theta < \theta_{c1}=0.85 \div 0.7$.Therefore, the developed technique should provide a measurement of the volume when liquid outflow or when the confined liquid evaporates.

The technique for measuring the volume of a non-wetting liquid system - a porous medium is analogous to the methods of mercury or water porosimetry. It is described in detail in [25]. This technique makes it possible to carry out investigations at temperatures from 243 to 393 K with an accuracy of 0.2 K.

IOP Conf. Series: Journal of Physics: Conf. Series **1099** (2018) 012022 doi:10.1088/1742-6596/1099/1/012022

2.3. Results of experiments

Figure 2 shows the experimentally determined values of θ for different relaxation times (t) for temperatures T = 279 K and 293 K. The experimental values of θ are obtained by a step – by – step method with two cycles of filling – outflow and a varying time t of observation of relaxation from t =10 s to t = 10^5 s at each step. The points in (Figure 2) reflect the results to six measurements θ for each time t. The spread of the values of θ does not exceed the measurement error. This is evidence that the initial state of the liquid – disordered porous medium system is reproduced when the memory is "erased" (after full filling in the first and second cycles) about the previous states. The coincidence of the values of θ can be attributed to the fact that after full filling and subsequent removal of excess pressure the system was in one of a large number of degenerate states with the formation of an "infinite" percolation cluster filled at $\theta > \theta_c$ or empty pores at $\theta < \theta_c$. All these states are characterized by the same degree of filling.

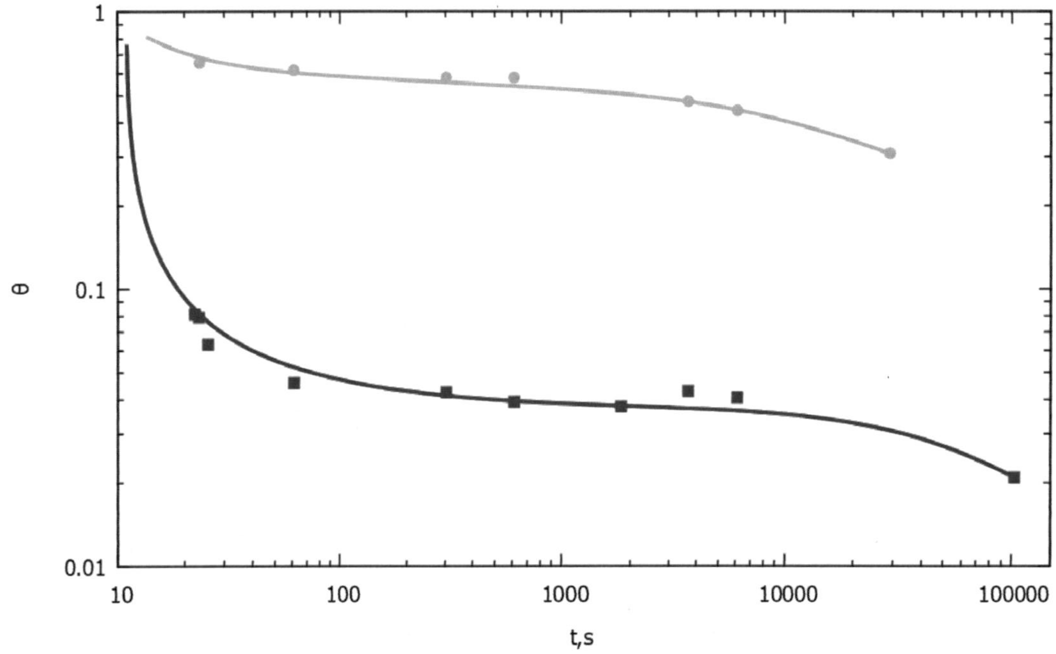

Figure 2. The values of θ for a different time t show a relaxation for temperatures T = 279 K and 289 K.

Between the temperatures T = 293 K and 279 K in the water – porous medium L23 system, a dispersion transition occurs when the liquid dispersed in the porous medium becomes metastable with decreasing temperature [22 – 24]. In a narrow region (ΔT ≈ 10 K) of temperatures in the vicinity of the critical temperature, the transition occurs adiabatically with increasing temperature from an almost complete non – flow of liquid at T = 279 K to an almost complete outflow at T = 293 K. It follows from (Figure 2) that in the time interval $60 < t < 6 \cdot 10^3$ s at T = 279 K a slow relaxation θ(t) is observed, and the dependence θ(t) in the double logarithmic scale within the error margin measurement is described by a straight line. A similar picture is also observed at T = 293 K for another gas mechanism of transport and time in the interval $60 < t < 6 \cdot 10^3$ s. However, the slope of the function θ(t) at this higher temperature is lower, and, consequently, the relaxation rate is smaller in accordance with T = 293 K. In Figure 3 shows the dependences of θ(t) at six temperature values from 277 K to 289 K in the time interval from 13 s to $3 \cdot 10^4$ s. The dependencies presented in Figure 3 can be described, within the error of measurements, by the law of a stretched exponent, in which the exponent (β) and the relaxation time (τ_0) are functions of temperature.

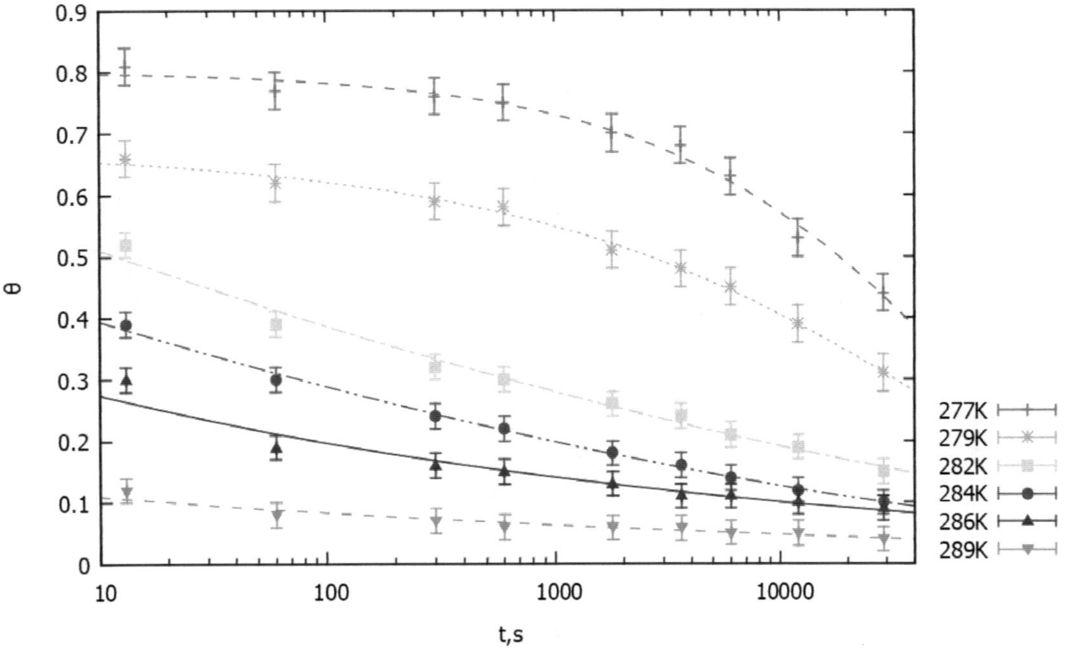

Figure 3. Dependences of θ(t) at six temperatures from 277 K to 293 K in the time interval from 13 to $4 \cdot 10^4$ s. (Points – experiment, solid lines – approximation by the function $\sim\exp[-(t/\tau_0)^{\beta}]$)

T = 277 K: $\beta = 0.55$, $\tau_0 = 7 \cdot 10^4$ s; T = 279 K: $\beta = 0.39$, $\tau_0 = 5.5 \cdot 10^4$ s; T = 282 K: $\beta = 0.07$, $\tau_0 = 1.4 \cdot 10^{-2}$ s; T = 284 K: $\beta = 0.065$, $\tau_0 = 4 \cdot 10^{-2}$ s; T = 286 K: $\beta = 0.0053$, $\tau_0 \ll 10^{-10}$ s; T = 289 K: $\beta = 0.005$, $\tau_0 \ll 10^{-10}$

From Figure 4 it follows that with an increase in temperature from 277 K to 279 K, the value of the exponent decreases from values β = 0.55 to β = 0.39. A further decrease in temperature by 3 K leads to a significant (by an order of magnitude) change in the parameter to values β = 0.07. In this case, at a temperature of 282 K, the characteristic relaxation time decreases sharply (by six orders of magnitude) (Figure 5).With a further increase in temperature to 286 K, the parameter β falls to values β = 0.0053 (Figure 4). The relaxation time also decreases by several orders of magnitude (Figure 5). Figure 4 and Figure 5 show the temperature dependence of the parameter and the characteristic relaxation time.

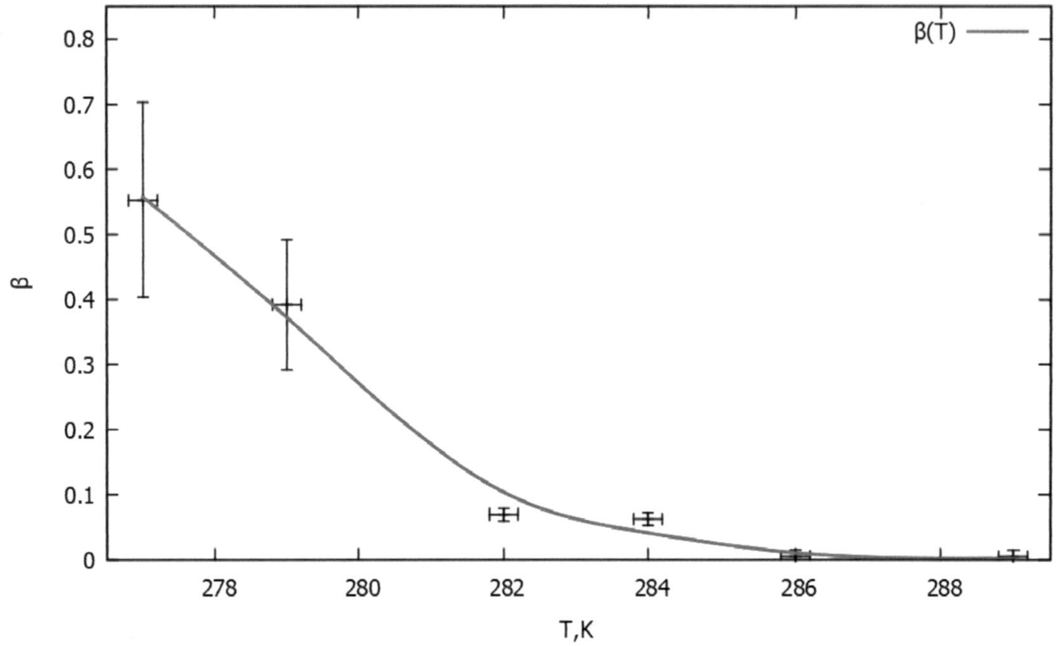

Figure 4. Temperature dependence of the parameter β in the law of a stretched exponent $\theta \sim \exp\left[-(t/\tau_0)^{\beta}\right]$.

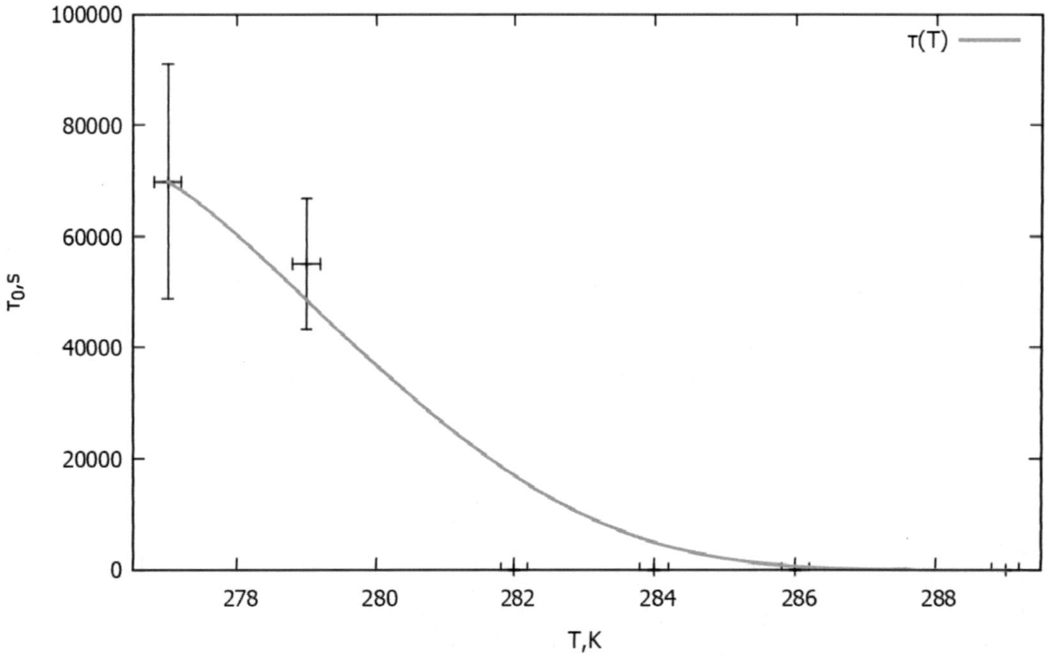

Figure 5. Dependence on the temperature of the relaxation time in the law of a stretched exponent $\theta \sim \exp\left[-(t/\tau_0)^{\beta}\right]$

3. Conclusion

Analyzing the results of the experiments (Figures 3-6), it can be stated that after filling the porous medium and then removing the excess pressure, it is possible to realize two different ways of relaxation. At "high" temperatures, T > 284 K, almost all the liquid (> 80%) is concentrated in unstable local configurations for which there are flow paths through an infinite cluster of filled pores. As a result, the leakage occurs during the characteristic time of reducing the excess pressure t ~ 10 s (Figure 6).

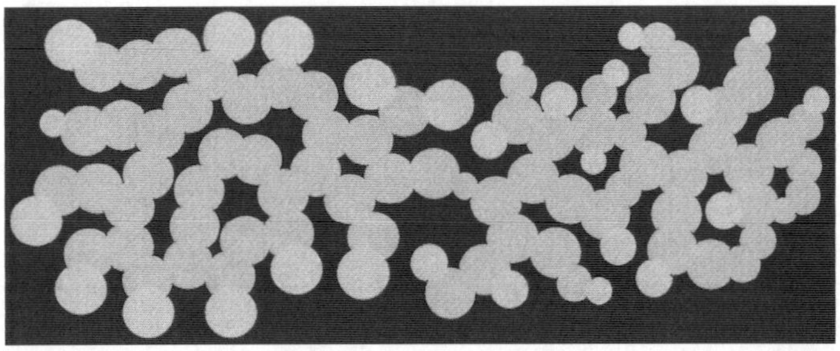

Figure 6. Schematic representation of a filled porous medium. (Purple – frame, blue – filled pores, green – pores loosely associated with a percolation cluster)

The confined liquid part (with a filling degree less than the percolation threshold $\theta < \theta_c$) slowly (at times $t \gg 10$ s) outflows according to the Kohlrausch law with a small index $\beta \leq 0.05$. With this degree of filling, a lower percolation threshold θ_c, the confined liquid is in unconnected local configurations (Figure 7).

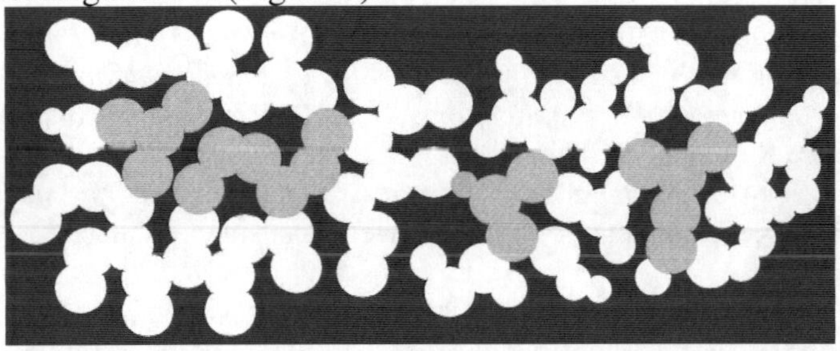

Figure 7. Schematic representation of a filled porous medium. (Purple – frame, blue – filled pores, white – empty pores)

These local configurations consist mainly of empty pores, since $z(1 - \theta_c) \sim z$ (for $\theta_c \sim 0.1$) empty pores are contained in pores of the local configuration (z – the number of nearest neighbors). Thus, the filled pores of the local configuration turn out to be jammed or "blockage". Therefore, the reduction in the degree of filling can occur according to the following scenario: 1). capillary evaporation 2). diffusion through unfilled pores 3). capillary condensation at the boundary of a porous medium. This leads to a significant slowing down of the relaxation. At temperatures T < 282 K, likewise, part of the liquid (< 40%) of the initial volume follows the characteristic time of removal of the excess pressure. However, the majority of the liquid (> 60%) at observation times of the order of $3 \cdot 10^4$ s, in accordance with [25], is metastable and slowly relaxes. Relaxation occurs according to the law of a stretched exponential with exponent depending on temperature. Thus, at temperatures T ~ 282 K, a crossover occurs in the behavior of the system studied, which is apparently due to the appearance of a dispersion transition in it. This is due to a change in the relaxation mechanism: due to the jamming of the outflow of liquid after the break of the percolation cluster of the filled pores after the dispersion transition.

4. Acknowledgements

The work was supported by the Russian Science Foundation grant No. 18-13-00398.

References

[1] Balitska V, Shpotyuk O, Brunner M and Hadzaman I 2018 *Chem. Phys.* **501** 121–7
[2] Phillips J C 1996 *Rep. Prog. Phys.* **59** 1133
[3] Langer J S 2014 *Rep. Prog. Phys.***77** 042501
[4] Cipelletti L, Ramos L, Manley S, Pitard E, Weitz D A, Pashkovski E E and Jahnsson M, 2003 *Faraday Discuss*. **123** 237–51
[5] Cipelletti L, Manley S, Ball R C and Weitz D A 2000 *Phys. Rev. Lett.* **84** 2275–78
[6] Ballesta P, Duri A and Cipelletti L 2008 *Nat. Phys.* **4** 550–4
[7] Narayanan R A, Thiyagarajan P, Lewis S, Bansal A, Schadler L S and Lurio L B, 2006 *Phys. Rev. Lett.* **97** 075505
[8] Srivastava S, Kandar A K, Basu J K, Mukhopathyay M K, Lurio L B, Narayanan S and Sinha S K, 2009 *Phys. Rev. E.* **79** 021408
[9] Caronna C, Chushkin Y, Madsen A and Cupane A 2008 *Phys. Rev. Lett.* **100** 055702
[10] Guo H, *et al*. 2009 *Phys. Rev. Lett.* **102** 075702
[11] Ruta B, Chushkin Y, Monaco G, Cipelletti L, Pineda E, Bruna P, Giordano V M and Gonzalez-Silveira M 2012 *Phys. Rev. Lett.* **109** 165701
[12] Borman V D, Belogorlov A A, Byrkin V A and Tronin V N 2013 *Phys. Rev. E.* **88** 052116
[13] Berthier L and Biroli G 2011 *Rev. Mod. Phys.* **83** 587
[14] Biroli G and Garrahan J 2013 *J. Chem. Phys.* **138** 12A301
[15] Stillinger F H and Debenedetti P G 2013 *Annu. Rev. Condens. Materr Phys.* **4** 263
[16] Borman V D and Tronin V N 2016 *Physica A: Statistical Mech. and its Appl.* **457** 391–405
[17] Vafai K 2011 *Porous media: applications in biological systems and biotechnology* ed.by K.Vafai (Boca Raton FL: RC Press)
[18] Lowell S *et al*. 2004 *Characterization of porous solids and powders : surface area, pore size, and density* (Dordrecht and Boston: Kluwer Academic Publishers) Vol. 16 of Particle technology series
[19] Fadeev A Y and Eroshenko V A 1997 *J. Colloid Interface Sci.* **187** 275–82
[20] Lisichkin G V 2003 *Chemistry of Grafter Surface Compaunds* (Moscow: Fizmatlit)
[21] Rigby S P and Edler K J 2002 *J. Colloid Interface Sci.* **250** 175–90
[22] Borman V D, Belogorlov A A, Byrkin V A, Tronin V N and Troyan V I 2012 *JETP Lett.* **95** 511–14
[23] Borman V D, Belogorlov A A, Byrkin V A, Tronin V N and Troyan V I 2013 *J. Exp. Theor. Phys.* **117** 1139
[24] Borman V D, Belogorlov A A, Grekhov A M and Tronin V N 2014 *Eur. Phys. J. B* **87** 249
[25] Borman V D, Belogorlov A A and Tronin V N 2016 *Phys. Rev. E.* **93** 022142

Study of local configurations in the systems "disordered nanoporouse medium – non-wetting liquid"

S A Bortnikova[1], A A Belogorlov[1,2], V D Borman[1], V N Tronin[1]

[1] National Research Nuclear University MEPhI (Moscow Engineering Physics Institute) Molecular Physics Department, 115409, Russia, Moscow, Kashirskoe highway, 31
[2] Topchiev Institute of Petrochemical Synthesis, Russian Academy of Sciences, 119991, Russia, Moscow,Leninsky prospect, 29

E-mail: svetlana.sweta7@yandex.ru, AABelogorlov@mephi.ru

Abstract. The analysis of redistribution functions of non-wetting liquid (water) dispersed in disordered nanoporous media of Libersorb 23, obtained in work [22]. Local configurations which describe the time relaxation of a non-wetting liquid dispersed in a disordered nanoporous medium are studied. It is shown that various possible local configurations can be used to describe the temporal relaxation of a liquid dispersed in a nanoporous medium. A qualitative picture is presented of the possible mutual arrangement of the pores in the porous medium Libersorb 23. The presented approach allows qualitatively to restore the spatial distribution of pores of a porous medium.

1. Introduction

The problem of the relaxation of metastable states in disordered media such as glasses, colloids, and polymers is one of the most urgent problems for both theoretical and experimental physics [1-15]. It was shown that the state of such media is nonergodic and their relaxation is described by the law of a prolonged exponential [3, 11, 15]. For the description of states and relaxation of glasses, colloids and polymers, the concept of local structures is used in such models as shear transformation zone (STZ), dynamic heterogeneity(DH),and random first order transition theory (RFOT) [4, 7, 8].

At the same time, the applied meaning of solving the relaxation problem is important for the industry, since porous media, mainly disordered media, are widely used in such processes as catalysis, filtration [16, 17]. In addition to the interest in the study of the relaxation of porous media, there is the problem of the spatial arrangement of pores in a porous medium, since knowledge of such structural characteristics of the medium as pore size distribution and connectivity is often insufficient for predicting processes in porous media [18, 19]. For example, for heat and mass transfer during the flow of gases and liquids in porous media, determination of flow velocity, temperature and pressure [20].
Anomalously slow relaxation of a non-wetting liquid in a disordered nanoporous medium [21] and method of restoring the distribution function of a non-wetting liquid dispersed in a nanoporous medium [22] can be used to create a technique for restoring the spatial distribution of pores in a porous medium. Anomalously slow relaxation implies that limited fluid clusters can be located inside the pore space for long periods of time, and can indicate the existence of certain local pore configurations (some spatial arrangement of the pores). The spectrum of the relaxation times of metastable states was

calculated in work [23]. It was also shown there that the relaxation of metastable states is described by a power law. In this case, a quantitative analysis of the dispersed non-wetting liquid will allow to determine the most probable configurations and estimate their number.

The presented results of the analysis of the redistribution functions of a non-wetting liquid dispersed in a disordered nanoporous medium of Libersorb 23 (L23), obtained in work [22] show, that various possible local configurations can be applied to describe the temporal relaxation of a liquid dispersed in a nanoporous medium. It is also shown, that the proposed analysis can be the basis for a technique for determining the spatial distribution of pores of a porous medium that can be described by various local configurations.

2. Methods and materials

The observation of local configurations in the system disordered nanoporous medium – non-wetting liquid was carried out according to the experimental data for the system L23 – distilled water [24]. The porous medium L23 is the hydrophobized silica gel KSK-G. This sample has characteristics: skeleton density $\rho = 1.7798\pm0.0016$ g/cm^3, specific pore volume $V_p = 0.62\pm0.02$ cm^3/g, porosity of the material $\varphi = 0.52$, surface area $S_p = 199\pm7$ m^2/g, average granules size ~ 10 μm, average pore radius $\langle R \rangle = 5.0\pm0.2$ nm. Method of obtaining this material and its characteristics described in detail in the work [21].

Results of research on the process filling porous medium by non-wetting liquid and process outflow of non-wetting liquid from the porous medium in quasistatic pressure change were obtained at the experimental installation. Data were obtained according to the method of measuring the time relaxation of non-wetting liquid dispersed in a nanoporous medium. Experimental installation and measuring techniques described in work [25]. The results were obtained for the following experimental conditions: nanoporous medium weight – 4g, volume of liquid – 55 cm^3, temperature – 6, 9, 16 ^{0}C and time interval between infiltration-defiltration cycles – 1, 10, 100 minutes.

The analysis of the experimental data was carried out in accordance with method of restoring the distribution function of non-wetting liquid dispersed in the pores of the porous medium, after reducing the excess pressure to zero [22]. The basis of this method is to compare the functions of empty pores for different time delay after the reduction of the excess pressure to zero. Main processing points:

1. Deduction from the data of the change in the volume of the system associated with elastic deformation;
2. Normalization of experimental data by volume on the total pore volume in the first experiment;
3. Smoothing of the obtained experimental data by averaging over five points;
4. Comparison of processing results with experimental data taking into account the measurement error;
5. Numerical differentiation of data after processing;
6. Approximation of the obtained data by the analytic function. This function is the sum of five Gauss distribution functions and ensures the reliability of the approximation $R^2 \geq 0{,}998$;
7. Integration of an analytic function and comparison with experimental data.

As a result, derivatives were obtained for the experimental set of points in the analytical form. The distribution function of a non-wetting liquid dispersed in porous of media in terms of pore sizes was determined as:

$$f_t^c(P) = f_0(P) - f_t(P),$$

were $f_0(P)$ - function of the distribution of the pore volume in terms of the sizes available for the liquid at the initial instant of time, $f_t(P)$ - function of the distribution of the pore volume in terms of the sizes available for liquid after the waiting time (t) after filling. The result of calculating the total integral for the obtained distribution function of a non-wetting liquid dispersed in a nanoporous medium gives a value equal to the fraction of the non-wetting liquid from the pore space.

3. Results

We consider in detail the results obtained for 6^0C. The results of calculating the distribution of non-wetting liquid dispersed in a nanoporous medium, according to the described procedure [22], shown in the figure 1a. The figure shows the presence of a peak in the region of small pores (4 nm) and pronounced peak in the region of large pores (~5 nm). According to the distribution of liquid dispersed in a nanoporous medium at full filling (figure 1a, solid line) will outline the main pore sizes: large pores (~5 nm), pores of medium size (~4.5 nm) and small pores (~4 nm). Their volume fraction accordingly is 70, 20 and 10 % of the total volume. In work [21] was shown, that for medium L23 the characteristic number of nearest neighbors for a given pore is 4-6 and porosity φ=0.52. To estimate the probability of neighborhood of pores of different sizes, we use formula [26]:

$$P(\theta) \sim \theta_1 * \theta_2$$

were θ_1 and θ_2 - fraction of pore volume with a given radius of the total pore volume. Probability of pore space "small-small" is 0.01; probability of "small- large" is 0.07; probability of "large - medium size" is 0.14; probability of "small - medium size" is 0.03; probability of "large - medium size" is 0.49. According to estimates, nanoporous medium L23 is a cluster of large pores surrounded around the perimeter of mostly small pores. On the figure 1b shown qualitatively section possible arrangement of clusters.

Let us consider the distribution of a liquid dispersed in a nanoporous medium 1 minute after full filling (see figure 1a) dot-dash line. The figure shows a decrease in volume over the entire range of pore sizes. Most of the liquid remains in large pores (~5 nm). Also, there is a seizure of liquid in pores of medium size (~4.5 nm) and small pores (~4 nm). Such a distribution may correspond to the picture in the figure 1b. Here the liquid filled pores are shown in blue, and white - empty pores. Cluster of non- filling 1 in the Figure 1b corresponds to a peak in the region of large pores (~5 nm). In Figure 9a, the dashed-dot line shows the distribution of the non- filling liquid 10 minutes after full filling and the reduction of the excess pressure to zero. It can be seen that a decrease in the volume of non- filling occurs in the pore area of medium size (~4.5 nm), slightly decreases in the region of large pores (~5 nm), and in the region of small pores (~ 4 nm) there is no change in comparison with the distribution after 1 minute. The possible distribution of the non- filling liquid is shown in Figure 9b. Here the blue color shows pores filled with liquid, and white - empty pores. The cluster of non- filling liquid 1 corresponds to a peak in the region of large pores (~5 nm). Cluster 2, shown in Figure 1b, corresponds to the pores of large (~ 5 nm) and medium (~ 4.5 nm) sizes disappear.

Consider the distribution of the non- filling liquid dispersed in a nanoporous medium 100 minutes after filling (see Figure 3a). The dotted line shows the distribution of the non- filling liquid dispersed in a nanoporous medium 100 minutes after full filling and reducing the excess pressure to 0. The figure shows that the liquid as well as for 10 minutes does non-outflow of large pores (~ 5 nm). It flows from pores of medium size (~ 4.5 nm), and partly from small pores (~ 4 nm). A possible picture of the location of clusters of non- filling liquid is shown in Figure 3b. Where blue is filled pores, and white - empty pores. Cluster 1 in Figure 3b corresponds to peaks in the region of large pores (~ 5 nm) and in the region of small pores (~ 4 nm).

Thus, the assumption of the arrangement of small pores (~ 4 nm) along the perimeter of the large-pore cluster is completely within the framework of the result obtained. The flow of medium pores (~ 4.5 nm) and small pores (~ 4 nm) observed after 100 minutes can be explained by the assumption that small pores located along the perimeter of the cluster are connected with a large-pore cluster through pores of medium size.

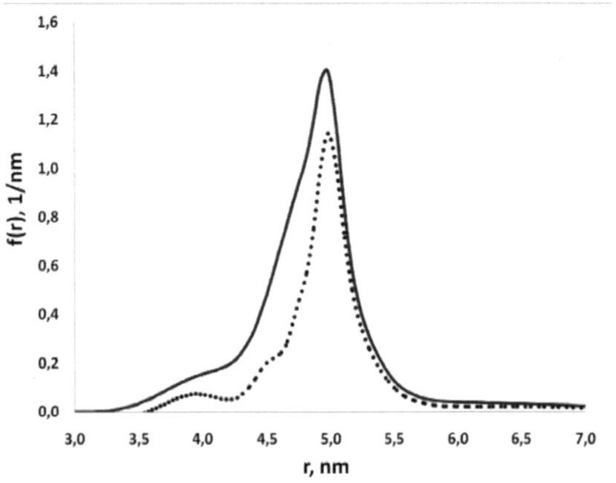

Figure 1a. The pore size distribution function of pores filled non-wetting liquid (solid line); distribution of pores filled with liquid for 1 minutes time delay (dot-dash line).

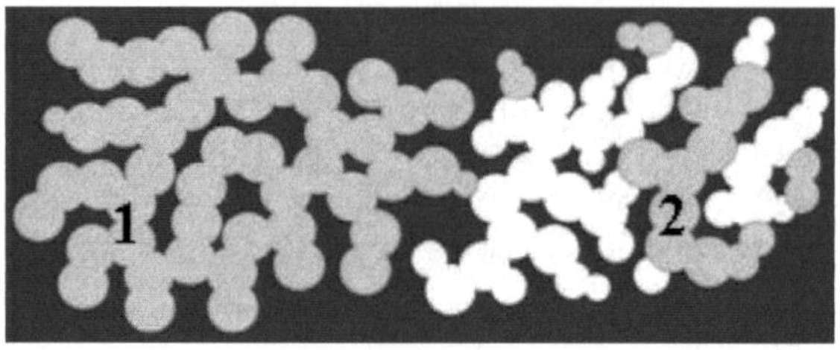

Figure 1b. The qualitative distribution of the liquid of the remaining porous medium in space after 1 minutes time delay. In blue - filled pores, white - empty pores.

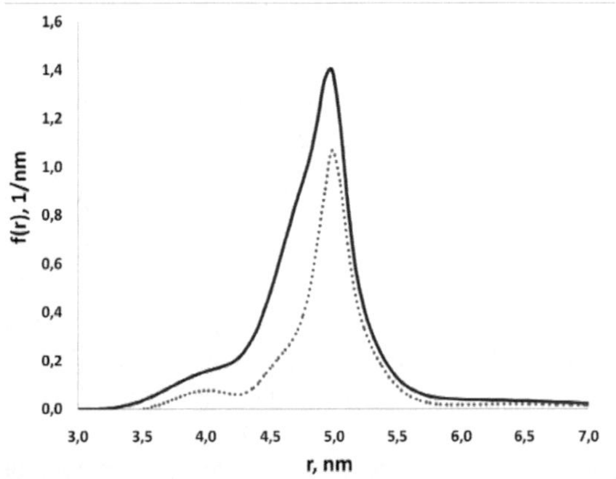

Figure 2a. The pore size distribution function of pores filled non-wetting liquid (solid line); distribution of pores filled with liquid for 10 minutes time delay (dot-dash line).

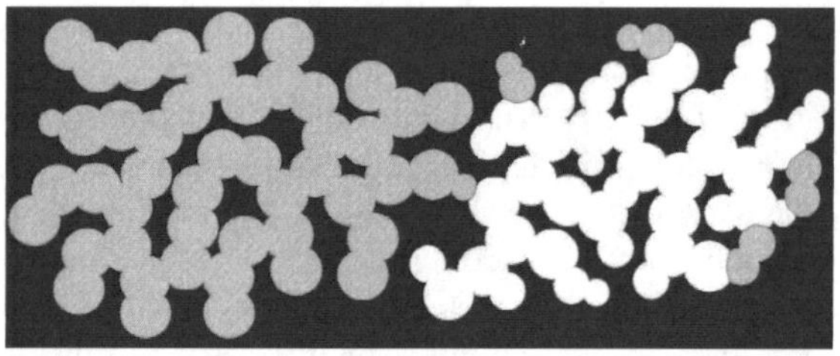

Figure 2b. The qualitative distribution of the liquid of the remaining porous medium in space after 10 minutes time delay. In blue - filled pores, white - empty pores.

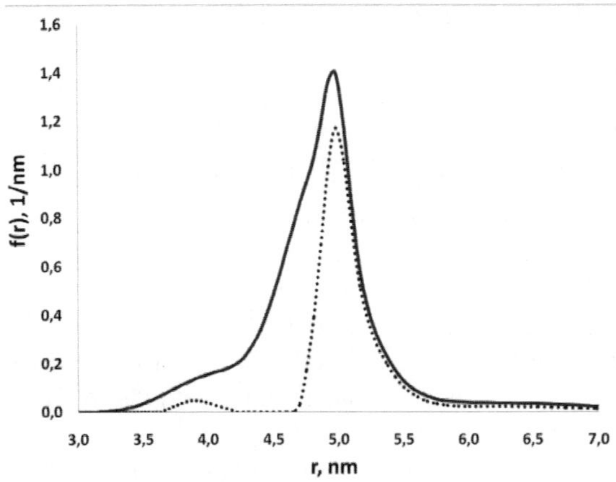

Figure 3a. The pore size distribution function of pores filled non-wetting liquid (solid line); distribution of pores filled with liquid for 100 minutes time delay (dot-dash line).

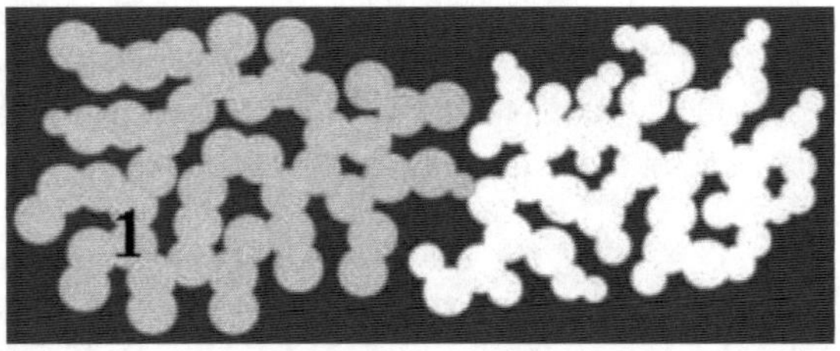

Figure 3b. The qualitative distribution of the liquid of the remaining porous medium in space after 100 minutes time delay. In blue - filled pores, white - empty pores.

4. The discussion of the results

Consider the redistribution of liquid at 6 ^{0}C. According to the figures 1b – 3b, the non-wetting liquid in the pore space forms certain local configurations. Examples of such possible local configurations are shown in Figure 4. Figure 4 is represent section of porous media. For given local configurations, we estimate the energy and time of fluid flow from a given pore.

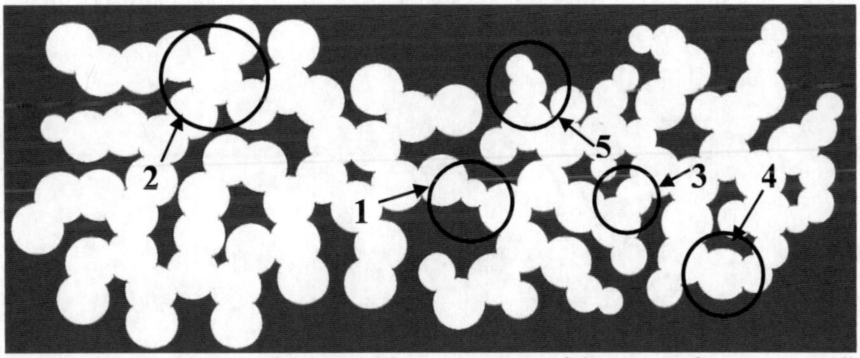

Figure 4. A qualitative picture of the mutual arrangement of the pores of a porous medium L23 with isolated local configurations.

Consider a possible local configuration representing a small pore surrounded by four large. In Figure 14, it is denoted by 1. We estimate the energy of a given small pore (~ 4 nm). According to the work [27], the change in the surface energy ΔE of the liquid in the pore is described as:

$$\Delta E = \Delta E_s + \Delta E_l \tag{1}$$

Were ΔE_s - is the change in the surface energy at the solid-liquid interface, and ΔE_l - is the change in the surface energy at the liquid-gas interface. According [27], connection between energy and characteristics of a porous medium and a liquid: $\delta\sigma$ - the change in the specific energy of the surface of a solid with the outflow of liquid, s - pore surface area, s_m - surface area of the meniscus, z - number of pores – neighbors, n - number of filled pores – neighbors, is defined by the equation:

$$\Delta E_s = -\delta\sigma(s - s_m z) \tag{2}$$
$$\Delta E_l = \sigma s_m (2n - z) \tag{3}$$

Flow is possible if $\Delta E<0$. For the given pore$\Delta E > 0$ if n=3. Then the pore is in the filled state. The time for the outflow of liquid from the pore is determined by the relation:

$$\tau = \tau_0 exp(\delta A/T) \quad [21]. \tag{4}$$

Were τ_0~10 $^{-1}$ s - characteristic hydrodynamic flow time of the fluid in the experiment, and $\delta A=pV+\Delta E$ - work that is needed to spend on outflow liquids from the pores. In the absence of excess pressure in the system (p = 0), $\delta A=\Delta E$. For a local configuration, a small pore with three filled neighbors of four $\tau\approx 3\times 10^5$ s, which exceeds the experimental observation time (6×10^3 s).

We also estimate the leak time for a possible local configuration indicated in Figure 4 the number 2. It is a large pore (~ 5 nm), surrounded by 5 neighbors. For a given pore, with 4 empty of 5 neighboring pores, it is estimated that the time $\tau \approx 4\times 10^8$ s, hat exceeds the time in the experiment. Consider the local configuration 3 (Fig.14), which is a given time of average size (~ 4.5 nm) with six possible neighbors. The time estimate, according to the formulas (1-4), with 4 occupied neighbors of 6 gives the value $\tau \approx$ 10s. That corresponds to a "fast" leakage (up to 20 s).

Another possible local configuration is shown in Figure 4 under number 4. It is a given large pore (~ 5 nm) with 2 possible neighbors. Provided that all 2 neighbors are in the filled state, $\tau \approx 4\times 10^2$ s.

Figure 4 shows the local configuration at number 3. It is a time of medium size (~ 4.5 nm), with 5 neighboring pores. In the case that 4 out of 5 neighbors are in a full state, the time $\tau \approx 5\times 10^3$ s, which corresponds to the observed outflow at times of 100 minutes.

The obtained estimates of the fluid outflow time for certain local pore configurations can be correlated with the theoretical description of the relaxation of a non-soaking non-wetting liquid obtained in [28], as a function of time. Figure 5 shows the specific local corresponding to the specific local pore configurations described above.

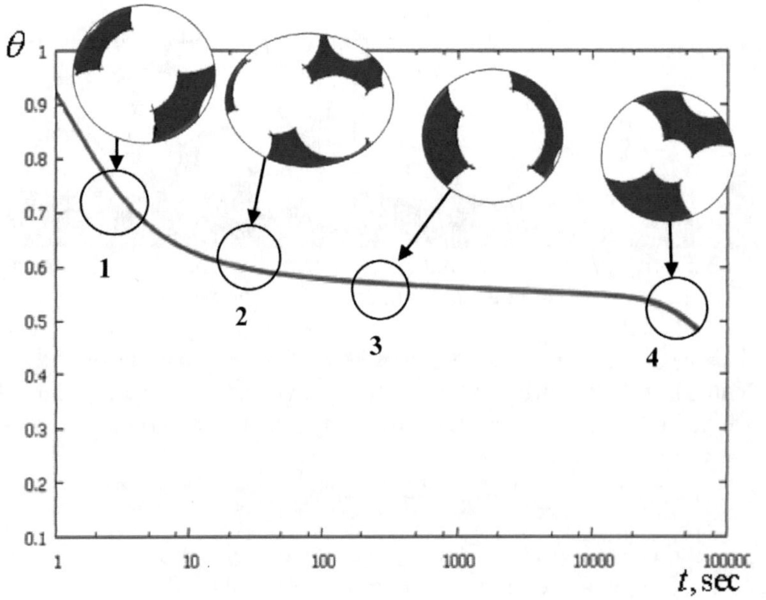

Figure 5. Qualitative depending of the volume fraction remaining liquid on time and local configurations.

For example, region 1 in Figure 5 can correspond to local configurations: a medium-sized (~ 4.5 nm) area with 6 possible neighbors with 4 occupied neighbors of 6. A region 2 can correspond to a large pore (~ 5 nm) with 2 possible neighbors with 2 filled. Area 3 corresponds to the pore of medium size (~ 4.5 nm), with 5 possible neighbors with 4 filled out of 5. Region 4 corresponds to a pore of small size (~ 4 nm) with 4 neighbors at 1 filled from 4. The obtained calculated values are qualitatively consistent with the results of the theoretical description.

5. Conclusion

On the basis of the functions of the distribution of a non-outflow liquid from pores by the size obtained in [22], an assumption was made about the possible mutual arrangement of the pores in a porous medium. Received qualitative the picture of the mutual arrangement of the pores. The evaluation of the times of non-wetting liquid from the given local configurations was presented. Experimental data of the dependence of a non-wetting liquid, dispersed in a nanoporous medium on time was correlated with theoretical description, presented in work [28].

6. Acknowledgments

This work was performed within the framework of the Center “Physics of non-equilibrium atomic systems and composites” supported by MEPhI Academic Excellence Project (Contract No. 02.a03.21.0005) and was supported by the Russian Foundation for Basic Research (project no. 18-08-01161).

References

[1] Falk M L and Langer J S 2011 *Ann. Rev. Cond. Matt. Phys.*2353.
[2] Bouchbinder E and Langer J S 2009 *Phys. Rev. E* **80** 031138.
[3] Langer J S 2012 *Phys. Rev. E* **85** 051507.
[4] Langer J S 2013 *arXiv*:1308,6544v2
[5] Bouchbinder E and Langer J S 2011 *Phys. Rev. Lett.***106** 148301.
[6] Borman V D, Belogorlov A A, Grekhov A M and Tronin V N 2014 *Eur. Phys. J.* B **87** 249.
[7] Biroli G and Garrahan J P *2013 J. Chem. Phys.***138** 12A301.
[8] *Dynamical Heterogeneities in Glasses*, Colloids, and Granular Media 2011 ed L Berthier, G Biroli, J P Bouchaud, L Cipelletti and W van Saarloos Int. Ser. Monogr. Phys.150
[9] Berthier L and Biroli G 2011 *Rev. Mod.Phys.*83587.
[10] Chander D and Garrham J P 2010 *Ann. Rev. Phys. Chem.***61** 191.
[11] Potuzak M, Welch R C and Mauro J 2011 *J. Chem. Phys.*135214502.
[12] Tanaka H, Kawasaki T, Shintani H and Watanabe K 2010 *Nat. Mater.*9324.
[13] Borman V D, Belogorlov A A and Tronin V N 2015 *Phys. Proc.*724-9
[14] Rigby S P and Edler K J 2002 *J. Coll. Int. Sci.*250175.
[15] Phillips J S 1996 *Rep. Prog. Phys.*591133.
[16] Steefel C, Depaolo D and Lichtner P. 2005 *Earth and Plan. S. Lett.,***240**(3-4):539–558,
[17] Wood B.D., Radakovich K. and Golfier F.. 2007 *Adv in Water R,***30**(6-7):1630–1647
[18] Iliev Oleg, Lakdawala Zahra, et. all 2017 *Math. Mod. and Analysis*, 671-694,
[19] Raoof A., Hassanizadeh S.M. and Leijnse A.. 2010 *Upscaling.Vadose Zone J* **9**(3):624–636,
[20] Meyers J.J., Nahar S., et. all, 2001 *J of Chr. A*, **907** 57–71.
[21] Borman V D et al 2015 *J. Exp. Theor. Phys.* **148**(6), 1169
[22] Belogorlov A A, Borman V D, et al., 2016 *J. Phys.: Conf. S.* **751**(1), 012030
[23] Borman V D, Belogorlov A A, Byrkin V A and Tronin V N 2013 *Phys. Rev.* E **88** 052116
[24] Borman V D, Belogorlov A A and Tronin V N 2016 *Phys. Rev.*E **93** 022142
[25] Borman V D, Belogorlov A A, Grekhov A M et. all , 2005 *J. Exp. Theor. Phys.* **127**(2), 431
[26] Grinchuk P and Rabinovich O 2003 *J. Exp. Theor. Phys.*96 301.
[27] Borman V D, Belogorlov A A, Byrkin V A, Tronin V N and Troyan 2013, *JETP* **144**, 1290
[28] Borman V D and Tronin V N, 2016 *J. Phys.: Conf. S.* **751**, 012034

IOP Conf. Series: Journal of Physics: Conf. Series **1099** (2018) 012024 doi:10.1088/1742-6596/1099/1/012024

The number of nearest neighbors in the model of a disordered porous medium

A S Mitrofanova, V A Byrkin

National Research Nuclear University MEPhI, Kashirskoe sh. 31, Moscow, 115409 Russia

E-mail: mitrofanova2194@yandex.ru

Abstract. The paper considers a model of a disordered porous medium consisting of chaotically located intersecting spheres of various sizes. Based on the model and the analytical approach, the number of nearest neighbors is calculated for each pore in from the pore radius. In conclusion, an analysis of this dependence is presented for packages with different pore size distributions.

1. Introduction

Currently, the "nanoporous medium – non-wetting liquid" systems are of great interest from a practical point of view, because of the large nonlinear response, such systems can be used in temperature sensitive sensors, passive protection systems, for example in fire extinguishing systems, in drug delivery systems [1], in systems with controlled permeability for liquids or gases.

Recent studies show that such a response may be connected with a multiparticle pore correlations in a disordered porous medium, collective phenomena and correlations in the mutual arrangement of clusters of liqiud in a random porous medium [2].

One of the important parameter in describing the local pore configuration is the number of nearest neighbors (coordination number). It is widely used in the evaluation of structural properties and bonds between pores, including strength characteristics [3, 4], heat and mass transfer [5] and phase formation [6]. Two methods are widely used for the investigation and determination of the number of nearest neighbors: the experiment [7, 8] and computer simulation [9, 10]. In works [15–18] devoted to modeling the motion of liquids/gases through a porous medium, the importance of understanding the pore configuration, the dependence of coordination numbers on pore size distribution, is noted.

This article is devoted to the theoretical method of investigation.

The study of the number of nearest neighbors, generally, is focused on two aspects:

- determination of the relationship between the number of nearest neighbors and porosity
- determination of the relationship between the number of nearest neighbors and the distribution of pores by size [11].

The current understanding of the relationship between the number of nearest neighbors and the pore size distribution is very limited. One of the most important characteristics of porous media is the pore size distribution function (PSDF). In the nanopore region, the experimental methods for the getting of PSDF are based on information on the hysteresis loop. In this case, the PSDF is calculated from its adsorption and desorption branches.

The calculation of the PSDF for mesoporous media is based on the capillary condensation theory. As a rule, the considered hysteresis loop is typical for porous medium with a corpuscular structure. The stochastic geometry of corpuscular porous medium is very complicated. For this reason, when analyzing the experimental data, the first approximation is the model of effective spherical or cylindrical pores [12].

In 1945, Wheeler, taking as a basis the model of an open cylindrical capillary, proposed a method for calculating the PSDF. Later his approach was developed in the works of Barrett Joyner-Halenda, Creston-Inkley and Dolimer Hill. On the basis of Wheeler's approach, a generalized equation was proposed that makes it possible to obtain a PSDF for pores of spherical, cylindrical, and slit forms.

In this paper, the model of a disordered porous medium is proposed consisting of chaotically located intersecting spheres of various sizes. The pore space represents the voids inside the spheres, the rest of the space represents a solid phase. This model allows to take into account multiparticle pore correlations and within the framework of the model it becomes possible to analytically calculate the number of nearest neighbors in each pore taking into account the pore size distribution [13]. In addition, the calculation of the number of nearest neighbors of such a system will allow to reconstruct the pore size distribution function in view of their correlations. This fact makes it possible to describe the thermodynamic states of a liquid in a nanoporous disordered medium as an ensemble of quasiparticles consisting of clusters of liquid in neighboring pores.

2. Main part

A model of chaotically located spheres was proposed by Weissberg to describe disperse and porous media. The model is a system of chaotically located intersecting particles of a spherical shape that do not correlate with each other. The probability of finding an arbitrarily chosen center of a particle at a certain point in space does not depend on the location of the remaining centers. This model is universal, because allows us to describe both the structure of the body skeleton and the structure of the pore space. The simplest case of this model is a system of spheres of the same radius [14]. Let us consider a system of chaotically located intersecting cavities of spherical shape with concentration n. In this model, by porosity is meant the ratio of the volume outside the spheres to the total volume of the medium. That is, the porosity is equal to the probability that an arbitrary point of space is in the space of pores; in other words, it is the probability that no center is in the sphere of radius R.

Let us find the probability that the volume V_0 ($V_0 \ll V$) does not contain any center, i.e. all the centers are located in the volume $V - V_0$. V is the total volume of the medium.

Then the probability that an arbitrarily chosen center is outside the volume V_0 is equal to:

$$P(V_0) = \left(\frac{V - V_0}{V}\right)^{nV} \tag{1}$$

In this case, the total number of particles nV is large, and the product nV_0 is finite, so we can use the definition of the number e as the limit of the expression $(1-\alpha)^{-1/\alpha}$ for $\alpha \to 0$.

Then the probability becomes:

$$P(V_0) = e^{-nV_0} \tag{2}$$

where $V_0 = \frac{4}{3}\pi R^3$ is a volume of one sphere.

From the expression for the probability we get the value of porosity:

$$\varphi = e^{-\alpha} \tag{3}$$

where $\alpha = \frac{4}{3}\pi R^3 n$ is a dimensionless parameter.

The pore structure is a random lattice, the nodes of which correspond to voids, and the connections to the throats. By the number of nearest neighbors (or coordination number) is meant the number of points of contact of one pore, in other words the number of links originating from the node. Too large and too small numbers of nearest neighbors are unlikely.

The probability P_z that an arbitrarily chosen cavity has Z throats is equal to the probability that exactly Z lies inside a sphere of radius $2R$ from $(n-1)$ centers of cavities located in a unit volume containing the chosen cavity, centered at the center of the selected cavity:

$$P_z = C_{n-1}^{z}\left(\frac{4}{3}\pi(2R)^3\right)^z\left(1-\frac{4}{3}\pi(2R)^3\right)^{n-1-Z} = \frac{(8\alpha)^Z e^{-8\alpha}}{Z!} \tag{4}$$

The average number of nearest neighbors is

$$\bar{Z} = \sum ZP_Z = 8\alpha \tag{5}$$

Consider a system of N sort of spheres.

$$\begin{cases} \text{1-sort} - a_1R, b_1n_1, \text{where } a_1 = 1, b_1 = 1 \\ \text{2-sort} - r_2 = a_2R, n_2 = b_2n_1 \\ \text{3-sort} - r_3 = a_3R, n_3 = b_3n_1 \\ \ldots \\ \text{N-sort} - r_N = a_NR, n_N = b_Nn_1 \end{cases} \tag{6}$$

Then the dimensionless parameter α takes the form:

$$\alpha_j = \frac{4}{3}\pi R^3\frac{n_j}{b_j}\sum_{i=1}^{N} a_i^3 b_i \tag{7}$$

The average number of nearest neighbors is

$$\bar{Z}_k = \sum_{k=1}^{N}\frac{(a_j+a_k)^3 b_j\alpha_j}{\sum\limits_{i=1}^{N} a_i^3 b_i} = \frac{4}{3}\pi n_j\sum_{k=1}^{N}(r_j+r_k) \tag{8}$$

3. Results and discussion

In the course of this study, a numerical experiment was conducted, the technique of which is as follows. The initial simulated region is a cube with a linear dimension $L = 32$. The cube is filled with spheres of various sizes that can overlap. Generation of a sphere occurs by assigning its coordinates to the center and radius. In turn, the center refers to an arbitrarily given point inside the cube, and the radii of the spheres obey the Gaussian distribution. As a result, the space outside the spheres will represent the skeleton of the porous body, and the voids inside - the pore space.

To verify the numerical experiment, at the start the monodisperse sphere approach was applied (Eq. (5)).

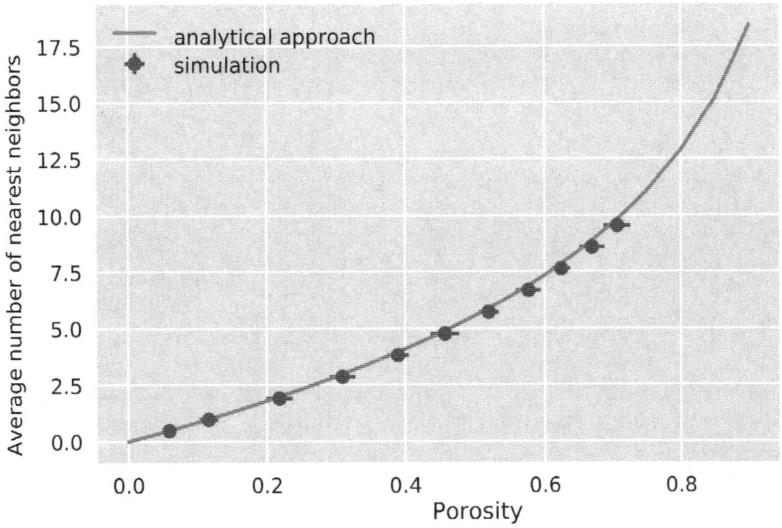

Figure 1. Dependence of the average number of nearest neighbors on the porosity of the system

In the case of polysperpic spheres, the dependence of the number of nearest neighbors for three variants of pore size distribution was studied:

- Narrow distribution with parameters: $\mu = 1, \sigma = 0,1$
- Wide distribution with parameters: $\mu = 1, \sigma = 0,4$
- Normal distribution with parameters: $\mu = 1, \sigma = 0,2$

where μ is the mean dimensionless radius and σ is variance.

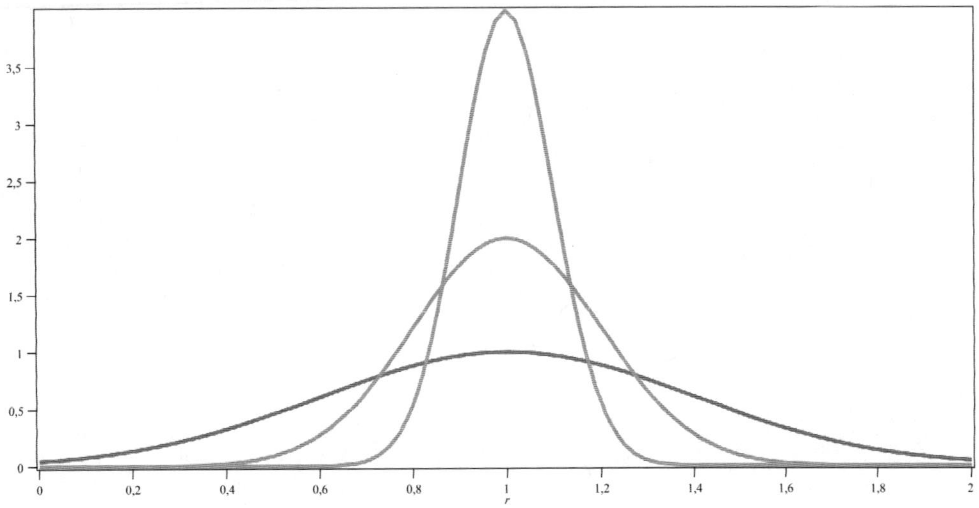

Figure 2. Three cases of pore size distribution

IOP Conf. Series: Journal of Physics: Conf. Series **1099** (2018) 012024 doi:10.1088/1742-6596/1099/1/012024

3.1. Narrow pore size distribution

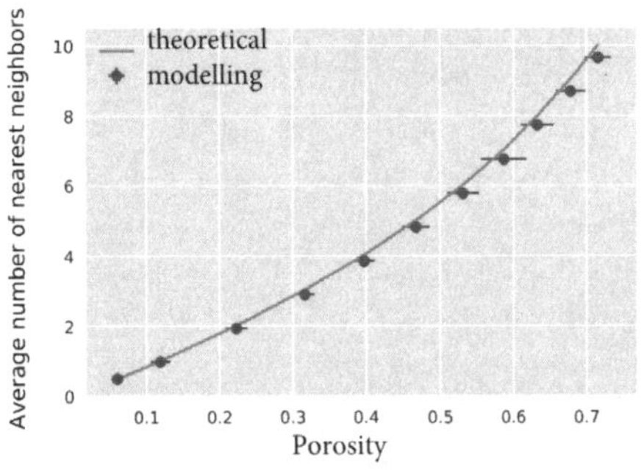

Figure 3. The average number of nearest neighbors, depending on the porosity of the system

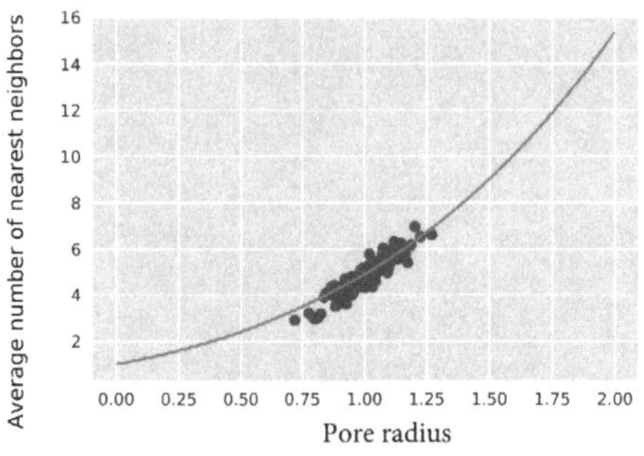

Figure 4. The average number of nearest neighbors at the pore radius r

3.2. Normal pore size distribution

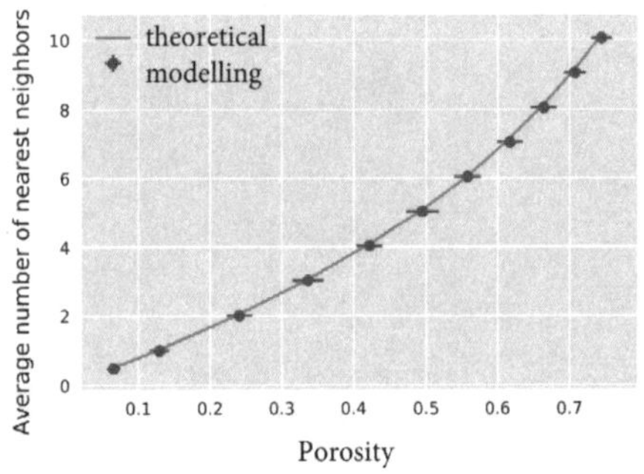

Figure 5. The average number of nearest neighbors, depending on the porosity of the system

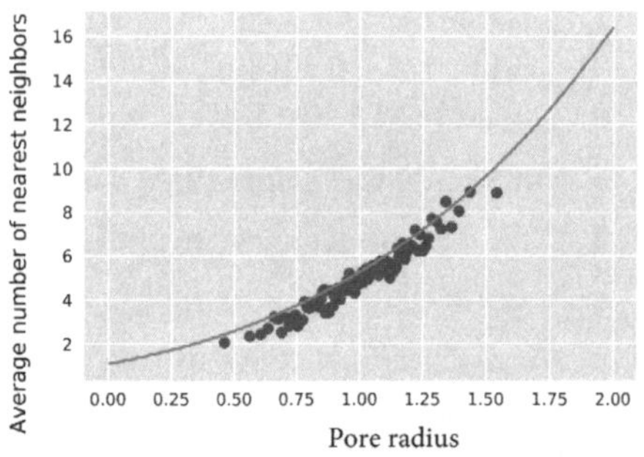

Figure 6. The average number of nearest neighbors at the pore radius r

3.3. Wide pore size distribution

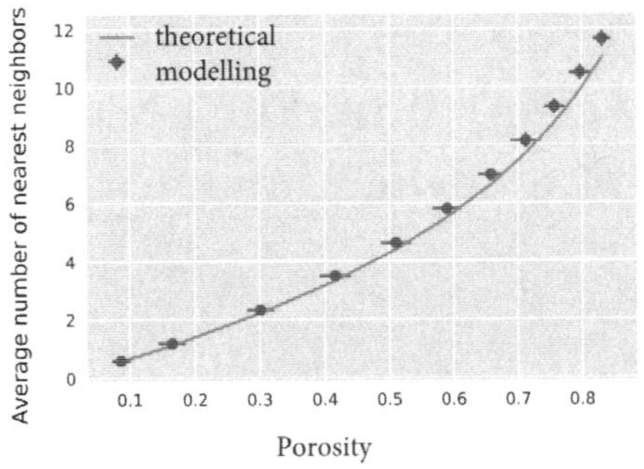

Figure 7. The average number of nearest neighbors, depending on the porosity of the system

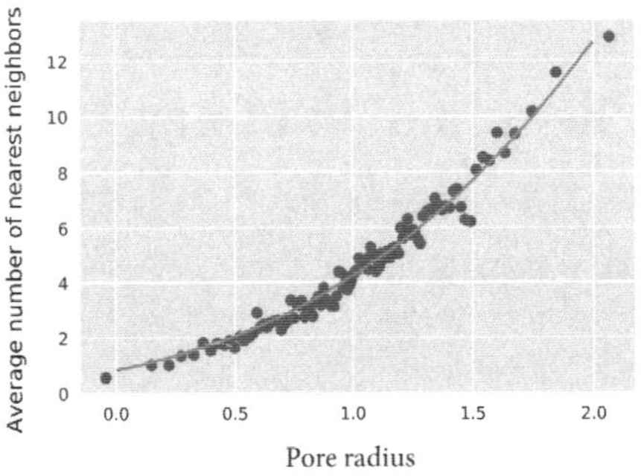

Figure 8. The average number of nearest neighbors at the pore radius r

In the course of the work, a model of chaotically located overlapping spheres of various sizes was considered. Within the framework of this model, a formula was obtained that describes the dependence of the number of nearest neighbors on the radius of spheres. The distribution of the nearest neighbor numbers was plotted as a function of the pore radius for different values of porosity and the dependence of the number of nearest neighbors on the pore radius r. The exact experiment was verified for the case of monodisperse spheres. It was shown that the results obtained from the formula are consistent with the data obtained in the course of the numerical experiment.

Acknowledgments
The present work was supported by Russian Foundation for Basic Research (project no. 18-08-01161).

References

[1] Ksenofontova O I, Vasin A V, Egorov V V, Bobyl A V, Soldatenkov F Y, Terukov E I, Ulin V P, Ulin N V, Kiselev O I 2014 *Technical Physics* **59** 66-77
[2] Borman V D, Belogorlov A A, Byrkin V A, Tronin V N 2013 *Phys. Rev. E* **88**
[3] Johnson K L, Kendall K, Roberts A D 1971 *Proc. of the Royal Soc. of London* **324** 301–313.
[4] Cundall P A, Strack O D L 1979 *Geotechnique* **29** 47
[5] Wakao N 1982 *Heat and Mass Transfer in Packed Beds* (New York:Gordon and Breach Sci. Pub)
[6] German R M 1989 *Particle Packing Characteristics* (New Jerse: Metal Powder Industries Federations)
[7] Zou R P, Bian X, Pinson D, Yang R Y, Yu A B, Zulli P 2003 *Part. Part. Syst. Charact* **20** 335–341
[8] Bernal J D, Mason J 1960 *Nature* **188** 910
[9] Suzuki M, Oshima T 1985 *Powder Technol* **44** 213
[10] Oger L, Troadec J P, Bideau D, Dodds J A, Powell M J 1986 *Powder Technol* **46** 121
[11] Schkolnikov E I, Sidorova E V 2007 *Doklady RAN* **412** 357-360
[12] Karnauhov A P 1999 *Texture of dispersed and porous materials [in Russian]* (Novosibirsk: Nauka)
[13] Byrkin V A 2015 *Phys. Proc.* **72** 14–21
[14] Kheifets L, Neimark A 1982 *Multiphase Processes in Porous Media (in Russian)* (Moscow: Khimja)
[15] Li M, Tang Y B, Bernab Y, Zhao J Z, Li X F, Li T 2017 *Journal of Geophysical Research: Solid Earth* **122** 4918–4930
[16] Veyskarami M, Hossein A, Ghazanfari H H 2018 *Comp. Geosci.* **22** 329-346
[17] Mahabadi N, Jang J 2017 *ICSMGE 2017 - 19th International Conference on Soil Mechanics and Geotechnical Engineering* 1201–1204
[18] Ding B, Li C, Zhang M, Ji F, Dong X 2015 *Chem. Eng. Sci.* **127** 40–51

Phenomenon of non-outflow of a non-wetting liquid dispersed in nanoporous medium. The influence of modification and size of granules.

A Yu Emelianova[1], S A Bortnikova[1] and A A Belogorlov[1,2]

[1] National Research Nuclear University MEPhI (Moscow Engineering Physics Institute) Molecular Physics Department, 115409, Russia, Moscow, Kashirskoe highway, 31
[2] Topchiev Institute of Petrochemical Synthesis, Russian Academy of Sciences, 119991, Russia, Moscow,Leninsky prospect, 29

E-mail: svetlana.sweta7@yandex.ru, AABelogorlov@mephi.ru

Abstract. In this paper was studied the non-outflow phenomenon of a non-wetting liquid dispersed in nanoporous medium and the influence of porous medium surface modification and size of porous medium granules. The study of this phenomenon is important for development of mechanical energy dampers and for extraction of medicinal substances under certain conditions devices. The presented results for fourcommercialhydrophobized porous media Fluka 100 C18 (60756), Fluka 100 C18 (60758), Fluka 100 C8 (60755) andFluka 100 C8 (60759) with the different thickness of the modifying layer and the size of the granules are show that the less of granules size and more narrow pore size distribution lead to decreasing of non-outflow phenomenon. The study of non-outflowphenomenon of non-wetting liquid from the pores of a nanoporous medium was carried out by time relaxation method and dispersed non-wetting liquid distribution recovery method at different temperatures.

1. Introduction

When the excess pressure is applied in a porous medium - nonwetting liquid system, the pores of the medium are filled with a liquid that passes into a dispersed state. When the excesspressure is reduced in the intrusion-extrusion cycle, the hysteresis is observed, and the filling pressure is greater than the outflow pressure [1]. Depending on various parameters of the system, both energy and geometric, different scenarios of this fluid behavior are possible with decreasing pressure. At the certain critical degree of filling and temperature, a dispersion transition occurs and part of the liquid can remain in a disordered porous medium [2]. It was assumed that the phenomenon of non-outflow of non-wetting liquid and the dispersion transition are determined by the filled pores arrangement, and the temperature dependence is related to the temperature dependence of the specific surface energy of the liquid-gas and the liquid-solidinterfaces([1,3]).

Parameters of the nanoporous media - non-wetting liquid system can significantly affect the amount of liquid remaining in the pores. Thus, it was shown that the volume of entrapment liquid depends, in addition to the observation time, on the temperature and size of the granules of the porous medium under investigation [4]. It was shown in [5, 6] that due to the increase in the length of the grafted alkylsilane, the contact angle of wetting of the modified surface also increases, while another group of researchers observed its insignificant decrease [7]. At the same time, it was shown in [9] that the effect of the pore geometry on the coating density is significantly manifested when the alkyl chain extends independently of the structure of the sample.

It was also described in [9, 10] that the non-outflow of liquid is observed when small pores surrounded by pores larger in size. The existence of such pore arrangements can also explain the dependence of the volume of entrapment liquid on the size of the silica gel granules. In [4], the link between the length of the modifier and the presence of large and small pores wasnoted using the example of the study of Fluka 100 C8 and Fluka 100 C18 silica gels, and the same was also obtained in [11]. Due to fluctuations in the pore size of the porous medium, which are determined from the pore size distribution function, at different times the liquid can flow out of pores of different sizes and the number of these pores also depends on the observation time.

The practical interest of observing the relaxation of the metastable state of a non-wetting liquid dispersed in a nanoporous medium is associated with the possible use of devices for the accumulation and dissipation of mechanical energy. Also, the results of these observations can be used in the study of time-dependent and temperature-dependent processes, such as extraction of drugs from porous structures.

2. Methods and materials

The investigated porous media were silica gel Fluka 100 C8 (catalog number 60755 and 60759) and silica gel Fluka 100 C18 (catalog number 60756 and 60758), the industrial samples of hydrophobized silica gel from Sigma-Aldrich, andcarriers of these samples after removal of the modifying layer.For the original samples and then for their carriers after removal of the modifying layer, the specific surface area (BET) [12], specific pore volume, density and pore size distribution function (BJH) [12] were determined. The measurements were carried out using a helium pycnometer ULTRAPYC 1200e and a gas sorption analyzer Autosorb IQ (Quantachrome Instruments). The resultsfor modified samples are shown in Tables 1 and compare of pore size distribution (BJH) for modified and non-modified samples in Figure 1. Samples Fluka 100C8 (60755) and Fluka 100C8 (60759) have the same length of the modifier, but different granule size, while in Fluka 100C8 (60755) and Fluka 100C18 (60756) the granules have the same size.

Table 1. Samples characteristics

Sample	$V_{por}, \frac{cm^3}{g}$	$S_{por}, \frac{m^2}{g}$	<R>, nm	$\rho, \frac{g}{cm^3}$	Granule size[1], μm
Fluka 100C8 (60755)	0.61	326	3.8	1.73	15-35
Fluka 100C8 (60759)	0.44	222	4.0	1.75	40-63
Fluka 100C18 (60756)	0.37	142	3.8	1.62	15-25
Fluka 100C18 (60758)	0.47	195	4.5	1.90	40-63

As can be seen from the presented table, the samples are similar in pore radius, but have different specific pore volume and surface area. At the same time, these values are thesmallest for the C18 samples, which is connected with the longer length of the modifier molecule and, correspondingly, its volume in the pores of the medium.

[1]Material safety data sheet

IOP Conf. Series: Journal of Physics: Conf. Series **1099** (2018) 012025 doi:10.1088/1742-6596/1099/1/012025

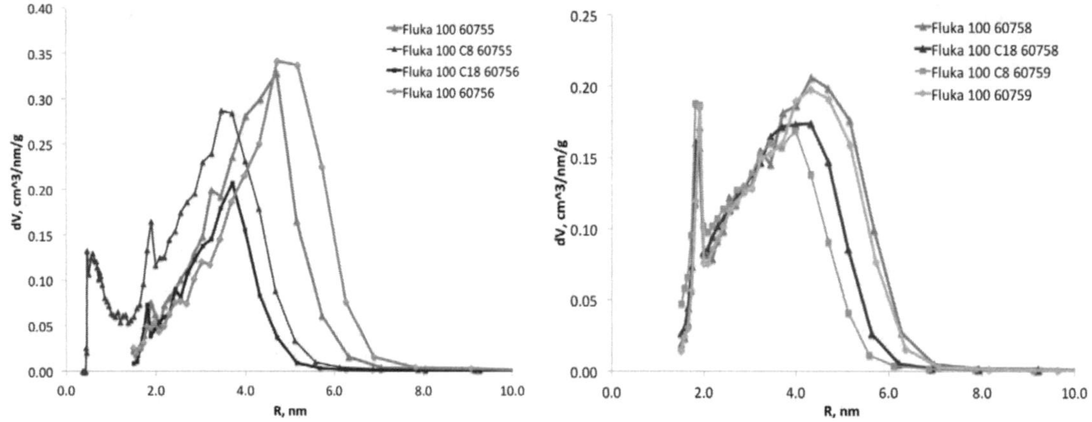

Figure 1. Pore size distribution of samples and samples carriers

From the obtained pore size distribution functions for carriers (Fig.1), it can be seen that they are similar in samples with the same granule size. This means that only the grafted layer of the modifier can affect the differences in the characteristics of these samples. The greatest difference in the characteristics of the modified sample and carrier are observed forFluka 100 C18 (60756), which are natural given the greater length of its modifier.

The study of the non-outflow phenomenon of a non-wetting liquid dispersed in nanoporous medium was carried out in accordance with [4] and [13].

3. Results

Figure 2 shows the intrusion-extrusion curves (with account of the compressibility of the system). As it can be seen from Fig. 2, investigated samples have the same intrusion pressure: $\mathbf{P_{in}} \approx (\mathbf{145} \pm \mathbf{5})$ atm.The pore size distribution functions of the samples were obtained according to the procedure described in [13]. Sample Fluka 100 C18 (60758) was excluded from the investigation because of modification layer destruction during experiments.

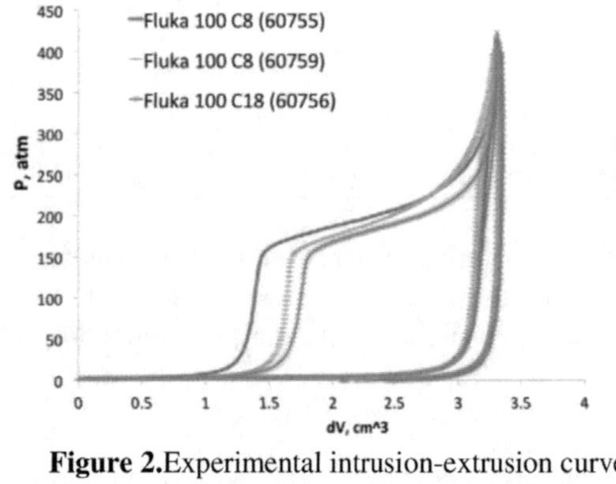

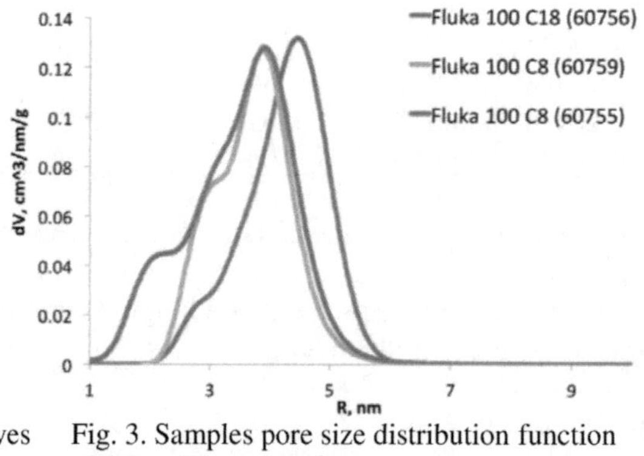

Figure 2. Experimental intrusion-extrusion curves (with account of the compressibility of the system)

Fig. 3. Samples pore size distribution function

The smoothed function $dV(p)$ is differentiated. The resulting function is approximated by the analytic function with high accuracy ($R^2 \geq 0.998$).The analytical form of the derivative $dV(p)/dp$ allows us to obtain a function of the pore size distribution, assuming pores are cylindrical and the contact angle $\theta = 115°$ (Fig. 3).It can be seen from the presented distributions that for an equal average pore radius, determined from the maximum value of the distribution functions, their half-width at half-height is different.

IOP Conf. Series: Journal of Physics: Conf. Series **1099** (2018) 012025 doi:10.1088/1742-6596/1099/1/012025

To observe the relaxation of the liquid in the space of media, they were filled at different time delay [4]. The fraction of the volume filled with liquid was defined as the ratio of the volume of liquid entering the pores to the total pore volume. The fraction of entrapment liquid was defined as the ratio of the volume of liquid remaining in the pores after removal of the excess pressure to the total pore volume. First, one full intrusion-extrusion cycle was done by loading and subsequent unloading of the system (t = 0c). Further the cycle was repeated after t = 1 s. Then, the cycle was repeated with various time delays.

The results of observing the relaxation of the liquid under study for three different temperatures (20, 40, 60 °C) are presented in Figures 4.1-4.3. It can be seen that the fractions of entrapment liquid remaining in the pores $\theta_{ent}$for the sample C8 (60755) remains approximately constant, but for samples with a smaller granule size C8 (60755) and C18 (60756) this fraction decreases significantly with temperature, especially for sample C18.

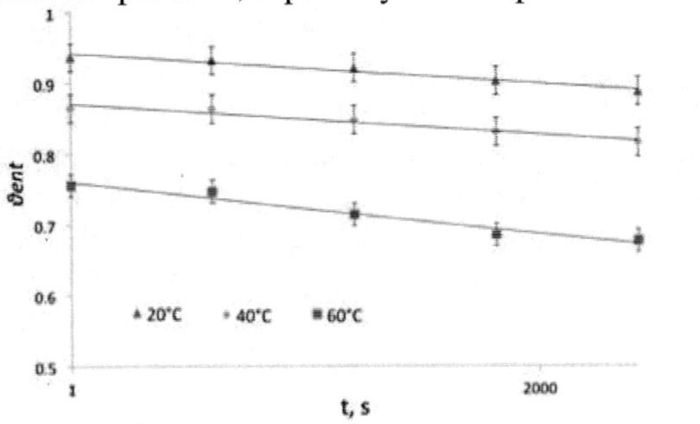

4.1.

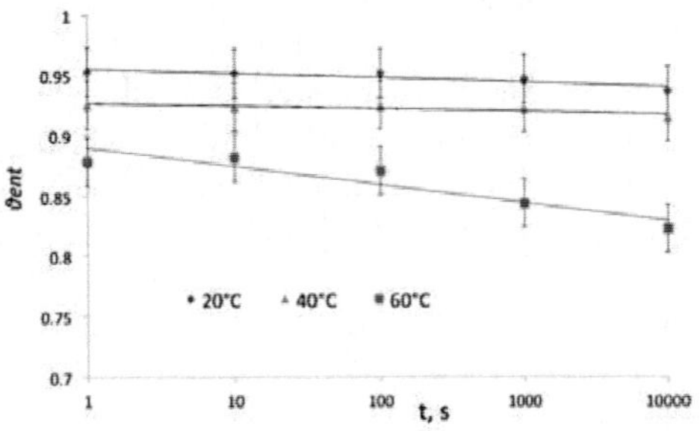

4.2.

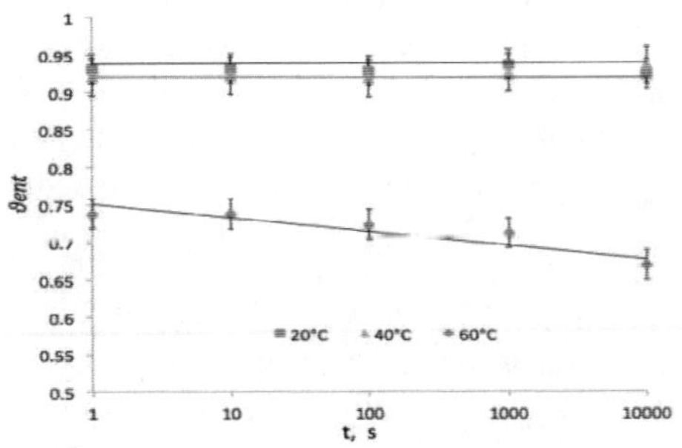

4.3

Figure 4. The time dependence for fraction of entrapment liquid for C8 (60755) (4.1), C8 (60759) (4.2), C18 (60756) (4.3) (logarithmic scale)

There may be a number of reasons for the observed differences. First, most of the liquid naturally flow from samples with a large granule size. Further, since the energy barrier of intrusion of fluid changes following the change in the surface tension of the liquid-porous interface, as the temperature increases to 60°C, it becomes negative and the liquid leave the porous medium. In turn, the interaction energy of the surface of silica gel and water depends on the modification of the surface, i.e. perhaps the thickness of the layer also has an effect. In addition, the reason can be the difference in the distribution functions of the investigated samples, and, consequently, the difference in the local configurations of empty and filled pores.

To study the size and number of pores that remain filled at different times, for a pair of samples with significant increase in temperature - Fluka 100 C8 (60755) and Fluka 100 C18 (60756) - pore size distribution functions were constructed (Fig. 6) for each intrusion cycle. The calculation method described in [13] was used. The results are presented for relaxation at 60 °C.

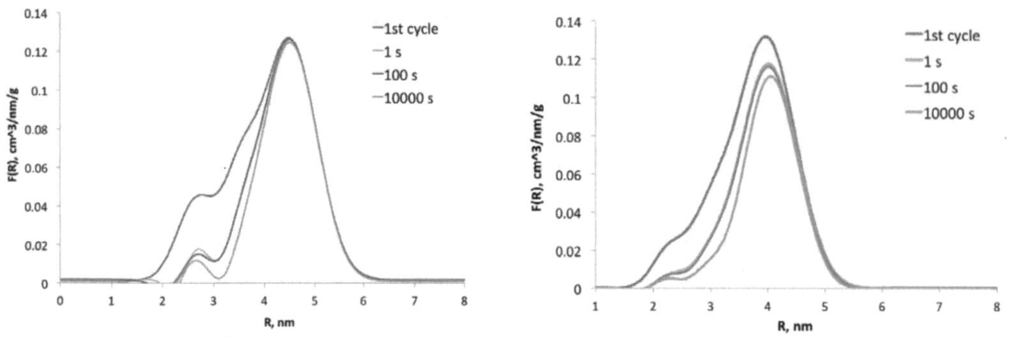

5.1 5.2

Figure 5. Pore size distribution for intrusion cyclesfor different time delaysfor Fluka 100 C8 (60755) (6.1) and Fluka 100 C18 (60756) (6.2) samples

As can be seen from Figs.5.1-5.2, the amount of liquid remaining in pores decreases with increasing observation time. In this case, the sizes of the pores remaining filled are different for the samples. In sample Fluka 100 C8 (60755), the liquid flows out only from small pores each time, in contrastto Fluka 100 C18(60756), for which in each cycle of intrusion the number of filled pores decreases among pores of different sizes. Perhaps this is a consequence of the difference in the width of the pore size distribution function of the samples.

4. Conclusion

The study of non-outflow phenomenon of non-wetting liquid from the pores of a nanoporous medium was carried out by time relaxation method and dispersed non-wetting liquid distribution recovery method at different temperatures.The presented results for four commercial hydrophobized porous media Fluka 100 C18 (60756), Fluka 100 C18 (60758), Fluka 100 C8 (60755) and Fluka 100 C8 (60759) with the different thickness of the modifying layer and the size of the granules are show that the less of granules size and more narrow pore size distribution lead to decreasing of non-outflow phenomenon.

Acknowledgments

This work was performed within the framework of the Center"Physics of nonequilibrium atomic systems and composites" supportedby MEPhI Academic Excellence Project (Contract No. 02.a03.21.0005)and was supported by the Russian Foundation for Basic Research(project no. 17-08-01424).

References

[1] Borman V D, Belogorlov A A, Grekhov A M, et al., 2005 *JETP* **139** 446.
[2] Borman V D, Belogorlov A A, Byrkin V A, et al., 2012 *JETP Lett* **95** 579.
[3] Jaroniec C P, Kruk M, Jaroniec M, 1998 *J. Phys. Chem.* **102** 5503.
[4] Borman V D, Belogorlov A A, Tronin V N, 2015 *Coll. Surf. A: Phys. Eng. Asp.* **496** 63.
[5] Turowski M, Morimoto T, Kimata K, Monde H, 2001 *Chrom. J.* **91** 177.
[6] Horr I J, Ralston J, Smart R, 1995 *Coll. Surf A: Phys. Eng. Asp.* 97
[7] Korosi G, Kovats E, 1981 *Colloid Surf*. **2** 315.
[8] Chabert J B, Chapel J P, Jerard J F, et al., 1997 *Langmuir* **13** 2271.
[9] Borman V D, Belogorlov A A, Byrkin V A, et al., 2013 *JETP Lett.* **114** 1139.
[10] Rigby S P, Edler K J, 2002 *J. Colloid Interface Sci.* **250** 175.
[11] Suciu C V, Iwatsubo T, Yaguchi K, et al., 2005 *J. Colloid Interface Sci.* **283** 196.
[12] Lowell S, Shields J E, Thomas M A, et al., 2006 *Characterization of Porous Solids and Powders: Surface Area, Pore Size and Density*, (Springer, New York) .
[13] Belogorlov A A, Borman V D, Bortnikova S A, et al., 2016 *J. Phys.: Conf. S.* **751**.

Intellectual properties of a granular nanoporous medium in a non-wetting liquid.

V D Borman, A A Belogorlov and V N Tronin,

National Research Nuclear University MEPhI (Moscow Engineering Physics Institute), Molecular Physics Department, 115409, Moscow, Kashirskoe highway, 31, Russia

E-mail: AABelogorlov@mephi.ru

Abstract. The intellectual properties of a suspension of a granular nanoporous medium in a non-wetting liquid under conditions of a pulsed pressure change are discussed in the work. The characteristic response time (0.1 ÷ 10) ms and the high specific energy intensity of the impact energy conversion (E) (10 ÷ 100) J / g make attractive the use of the such type suspensions in shock / explosion absorption devices. In the process of rapid compression upon impact and when the critical pressure (P_0) is reached, the suspension is transformed into a new state with a change in the compressibility ($\chi = dV / dP$) by (2 ÷ 4) of the order. In the new state, the process of volume reduction does not depend on the temperature and viscosity of the liquid, it is practically non-dissipative and at constant pressure. A microscopic mechanism of such a nontrivial nonlinear response is proposed. The intellectual properties of a suspended granular nanoporous medium in a non-wetting liquid under conditions of pulsed pressure changes are considered using of the Libersorb 23-water system as an example.

1. Introduction

According to [1] the intellectual materials it is materials satisfying the following conditions: 1. Intellectual material is created by synthesizing various components to obtain the required response; 2. Strong response to external impact; 3. The macroscopic properties of the response and the change in state must be determined by a microscopic mechanism; 4. After the response, the intellectual material must return to its initial state.

At the present time, intelligent materials are known with the use of the effects of change under the external influence of the properties of a flat or rough surface of materials [1]. For example, when the surface is chemically modified by polymer molecules, the self-cleaning effect of the surface is observed. This is achieved due to the mechanism of changing the conformation of grafted polymers to the surface with increasing temperature [2]. As a result, the surface changes from a hydrophilic state to a hydrophobic state at a low temperature as the temperature rises and self-cleaning of the surface is observed.

As it follows from the results of a study of the properties of three materials of granular nanoporous media, different in nature, in a non-wetting liquid [3] in its design, response to impact and microscopic mechanism of state change during impact, and these materials can be attributed to new intellectual materials. It should be noted that the characteristic time for the outflow of non-wetting liquid from a granular nanoporous materials can be ($10^4 \div 10^7$) s [4, 5], but also and ($10^{-2} \div 10^{-1}$) s at the characteristic filling time ($10^{-4} \div 10^{-3}$) s [3, 6].

Another attractive feature of such materials is the high (up to 100 J/g) specific energy consumption of the process of filling nanopores of granules at the excess pressure. It follows that the suspension of a granular nanoporous medium in non-wetting liquid can be used for multiple impact absorption as a material with a response time comparable to the impact time. However, questions about the choice of properties and under what conditions the material under discussion can demonstrate intellectual properties that differ fundamentally from the properties of irreversibly (or viscoplastic) deformable materials [7]. First of all, this is a response to a shock pre-programmed with a material that allows providing the force (pressure) of an impact on the protected object that is smaller than the energy (impulse) impact, without reference to its magnitude. This possibility is indicated by the results of [3, 6, 8]. However, in experiments [9, 10] under different conditions, a strong pressure-force dependence on the time under impact and on the characteristics of the liquid and the mass of the porous substance was observed for the granular nanoporous substance in the various liquids studied. Therefore, the question of the properties of the response of a granular porous medium suspended in a non-wetting liquid remains open.

2. The developing for providing of necessary response

The basic and necessary property for a nanoporous granular material is lyophobicity. This property can be both natural and obtained by modifying the surface of the nanoporous granular material. Thus, the investigated system Libersorb 23 (L23) - water consists of a porous material L23, which is a silica gel KSK-G with a SiO_2 frame material, the surface of which was chemically modified by alkylsilanes according to the procedure described in [11], to impart a hydrophobic property to the pore surface. Lyophobicity of this material to the water was confirmed in [12]. In addition to the necessary condition for the surface to liquid to be lyophobic when creating a material with predetermined properties, it is also necessary to know the specific pore volume of the porous medium (V_{por}), the compressibility of the porous medium (χ_{pm}) and the liquid (χ_{liq}), the characteristic of the pores filling pressure (P_{c0}) and the evidence of returning the system to its initial state after an external impact. Characteristics of the L23 - water system are as follows: $V_{por} = 0.56 \pm 0.02$ cm^3/g, $\chi_{pm} = 0.41\ 10^{-8}$ 1/MPa, $\chi_{liq} = 4.4\ 10^{-8}$ 1/MPa [13], $P_{c0} = 16 \pm 1$ MPa. It was also shown in Ref. [12] that the system returns to its initial state after pores filling with a non-wetting liquid, and the fraction of non-liquid is less than 5% at a temperature of 20 °C. Subsequent drying of the L23 samples showed their complete recovery to the initial state. Also in work [6] for this system it was found that when the impact is applied and the mass of the porous medium is sufficient, the pores filling pressure of the non-wetting liquid remains, up to the complete absorption of the impact energy, constant within the measurement error. Thus, the correct choice of a porous material and a liquid is an important step in the creation of a material, namely a suspension of a porous substance in a non-wetting liquid having a strong response to an external action.

3. Strong response to external influences and return to initial state

An investigation of the response to an external impact during the filling of porous systems with non-wetting fluids was performed on an experimental bench and using the procedure described in [6]. A typical record of the change in pressure time (P) and a decrease in the volume ($-\Delta V$) of a suspension of granular nanoporous material L23 (4 g) in a non-wetting liquid, water (55 cm^3), as a result of a 10 kg load impact with an energy of E = 30 J is shown in Fig. 1a and 1b. It can be seen that as the time increases in the process of impact due to the movement of the rod inside the chamber, volume decreases with a continuous increase in pressure to point 2 in Figs. 1a and 1b. In point 2, the mode of transformation of the impact energy is changed. Continuous growth of pressure with a decrease in volume is replaced by a decrease in volume at a constant, within the error of measurement, independent of time and pressure. Therefore, point 2 is the critical point at which the "compressibility" ($\chi = d\Delta V/dP$) (receptivity) of the material suspension in the liquid changes abruptly with a continuous decrease in volume. The volume of suspension of the granular material and liquid continues to decrease at constant pressure with an increase in time to point 3. At this point, the volume decrease volume reaches a maximum. This corresponds to stopping the stock and the load. In the time interval from point 3 to point 4 the stock moves from the chamber so that to point 4 the volume of material and liquid tends to the initial state.

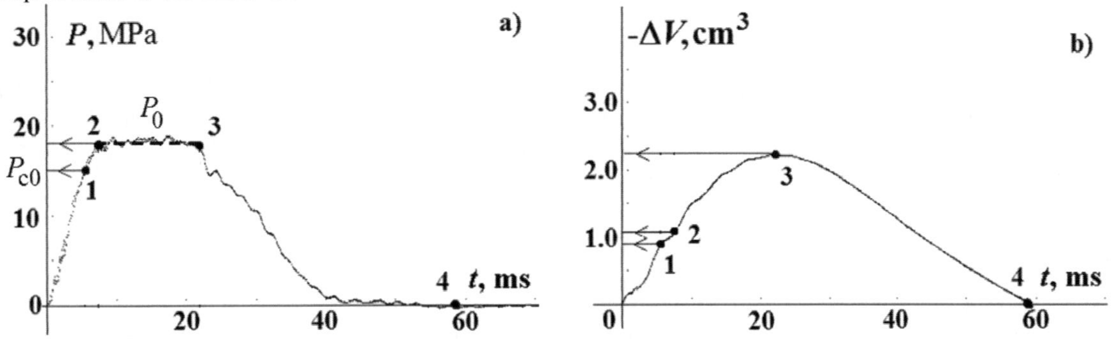

Figure 1. The experimental data of the pressure change in the chamber (a), and the internal volume of the chamber (b) versus time. The chamber is filled with water (liquid volume 55 cm^3) and a porous medium L23 with a mass of 4 g. The impact energy E = 30 ± 2 J.

As it follows from Fig. 1b in the time interval from point 0 to point 2, with a pressure increase to the value $P_2 \approx 18$ MPa, the volume decrease was $\Delta V_{0\text{-}2} = 1.104 \pm 0.05$ cm^3. Estimates show that the volume change in the time interval from point 0 to point 2 due to the elastic deformation is equal to $\Delta V_{el} = 1.05 \pm 0.05$ cm^3, and within the measurement error coincides with the measured value. From this it can be concluded that in the time interval from point 0 to point 2 the possible value of the volume of filled pores lies in the range (2 ÷ 3) % in accordance with the measurement error.

In the time interval from pint 2 to point 3 the pressure does not change. Therefore, the observed volume change cannot be related to elastic deformation and the volume decrease by $\Delta V_{2\text{-}3} = 1.14 +$ 0.05 cm^3 is due to the filling of nanopores granules with liquid. Knowing the value of $\Delta V_{2\text{-}3}$ makes it possible to estimate the fraction of the impact energy transformed into the energy expended for the pores filling $E_{2\text{-}3} = P_{2\text{-}3} \cdot \Delta V_{2\text{-}3} \approx 20.5$ J. This value, together with the estimates of the energy of elastic deformation on the time interval from point 0 to point 2, equal to $E_{el} \approx 10$ J, give, within the error of measurement, the value of the impact energy E = 30 ± 2 J.

The filling pores energy $E_{2\text{-}3}$ spends to creation of the nonwetting liquid - the framework of the porous medium surface in the filled pores, the meniscus liquid-gas surface, and the liquid flow $J = d\Delta V/dt$. As the filling increases, the surface of the frame-liquid interface increases and the kinetic energy of the load decreases. When the filled volume is increased, pores of smaller size should be filled and at a higher pressure in accordance with the Laplace formula. Therefore, one could expect an increase in the filling pressure in the interval from point 2 to point 3 with increasing time, rather than a constant $P = P_{2\text{-}3}$. Alternatively, according to [6], transport channels from such multiparticle local pore configurations are filled throughout this time interval, which consist of all pores of different sizes in accordance with the pore size distribution. The constant value of the pressure corresponds to the constant value of the total filling flux $J = d\Delta V/dt$ under the assumption of a constant hydrodynamic resistance over the entire time interval from point 2 to point 3 with the exception of the vicinity of point 3 of stopping filling (stopping the stock), see Fig. 1. Another part of the energy is expended, apparently, on the provision of liquid transport and on heat release as a result of the formation of a liquid-frame surface of a porous medium and liquid-gas [14] and dissipation during a viscous fluid flow. Also, in order to understand the nature of this phenomenon, additional experiments were performed to fill the granules at a various temperatures.

It is known that in the temperature range (5 ÷ 60) °C [13] the coefficient of water viscosity varies three times. Therefore, we could expect a change in the pressure-time dependence upon impact with a decrease in temperature from 55 to 7 C. However, the P(t) dependences for the water of the granular material L23 within the measurement error at T = 7 C and T = 55 C were the same (see Fig. 2a). which coincides with the result for the system of an aqueous $CaCl_2$-L23 solution [6]. On this basis, it can be assumed that water transport in nanopores with an average radius of 5 nm does not depend on the viscosity of the liquid in the regime of rapid compression upon impact. Similar results were observed for nanoporous hydrophobic zeolite granules and Silica gel in water and glycerol in [9, 15]. Since the flow of liquid when filling the pores with rapid compression is independent of temperature and therefore of viscosity, the main source of thermal effects can be heat release and heat absorption when forming the liquid-frame boundary of the porous medium and liquid-gas during filling [14]. It was shown in [16] that the energy of heat release when the number of nanoporous media is slowly filled makes up several percent of the filling energy.

Figure 2b shows the results of the pressure change from the time of the investigated system at T = 55 °C without replacing the porous medium and the fluid for impact No.1 and impact No. 5. As the results coincide within the measurement error, we can conclude that the system returns to its original state after the response to external influence.

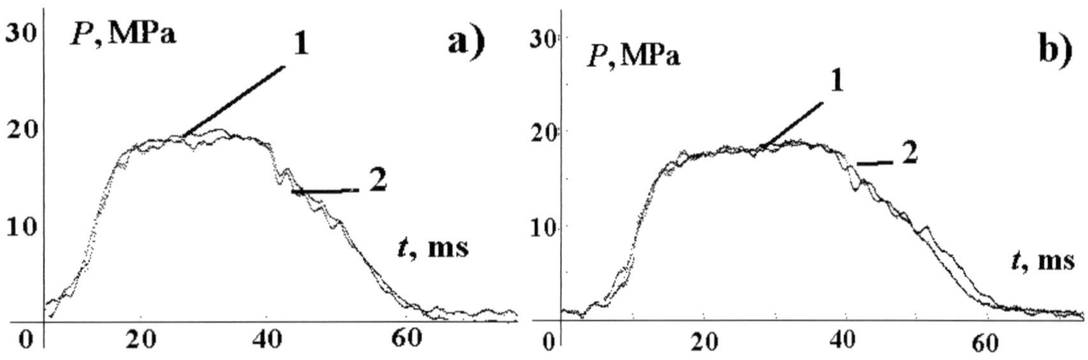

Figure 2. Dependences of P(t) for the system L23 (4 g) - water (55 cm^3) with an impact energy of 40 J a) at temperatures of (1) 7 °C, (2) 55 C and b) at a temperature of 55 °C (1) impact No.1 (2) impact No. 5.

From the dependences presented on Fig. 2 and additional experiments it follows that the critical filling pressure P_0 for the system under investigation does not depend on the temperature, fluid viscosity and the number of impacts with a constant impact energy. The transformation of impact energy by a granular nanoporous material in a non-wetting liquid occurs at constant pressure by an essentially non-dissipative filling process that begins at the critical pressure (P_0 in Fig. 1). These pressures determine the pores filling threshold, at the percolation cluster formation pressure, described in percolation theory [17].

In the quasistatic regime, in accordance with [18], pores filling with a non-wetting liquid occurs at $P \approx P_{c0}$ at the percolation threshold and under the condition that the characteristic time (τ_P) of the pressure increase is close to the characteristic filling time (τ_V). Under these conditions, the increase in pressure does not depend on the speed of the faster process of elastic compression and is determined by the slower process of volume reduction due to filling. Another situation with rapid compression is when, in the time interval t_0 - t_2, the pressure increase may outstrip the filling. Therefore, we can assume that the liquid-granular material system skips the quasistatic filling point and is "thrown" into the new state at a new critical pressure $P_0 > P_{c0}$ when the bound nanopores of the granules are filled without changing P, i.e., at constant pressure. This filling is analogous to the spontaneous filling of the pores of the medium with a wetting liquid.

The additional experiments carried out on the L23-water system at different impact energies and the mass of the porous material showed that the pressure P_0 within the measurement error does not depend on the impact energy E or on the mass L23. As a result, the independence of P_0 from the impact energy and the sample mass implies that the critical pressure P_0 is constant and the filling rate in the vicinity of P_0 is not limited by the transport of liquid through the surface of the granules. Therefore, the critical point at P_0 is the point of transition to spontaneous filling when the filling rate is determined by the hydrodynamic resistance of the nanopores system. At a time $t > t_2$ and at $P \approx P_0$ the rate of such filling becomes higher than the elastic compression rate and does not depend on the viscosity of the liquid.

In Fig. 3 shows the time dependence of the pressure P(t) and the volume change -ΔV(t) for the system, the granular material L23 (4 g) in water (55 cm^3) at T = 20 °C for the impact energy E from 5 to 70 J. It is shown, that the -ΔV(t) dependences differ by different time t_3 of reaching the maximum of the values (ΔV_{max}), the degree of symmetry at the maximum, and different derivatives of the volume dependence on time. When the energy of impact E increases, the transition from the P = const dependence to the P(t) dependence is close to linear over the time interval from point 1 to point 2. For the energy E = 5 J, the dependences P(t) and -ΔV(t) are symmetric at the maximum t_3 = 18 ms. The magnitude of the maximum pressure $P(t_3)$ = 12.5 MPa is less than the pressure of the start of pores filling in the L23-water system equal to P_0 = 18 MPa and the pressure of the pores filling start in the quasistatic mode P_{c0} = 15 MPa. Therefore, at E = 5 J, the filling conditions do not arise. Close to the symmetric dependence -ΔV(t) is also characteristic for the transformation of the impact energy into the elastic energy of the system.

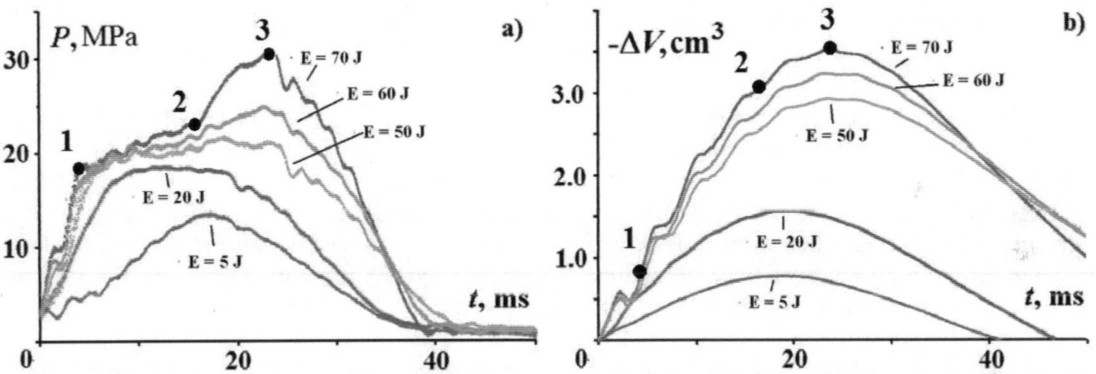

Figure 3. Dependences a) P(t) and b) -ΔV(t) for the system L23 (4 g) - water (55 cm^3) with impact energies of 5, 20, 50, 60 and 70 J at a temperature of 20 °C.

At an energy E from 20 to 50 J, pores are filled at constant pressure. The filling at P = const at the time t_3 reaches the maximum -ΔV(t). This is the moment of stopping the movement of the stock in the chamber and its subsequent return motion. Also a third qualitatively different regime for the transformation of the impact energy can be observed. At an energy E = 70 J, the compressibility of the system changes abruptly in point 1 at P = 18 ± 1 MPa. This value coincides with the value of the critical pressure P_0 at lower energies. As the time increases after point 1, the pressure does not remain constant, but increases according to a law close to linear to a value P = 22 MPa in point 2. This increase in pressure is accompanied by a change (decrease) in the volume $\Delta V_{1\text{-}2}$ = 2.2 ± 0.1 cm^3, the value of which, within the error of measurement, is close to the value of the total pore volume of the sample (~ 2 cm^3), taking into account the volume change due to elastic compression (~ 0.2 cm^3). After point 2, the pressure and volume changes increase, reaching simultaneously a maximum in point 3 and then decrease to the original values. The magnitude of the volume change and the symmetric form of the P(t) and -ΔV(t) dependences near the maximum in the time interval from point 2 to point 3 correspond to elastic compression. The total energy of transformation at the section 0 - 1, 1 - 2, 2 - 3 to the point of load stopping within the measurement error coincides with the kinetic energy of the load.

In the energy conversion mode, when filling nanopores at constant pressure and increasing the impact energy at E <50 J, the pores filling rate, J = dΔV/dt, increases. This is due to the fact that with increasing impact energy, the speed of stock movement into the chamber increases and, consequently, the rate of decreasing of the volume of the system. Therefore, at a constant pressure and, consequently, a zero volume change due to compression, the pores filling rate of the pores of the granules is determined experimentally as the rate of decrease in the volume of the system. In accordance with Fig. 3b, the value of J(t) as a function of time is close to linear in the entire time interval 0 - 1 except for the vicinity of the maximum -ΔV. Since the pressure difference that initiates the filling of empty pores can be assumed equal to the measured pressure, the relation P = J · ρ should be satisfied, where ρ is the hydrodynamic resistance to filling the pores system in all granules. Then for ρ = const, the increase in the flux J should lead not to the observed constancy of the pressure, but to its increase. Either the increase in J should cause a nonlinear response in the form of a decrease in the pores filling resistance ρ, and then P = const, taking into account the independence of the flow from the viscosity of the liquid. This response was proposed in [5] and will be discussed in Section 4.

Thus, it follows from the foregoing that there exists a maximum flux J in the spontaneous-filling mode at P = const. We note that when the impact energy E increases, the value of J tends toward the maximum value determined by the maximum value of the slope angle -ΔV(t).

4. Microscopic mechanism

The dynamics of filling with a non-wetting liquid nanopores granules is described in [6] with the aid of a system of kinetic equations for the distribution functions of clusters in terms of the number of available pores in them and clusters of filled pores. By available pores, by definition, are pores that are formed without delay (instantaneously) in accordance with the magnitude of the pressure. In contrast to them, the filled pores are formed after a finite time, necessary for pouring in and filling the pores with liquid.

As the hydrodynamic pressure in the nanopores system increases, the number and size of clusters from the available pores increase, and the process of their growth is described as the fusion ("interaction") of such clusters. This is a process with a characteristic time that coincides with the time of pressure increase. The process of the disappearance of clusters of accessible pores and the formation of clusters of filled pores is described as the process of filling the liquid with available pores through neighboring clusters of filled pores. The characteristic time of these processes is the hydrodynamic time of fluid motion and the filling of a cluster of accessible pores.

There are three processes with different characteristic time scales. This characteristic time τ_P is inverse to the value of the pressure growth rate, the characteristic time τ_k is the growth in the number of clusters from a finite number of filled nanopores and the characteristic time τ_V of the change in the macroscopic volume. In view of the relative smallness of the volume of clusters from a finite number of filled pores, $\tau_k << \tau_V$. The ratio of the times τ_P and τ_V can be controlled in experiments, since these quantities are related by the relation:

$$\tau_P = P/dP/dt = V/dV/dt \cdot P/V \cdot dV/dP = \tau_V \cdot \chi.$$

Here, χ is the compressibility of the suspension with a volume V at a pressure P. The reason for the compressibility of the suspension can be either an elastic compression of the liquid and granules, or a change in the volume of liquid in the pores.

In experiments with the studied systems (see Fig. 3), when the impact is absorbed in the time interval 0 - 1, a linear dependence of pressure on time and changes in the volume of suspended matter is observed. This dependence corresponds to the elastic compression of liquid and granules with unfilled pores. For the systems studied in the experiments, $\chi = (10^{-4} \div 10^{-2})$. Therefore, the characteristic growth time of the pressure $\tau_P << \tau_V$ is the filling time, and the rate of pressure growth is higher than the filling rate. As a result, the liquid-nanopores systems of the granules "slip" through the percolation threshold (Fig. 1a) in the quasistatic filling mode at a pressure of $P_{c0} < P_2$.

In the vicinity of point 2 (Fig. 1a) when the pressure value $P_0 \geq P_c$ should be observed dynamic percolation transition [6] (P_c - dynamic pressure percolation threshold). In the vicinity of the dynamic percolation P_0 transition at $\geq P_c$ clusters arise from filled pores. When passing through the point of dynamic percolation threshold in the neighborhood of this point sharply changes (Fig. 1a) the slope of P(t) at constant dV/dt, and changes by many orders effective compressibility χ suspensions. Therefore investigated in the time interval (Fig. 1, 2 - 3) the values of the characteristic times of change in pressure and volume of the same order: $\tau_P \sim \tau_V$ and the rate of change of pressure follows the rate of change in volume when nanopores are filled. This time relationship also means that the filling rate exceeds the deformation rate, and if the pressure $P_2 = P_3 =$ const is unchanged, deformation does not take place. Thus, when passing through the point of the dynamic percolation transition, the suspension is in a state where the transformation of the energy and momentum of the impact due to the filling of nanopores can be effective at high impact rates than with deformation.

In accordance with [6], the state of suspension of nanoporous granules beyond the threshold of dynamic percolation can be called a state of spontaneous filling. In this state, the filling rate exponentially depends on time with an exponent proportional to the number of clusters of filled pores. Such a mode in the dynamics of a percolating system is formed because the more the number of clusters of filled pores and the faster they are formed, the faster the "infinite" percolation cluster of available but empty pores is filled. Qualitatively, it can be said that as the rate of decrease of the volume $J = d\text{-}\Delta V(t)/dt$ increases and, consequently, the filling rate with increasing impact energy/momentum, the "number of filling channels" starting from clusters of filled pores increases. Therefore, we can also say that an increase in $J = d\text{-}\Delta V(t)/dt$ should lead to a decrease in the hydrodynamic resistance to motion of the non-wetting liquid in the disordered system of nanochannels of the granules. This apparent constancy of the resistance is obviously related not to a change in the viscosity of the liquid, but to an increase in the number of "filling channels". As a consequence, as the pores filling rate $J = d\text{-}\Delta V(t)/dt$ increases, the hydrodynamic (not viscous) resistance decreases, and the driving fluid the hydrodynamic pressure drop $\delta P = P_2$ - O, O - hydrodynamic pressure in empty pores, remains constant. This is precisely what is observed in the experiments carried out, if the maximum possible for suspension flow rate for spontaneous pores filling is not attained at impact. The maximum filling rate is determined by the maximum number of channels beginning at the surface of the granules and depends on the porosity of the granules.

5. Conclusion

Proceeding from the foregoing, it can be concluded that the suspension of a granular nanoporous medium in a non-wetting liquid under conditions of a pulsed pressure change exhibits intellectual properties and can be considered an intellectual material.

Acknowledgments
This work is supported by the Russian Science Foundation under grant 18-13-00398.

References

[1] Wang P (Ed.). (2016). Smart materials for advanced environmental applications. Royal Society of Chemistry.
[2] Ou R., Wang Y, Wang H and Xu T 2013 *Desalination* **318** 48-55
[3] Borman V D, Belogorlov A A, Tronin V N 2018 *Colloids and Surfaces A: Physicochemical and Engineering Aspects* **537** 540-548
[4] Borman V D, Belogorlov A A, Tronin V N 2016 *Colloids and Surfaces A: Physicochemical and Engineering Aspects* **496** 63-68
[5] Borman V D, Belogorlov A A, Tronin V N 2016 *Physical Review E* **93**(2) 022142
[6] Borman V D, Belogorlov A A, Lisichkin G V, Tronin V N, Troyan V I 2009 *Journal of Experimental and Theoretical Physics* **108**(3) 389-410

IOP Conf. Series: Journal of Physics: Conf. Series **1099** (2018) 012026 doi:10.1088/1742-6596/1099/1/012026

[7] Evans A G, He M, Deshpande V S, Hutchinson J W, Jacobsen A J, Carter W B 2010 *International Journal of Impact Engineering* **37** 947-959

[8] Borman V D, Belogorlov A A, Grekhov A M, Tronin V N, Troyan V I 2001 *Journal of Experimental and Theoretical Physics Letters* **74** 258-261

[9] Xu B, Qiao Y, Chen X 2014 *Journal of the Mechanics and Physics of Solids* **62** 194-208

[10] Surani F B, Kong X, Panchal D B, Qiao Y 2005 *Applied Physics Letters* **87** 163111

[11] Lisichkin G V, Fadeev A, Nesterenko P N, Serdan A A, Mingalev P G, Furman D B 2003 Khimiya privitykh poverkhnostnykh soedinenii (Chemistry of surface grafted compounds, in Russian).

[12] Borman V D, Belogorlov A A, Byrkin V A, Tronin V N, Troyan V I 2012 *JETP letters* **95** 511-514

[13] Haynes W M (Ed.) 2014 CRC handbook of chemistry and physics. CRC press.

[14] Borman V D, Belogorlov A A, Byrkin V A, Lisichkin G V, Tronin V N, Troyan V I 2011 *Journal of Experimental and Theoretical Physics* **112**(3) 385-400

[15] Han A, Lu W, Punyamurtula V K, Chen X, Surani F B, Kim T, Qiao Y 2008 *Journal of Applied Physics* **104**(12) 124908

[16] Gomez F, Denoyel R, Rouquerol J 2000 *Langmuir* **16** 4374-4379

[17] Sahimi M. 1993 *Reviews of modern physics* **65** 1393

[18] Borman V D, Belogorlov A A, Grekhov A M, Lisichkin G V, Tronin V N, Troyan V I 2005 *Journal of Experimental and Theoretical Physics* **100** 385-397

Activation of gold due to the size effect for the isotopic reactions in molecular hydrogen

O.A. Boeva*, A.A. Odintsov, K.N. Zhavoronkova
Mendeleev University of Chemical Technology of Russia, Moscow
*E-mail: olga_boeva@mail.ru

Abstract. In this paper, it shown that due to the size effect, we can create a catalytically active system based on gold nanoparticles. Gold, considered not so long ago, an inert metal, the transition in nanodispersed condition begins to show high catalytic activity in the reaction of ortho-para hydrogen conversion. It is shown that gold particle size from 1 nm to 40 nm possess high specific catalytic activity, which does not depend on the particle diameter in the size range. The assumption made that the hydrogen conversion proceed on magnetic mechanism, and, consequently, the gold particles become the new magnetic properties in comparison with diamagnetic massive metal.

1. Introduction

To test the catalytic properties of the various systems in this work uses two reactions involving molecular hydrogen:

$$H_2 + D_2 \leftrightarrow 2HD \quad (1)$$

$$\text{ortho-}H_2 \leftrightarrow \text{para-}H_2 \quad (2)$$

Both reactions have practical applications in the industrial process of cryogenic distillation of hydrogen and the production of liquid para-hydrogen used as rocket fuel in the boosters and space launch complexes, super-heavy class. These reactions in the gas phase do not occur without catalysts, and therefore requires catalytic reactors with highly efficient catalytic systems for the implementation of the above processes. Therefore, the development of highly efficient and affordable catalysts is relevant and timely task. One of perspective directions of creation of highly active catalysts is the synthesis of nanostructured systems based on nanoparticles of transition metals (Pt, Pd, Rh, Ru, etc.). This paper presents part of the research of catalytic activity of gold nanoparticles activated by the dimensional effect, as it is well known that the massive metal is inert and has no catalytic activity in these reactions.

2. Results

2.1. Synthesis of gold nanoparticles

Gold nanoparticles were synthesized in solutions by six methods:

1. chemical reduction in reverse-micellar solutions (RMS);
2. radiation-chemical reduction in RMS;
3. chemical reduction by Tannin;
4. chemical reduction by citrate ions with heating by the method of Turkevich;
5. chemical reduction by citrate ions under UV light;
6. chemical reduction by citrate ions with the addition of an excess precursor and further reduction by hydrogen at room temperature.

1) $NPs^{RMS\text{-}Chem}$ *synthesis*. Nanoparticles were synthesized by reduction of gold in RMS, which are an aqueous salt solution/surfactant/nonpolar solvent. As salt solutions 0.03 M, 1.1 M $AuCl_3$ solutions, 0.3 M $HAuCl_4$ solution were taken. As the surfactant used was bis(2-ethylhexyl) sulfosuccinate (AOT, (Sigma-Aldrich). Isooctan was used as dispersion medium.

Chemical gold reduction was carried out under aerobic conditions with quercetin solution (Qr, 2-(3,4-dihydroxyphenyl)-3,5,7,-trihydroxy-4H-1-benzopyran-4-one dihydrate, Merk). The reduction of gold ions occurs through a triple complex:

$$Qr...O_2...Au^+.$$

2) *NPs$^{RMS\text{-}RadChem}$ synthesis*. Radiation-chemical reduction of gold was carried out in anaerobic conditions under the influence of γ-radiation of cobalt ^{60}Co source. A dose of 30 kGy. Solvent electrons and radicals on water-organic micellar solutions of gold salts act as a reduction agent. In the process of irradiation, the reaction of gold ions reduction initiates the process of metal nanoparticles formation:

$$Au^+ + e^-_{aq} \rightarrow Au^0.$$

The size of gold nanoparticles depends on the diameter of the micelle pool, which characterized by coefficient of solubilisation $\omega=[H_2O]/[AOT]$. In this paper, it is: 3.0, 4.0, 5.0, 8.0 – for the chemical reduction ($NPs^{RMS\text{-}Chem}$), and: 1.5, 3.0, 5.0 – for the radiation-chemical reduction ($NPs^{RMS\text{-}RadChem}$).

3) *NPs^{Tannin} synthesis.* Gold nanoparticles were obtained by reduction of tannin ($C_{76}H_{52}O_{46}$, «Purified» Germany). Gold hydrosol was preparated at room temperature by reduction of the 0.025 M 0.025 M precursor ($AuCl_3$, "Aldrich") solution with tannin. Tannin in the reduction of gold goes into phlobaphene $C_{76}H_{52}O_{49}$. The reaction takes place in an excess of tannin, so it can be assumed that the free metal ions in the colloidal solution are absent.

4) *NPs$^{Cit\text{-}T}$ synthesis.* The required amount of sodium citrate $HOC_3H_4(COONa)_3$ ($CitNa_3$) added to the hot solution $Na[AuCl_4]$ (T = 368 ÷ 373 K). The resulting solutions were quickly brought to a boil in a water bath and boiled for 15 minutes with constant stirring (stirring speed – 800 rpm). Then the solutions were left in a water bath to cool to room temperature.

5) *NPs$^{Cit\text{-}h\nu}$ synthesis.* The solution containing $2.0{\cdot}10^{-4}$ M $Na[AuCl_4]$ (Aldrich) and sodium, citrate $CitNa_3$ with a concentration of $1.0{\cdot}10^{-3}$ M, was deaerated using a vacuum rotary vane pump NVR-4,5D «Vacma», equipped with a nitrogen trap with a residual pressure of $P = 1.5{\cdot}10^{-2}$ Torr = 2.0 Pa. After that, the solution in the quartz cell was irradiated with UV light of a low-pressure xenon pulse lamp with a total flow rate of $I_{UV} = 6.0{\cdot}10^{20}$ Kvant/s = $1.0{\cdot}10^{-3}$ Einstein/s. As result, there was a complete reduction of Au(III) ions.

6) *Reduction of gold NPs$^{Cit\text{-}H2}$* was carried out in several stages: obtaining priming in the form of metal particles of gold by citrate ions; adding an excess of ions Au^{3+} followed by hydrogen reduction at room temperature.

All solutions were prepared using deinonized water ($\varkappa = 17.0\ \Omega{\cdot}m$).

Control of nanoparticle synthesis, and obtaining composite catalytic systems by sorption of gold nanoparticles on support, was carried out on the optical absorbance spectras obtained on spectrophotometers Hitachi U-3010, Varian Cary 100 Scan UV-Vis.
Fig. 1 shows the absorbance spectrum in the process of gold nanoparticles synthesis by various methods , Fig. 2 shows the absorbance spectrum in the process of gold nanoparticles adsorption on the support, which are γ-aluminum oxide (brand «Trilistnik», Redkino Catalyst Company).

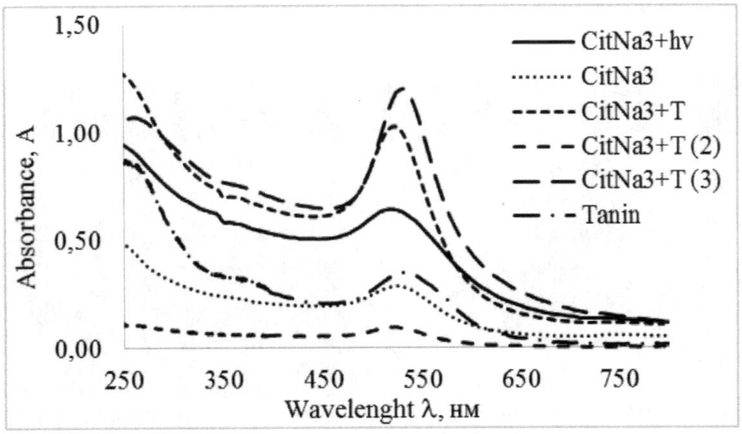

Figure 1. Absorbance spectrum of synthesis of gold nanoparticles.

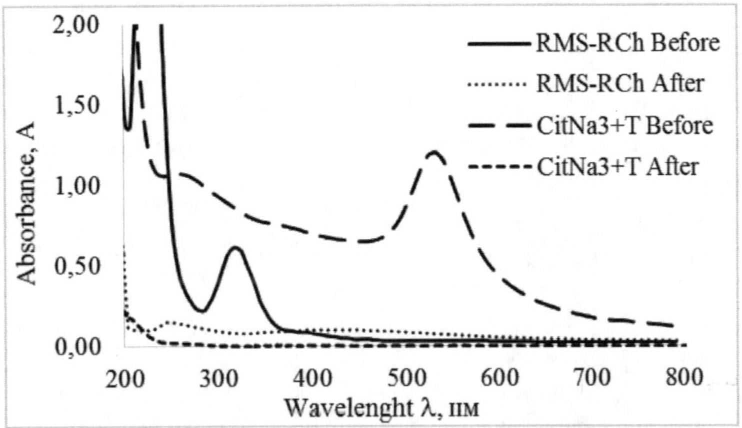

Figure 2. Absorbance spectrum of process of gold nanoparticles adsorption on support.

2.2. Microscopy investigations

Sizes of the nanoparticles NPs^{Cit-hv}, NPs^{Cit-T}, NPs^{Cit-H2} were determined on an Leo-912 AB Omega; C. Zeiss transmission electron microscope. Sizes of the nanoparticles $NPs^{RMS-Chem}$, $NPs^{RMS-RadChem}$, NPs^{Tannin} were determined on an EnviroScope 5.30 atomic force microscope.

Fig. 3 shows AFM-images of gold nanoparticles and histogram of particle size. Fig. 4 shows TEM-images of gold nanoparticles in basic colloidal solution and nanoparticles supported on γ-Al_2O_3.

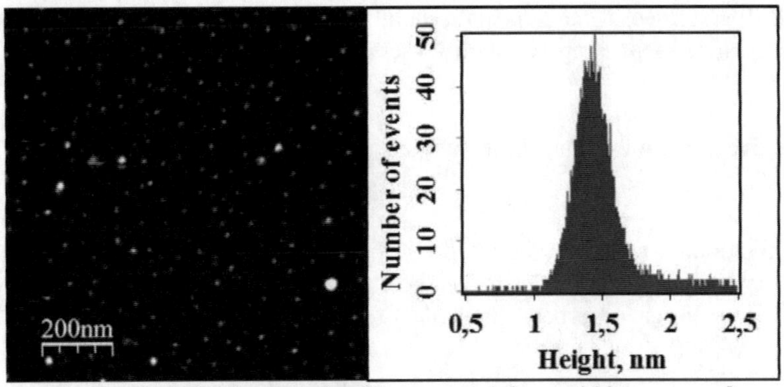

Figure 3. AFM-image of gold nanoperticles and histogram of particle size distribution.

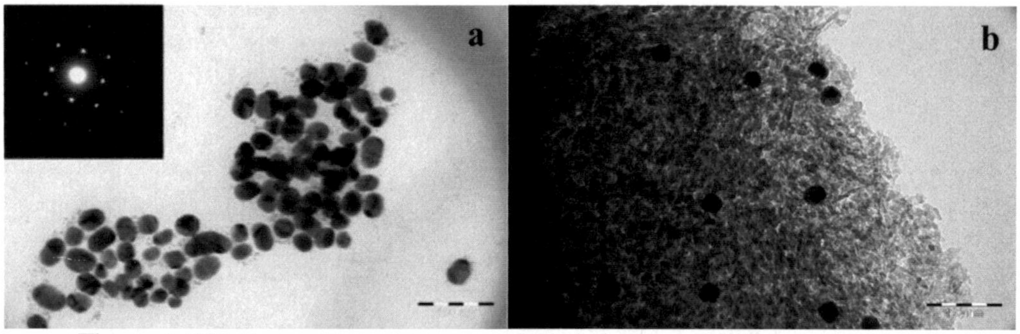

Figure 4. TEM-images of gold nanoparticles in basic colloidal solution (a) and nanoparticles deposited on γ-Al_2O_3 (b).

The synthesized nanoparticles have spherical form and mainly narrow unimodal size distribution. Changes in the shape or size of nanoparticles during sample preparation and research of catalyst samples do not occur. This is confirmed by TEM-images (Fig. 4) and indicates that the particles are firmly fixed on the surface of the support. The results of microscopic studies of all samples are presented in Table 1.

Table 1. Microscopy investigations results.

Synthesis method	Dominating size, nm
$НЧ^{RMS\text{-}Chem}$	0.7±0.2
$НЧ^{RMS\text{-}Chem}$	1.4±0.3
$НЧ^{RMS\text{-}Chem}$	1.7±0.2
$НЧ^{RMS\text{-}Chem}$	4.2±0.7
$НЧ^{RMS\text{-}RadChem}$	0.9±0.2
$НЧ^{RMS\text{-}RadChem}$	1.1±0.3
$НЧ^{RMS\text{-}RadChem}$	1.0±0.2
$НЧ^{Cit\text{-}hv}$	4.6±0.8
$НЧ^{Cit\text{-}T}$	19.4±4.2
$НЧ^{Cit\text{-}T}$	28.3±3.1
$НЧ^{Cit\text{-}T}$	40.1±5.4
$НЧ^{Cit\text{-}H2}$	7.4±1.1
$НЧ^{Cit}$	14.4±2.2
$НЧ^{Танин}$	20.5±4.5

2.3. *Hydrogen adsorption*

The active surface (the surface of the deposited metal) was determined by hydrogen adsorption. The measurements were carried out using volumetric method in a pressure range of from $1\cdot10^{-2}$ to $2\cdot10^{-1}$ Torr at T = 77 K.

The value of the active surface was calculated by the formula:

$$S_{H_2} = 2n_m \cdot N_A \cdot \sigma_{Me},$$

where n_m – is is an amount of chemisorbed hydrogen corresponding to the plateau on the adsorption isotherm, N_A – is the Avogadro number, σ_{Me} – is the average area occupied by one surface metal atom on which one hydrogen atom is adsorbed. The value of σ_{Me} for the gold is taken to be 6.51 Å^2.

Gold nanoparticles of all sizes from 0.7 nm to 40 nm are able to adsorb molecular hydrogen at 77 K. Based on the obtained experimental data, adsorption isotherms of a similar kind are constructed: a pronounced plateau is observed which is taken as a monolayer of chemisorbed hydrogen (Fig. 5).

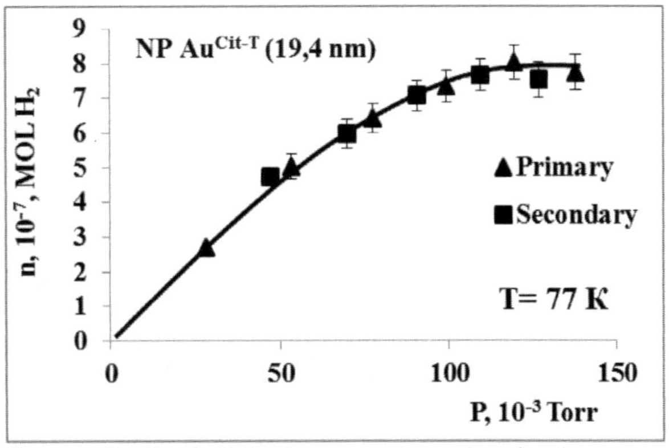

Figure 5. Typical hydrogen adsorption isotherm on gold nanoparticles

Experiments to determine the fraction of weakly bound hydrogen on the surface of gold nanoparticles showed that all hydrogen is weakly bound and easily removed from the surface by pumping at the adsorption temperature. This is evidenced by the coincidence of primary and secondary adsorption isotherms (Fig. 5).

During the research it was revealed that hydrogen adsorption on gold nanoparticles is observed at higher temperatures, up to room temperature. However, when the temperature increases from 77 K to 110 K, there is a sharp decrease in the adsorption value.

On the example of a sample with a nanoparticle diameter of 19.4 nm, the ratio of the amount of adsorbed hydrogen and the amount of metal in the catalytic system was estimated.

The concentration of gold in the colloidal solution of nanoparticles was $C_{Au} = 3 \cdot 10^{-4}$ M. According to the optical absorbance spectrums almost all gold nanoparticles were adsorbed onto the γ-Al_2O_3 (Fig. 2). The nanoparticles deposition was carried out on the basis of the ratio of 5 ml of solution to 1 g of support. Then, the amount of metal on the surface of 1 g of the catalytic system is $n_{Me} = 15 \cdot 10^{-7}$ mol. Taking into account the proportion of surface atoms $\sigma \approx 8$ %, the number of surface gold atoms in 1 g of the catalytic system is $n_{Me}^{surf} \approx 1.2 \cdot 10^{-7}$ mol.

From the experimental data on hydrogen adsorption it follows that the amount of adsorbed molecular hydrogen on the sample catalytic system (weight m = 0.7122 g) was $n_m = 7.7 \cdot 10^{-7}$ mol H_2 (Fig. 5). Thus, the catalytic system weighing 1 g is able to adsorb $10.8 \cdot 10^{-7}$ mol H_2, which is 9 times the number of surface gold atoms. Assuming that all hydrogen (fully or partically) is adsorbed with dissociation of molecules, the amount of adsorbed hydrogen in the form of atoms and molecules exceeds the total amount of metal, which, it would seem, can not be. However, the obtained results can be explained both by the dissolution of partially dissociated hydrogen in the volume of nanoparticles and, more likely, by the manifestation of the effect of the hydrogen spillover to the support. Estimates for other samples showed that the amount of adsorbed hydrogen exceeds the number of surface gold atoms in particles from 4 to 17 times. Similar results are comparable with the results of the study of hydrogen spillover on catalytic systems Pt/Al_2O_3 и Pd/Al_2O_3 [1].

2.4. Catalytic activity

The reactions were studied under static conditions without gas circulation at an operation pressure of 0.5 Torr. The analysis of the gas mixture was carried out continuously by the method of thermal conductivity based on various rotational thermal conductivity of gas mixtures of hydrogen isotopes and protium ortho-para modifications heat transfer of the platinum filament heated by a current in a cylindrical vessel (cell). The working cell (1 on Fig. 6) is part of reaction volume, and the comparison cell (2 on Fig. 6) is filled with a mixture of the original composition. The volumes of both cells are equal, are in identical conditions, and are included in the Wheatstone bridge circuit (Fig. 6). Copper coils located next to each other, which prevents temperature drift during operation are used in the circuit to stabilize the microvoltmeter (3 on Fig. 6).

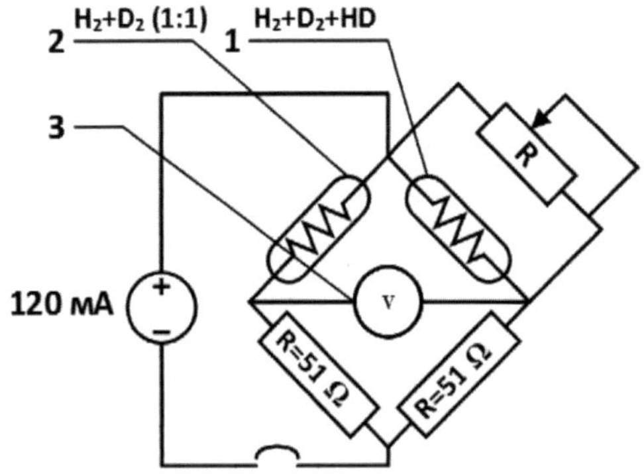

Figure 6. Measuring system circuit.

Specific catalytic activity (K_{sp}) is defined as the first-order rate constant (k_0) taking into account the number of molecules in the reaction volume at the reaction temperature (N_T) and divided by the catalyst active surface area (S_{H2}):

$$K_{sp} = k_0 \cdot \frac{N_T}{S_{H2}},$$

2.4.1. H_2-D_2 exchange

Protium-deuterium exchange reaction was studied in the wide range of temperature from 77 K to 530 K. All samples of catalytic systems based on gold nanoparticles ranging in size from 0.7 nm to 40 nm are shown high catalytic activity and have positive size-effect in H_2-D_2 exchange reaction [2]. Based on experimental data dependences of specific catalytic activity values on temperature are made in coordinates of the Arrhenius equation. (Fig. 7). Characteristic fractures conditionally divide the dependence graphs into two temperature regions, the reaction of which flows through various mechanisms with different values of activation energies. Because of experiments, it is shown that regardless of the method of obtaining nanoparticles, the reaction rate and the character of the kinetic curve coincide.

IOP Conf. Series: Journal of Physics: Conf. Series **1099** (2018) 012027 doi:10.1088/1742-6596/1099/1/012027

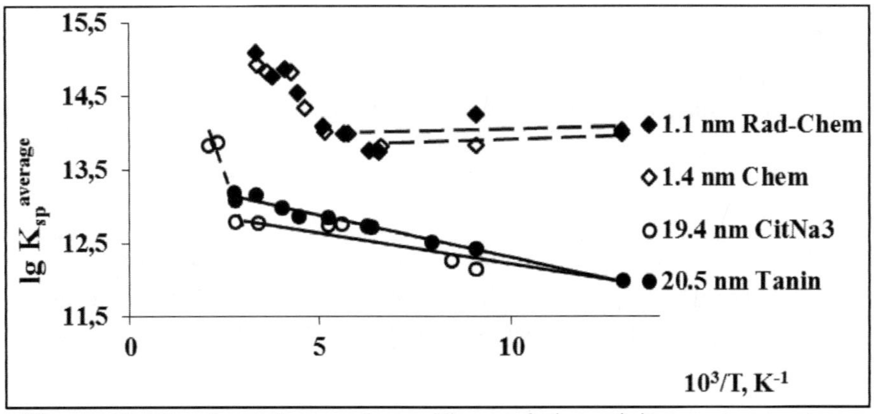

Figure 7. The dependence of specific catalytic activity on temperature in coordinates of the Arrhenius equation.

2.4.2. Ortho-para protium conversion

The ortho-para protium conversion reaction can proceed both by chemical and by magnetic mechanism, in which the change of the nuclear spin occurs under the influence of an inhomogeneous magnetic field.

The reaction was carried out at 77 K and 110 K. The results of experiments showed that all samples of catalytic systems showed extremely high catalytic activity, the value of which does not depend on the particle size and averages $K_{sp} = (3.9 \pm 1.2) \cdot 10^{14}$ molecules/cm²s, which exceeds the value of catalytic activity in the reaction of isotope exchange for small nanoparticles by 2-3 times, and for large nanoparticles with a diameter of 20÷40 nm to 400 times (Fig. 8).

Such a high value of catalytic activity of composite system based on gold nanoparticles is comparable with the most effective catalysts in the reaction of ortho-para protium conversion based on rare earth metals (Gd, Dy, Ho etc.), which by their nature have high magnetic moment values due to the presence of unpaired 4f-electrons. Thus, such a significant difference in the rates of reactions protium-deuterium exchange and ortho-para conversion of protium may indicate fundamentally different mechanisms of reaction. This proves that gold nanoparticles in contrast to massive metal have magnetic moment, which allows proceeding the reaction of o-p protium conversion on magnetic mechanism [3].

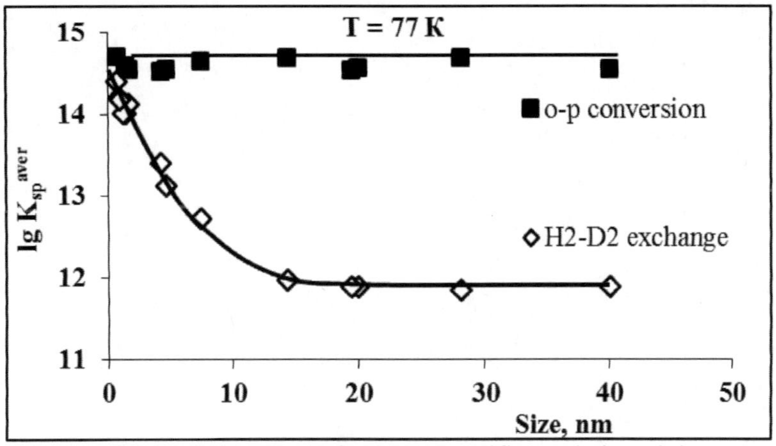

Figure 8. The dependence of average specific catalytic activity on size nanoparticles.

3. Conclusion

1) Due to the dimensional effect, during the transition from the catalytic inert massive gold to the nanostructured state, high-active composite catalytic systems in reactions involving molecular hydrogen were obtained, comparable in activity with catalysts based on transition metals;
2) It is shown that the specific catalytic activity of gold nanoparticles in the H_2-D_2 exchange reaction significantly depends on the particle size;
3) The specific catalytic activity of gold nanoparticles in the o-p protium conversion reaction does not depend on the particle size on the range of size from 1 nm to 40;
4) It is shown that the o-p protium conversion reaction flows through the magnetic mechanism, and, therefore, gold particles acquire magnetic properties in contrast to diamagnetic massive metal.

4. References

[1] Rozanov V V, Krylov O V. 1997. *Russian Chemical Reviews* **66** (2) 107-119.
[2] Boeva O A, Ershov B G, Zhavoronkova K N, Odintsov A A, Solovov R D, Abkhalimov E V, and Evdokimenko N D. 2015. *Doklady Phys. Chem.* **463** 165-167.
[3] Boeva O A, Odintzov A A, Solovov R D, Abkhalimov E V, Zhavoronkova K N, Ershov B G 2017. International Journal of Hydrogen Energy 42 22897-22902

Acknowledgements
We Express gratitude for the synthesis of nanoparticles in colloidal solutions to group employees, IPCE RAS: B.G. Ershov, R.D. Solovov, E.V. Abkhalimov.

IOP Conf. Series: Journal of Physics: Conf. Series **1099** (2018) 012028 doi:10.1088/1742-6596/1099/1/012028

Conceptual design of water supply systems based on low-power nuclear desalination complex

N I Laguntsov and R A Alexandrov

National Research Nuclear University MEPhI (Moscow Engineering Physics Institute), Kashirskoye sh., 31, Moscow, 115409, Russia

E-mail: NILaguntsov@mephi.ru

Abstract. The concept of environmentally friendly and energy-efficient water supply systems development on the basis of a desalination complex is considered to solve the problem of water deficiency, reduce the ecological load on the environment and provide fresh water to regions in which a lack of fresh water is found out. In the paper, a brand-new hybrid scheme of the water supply system is proposed, with the production of water containing reduced deuterium, based on desalination technology and recirculated water supply (recycling) with the help of membrane and distillation methods, as well as improved pretreatment and water purification technology.

1. Introduction

Nowadays water consumption is growing much faster than population. Russia is ranked 2th in the world for content of surface fresh water resources. However, only 20 percentages of fresh water sources are located in central and southern regions with a high population density, well-developed industry and agriculture. The lack of fresh water in these areas can be compensated by the desalination of salt seawater and brackish groundwater.

In order to solve the problem of water deficiency and to develop the desalination market in Russia, it is proposed to solve the global scientific and technical problem, in particular, to create environmentally-friendly water supply system which allows to obtain light water on the basis of desalination and recycling water technologies using membrane and distillation methods. This approach, from one hand, can solve the problem of water deficiency, and from another hand, will reduce the overall costs of water sector system operation in whole.

2. Conceptual design of water supply systems

Due to the high cost of desalination technology, technological scheme which will ensure the highest possible extraction by cheaper methods of purification before the desalination stage, should possess the significant advantage cause of reduction load on the desalination stage and also making it more efficient. Taking into account various groups of consumers of desalinated water (population, industry, agriculture) consuming water of different quality, it is advisable to use combined desalination schemes applying distillation, electrodialysis and low-pressure reverse osmosis (nanofiltration) technologies. In respond to different unit costs for the removal of various pollutant groups, the highest costs of which are salts, it's advisable to use recycling methods – reuse of purified waste water produced by activities of water consumers and characterized by a lower salt content. The use of recycling methods allows to reduce significantly the unit cost of different pollutants groups removal, satisfying the needs in the technical water for a wide range of consumers.

The daily human need for drinking water is 2-3 liters. Domestic water demand averages 100 liters per person daily. The proportion of fresh water used for industrial and agricultural needs is somewhat larger and averages 300 liters per person per day. Thus, for the water supply of a city with a population of 100 thousand people, about 0.85 m^3/s of water is needed, which is included in the most demanded range for the productivity of desalination plants, ranging from 50 to 200 thousand cubic meters per day (according to the IAEA) [1]. Table 1 shows the required amount of water used in various spheres of activity of the city, as well as the requirements for their quality and the recycling percentage, which reduces the required plant productivity.

Table 1. Calculation of the city water supply with a population of 100 thousand people.

Water assignment	Quantity, m^3/s	Degree of purification from salts, %	Degree of purification from toxic substances,%	Deuterium content, ppm	% of Recycling
Drinking	0,0035	94-96	in accordance with maximum permissible concentration (MPC)	120	–
Agricultural	0,35	94-96	in accordance with maximum permissible concentration (MPC)	140	50
Domestic	0,116	99,9	99,9	140	80-90
Own needs	0,00011	99,9	99,9	155	80-90
Industrial	0,35	99,9	99,9	155	80-90

The total heat and electric power for obtaining of the water required amount for a city with a population of 100 thousand people is about 500 and 10.0 MW, respectively. Taking into account the recycling, this value is almost halved and amounts to 300 MW of thermal energy and 4 MW of electricity. Thereby, while developing a hybrid water supply system for the desalination complex, it is proposed to use low-power reactors [1] or power-generating gas turbines in which exhaust gases have significant thermal power.

The use of gas turbines as a source of energy in combination with recycling technology can successfully solve the problem of water deficit in Crimea. In addition, taking into account the relatively low salinity of the Black Sea (18 kg/m^3), there is a principal possibility of a significant reduction in energy consumption for desalination due to a partial recession from distillation technology and an increase of desalination influence on the process by the method of low-pressure reverse osmosis.

On the figure below one of the variable configurations of the technological scheme is presented.

IOP Conf. Series: Journal of Physics: Conf. Series **1099** (2018) 012028 doi:10.1088/1742-6596/1099/1/012028

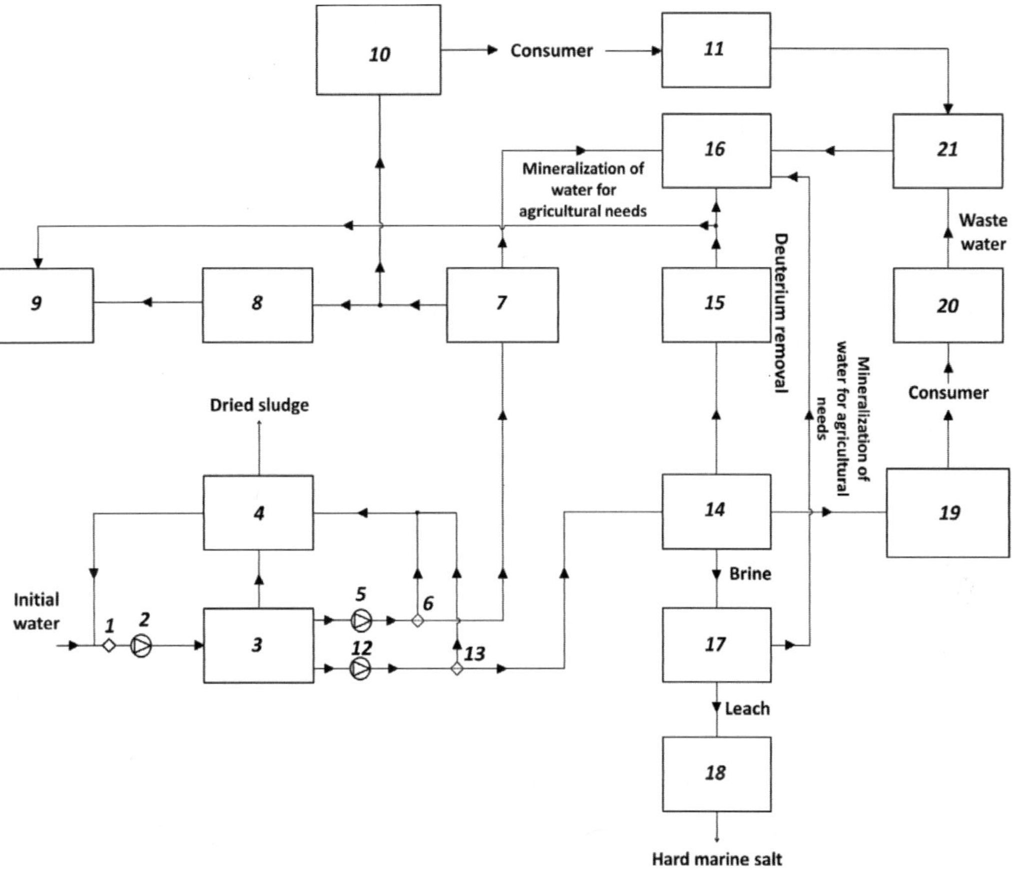

Figure 1. Principle technological diagram of the water supply system on the basis on desalination complex: 1 - coarse filter; 2, 5, 12 – pump; 3 – pretreatment unit; 4 - sewage-sludge treatment unit; 6, 13 - tangential filter; 7 – nanofiltration unit; 8 - UV –treatment; 9 - drinking water tank; 10 – system of domestic water supply; 11, 20 – sewage system; 14 - distillation unit; 15 - light water obtainment unit; 16 - agricultural water supply system; 17 – electrodialysis unit; 18 - vapour source; 19 - industrial water supply system; 21 - water prepurification.

For the long-term operation of desalination plants, the quality of the source water should meet certain rigid requirements. So, there is the need to develop a pretreatment system, taking into account the data on the pollutants composition, salinity and water preparation system performance. It seems promising to use modules of reagent treatment and flotation dividing of phases in the pretreatment unit [2].

This method was used in the development of technology for surface water treatment facilities of The Moscow International Business Centre (MIBC), also known as "Moscow City", with a capacity of 1.39 m^3/s. As the main reagent is used a low selective (removing a wide range of contaminants) mineral composite reagent, obtained by sulfuric acid decomposition of cheap aluminosilicate raw materials [3]. As a result of the use of this reagent, the concentration of heavy metals and oil products in water is reduced by 99%, the concentration of the hardness salts is 96%, and the residual aluminum content in the purified water is one order less than in the initial.

After the pretreatment unit, the salt water enters the desalination unit, which, as a rule, is very energy-intensive. In order to reduce the energy intensity of the desalination process, it is assumed to develop hybrid desalination schemes (collaborative use of distillation and membrane desalination methods) based on nuclear desalination, and these schemes make it possible to obtain drinking, technical and light water.

The part of the water from the pretreatment unit is directed to the distillation unit. The salt content of the desalinated water at the outlet (in the permeate flow) is minimal (<25 mg / l), so it is possible to use this water as a technical one for various industrial, domestic and own needs. The retentate section with high salt content is directed to the electrodialysis unit, afterwards it is entered the special evaporator in a form of brine in order to obtain leach, or is used for mineralization of agricultural water. In addition, the developed scheme has provision for the recycling of waste water, which after preliminary cleaning can enter to the input of the system or be used for agricultural purposes (Figure 1). The share of industrial water recycling is about 80-90%. The usage of the recycling scheme allows to reduce the salt content and consumption of the initial water and, consequently, to reduce the load on the desalination stage, which makes it more efficient.

Among other things, the scheme provides for the production of light water depleted heavy hydrogen isotope - deuterium. As known, the content of deuterium in seawater is about 155 ppm. For the production of light water, some of the desalted (technical) water is fed to the input of the deuterium removal unit. Depending on the application, there are various restrictions on the permissible content of deuterium in the water. For example, for drinking, it is sufficient to reduce the deuterium content to 120 ppm, which is somewhat lower than the corresponding level for fresh water reservoirs (Lake Baikal - 135 ppm). For irrigation waters of agricultural and household purposes, it is sufficient to reduce the deuterium content up to 140 ppm.

As known, total mineralization of drinking water should be 0.5-1.0 kg/m^3. In this case, the content of toxic ions in water should not exceed the MPC level. In order to obtain high-quality drinking water, another part of the water from the pretreatment unit, if necessary, is additionally purified from the toxic elements remaining in the water, after which it is fed to the nanofiltration and UV-treatment units. It is commonly known that nanofiltration membranes have reduced selectivity and increased permeability in comparison with reverse osmosis membranes. The operating pressure in nanofiltration processes is usually in the range from 0.3 to 1.0 MPa, which is much less than the operating pressure of the reverse osmosis process. In this case, the selectivity of nanofiltration membranes to the cations of Ca^{2+} and Mg^{2+} is different and depends on the composition of the water. In any case, the extraction degree of salts with nanofiltration is lower than with reverse osmosis. It affords an opportunity to obtain mineralized water used for drinking purposes at the outlet of the permeate flow. It is important to notice that it is necessary to dilute this water additionally with a depleted flow, entered from the deuterium removal unit, in which the deuterium content was reduced to a level of 100 ppm. The total content of deuterium in drinking water will be approximately 120 ppm. The concentrate from the nanofiltration membrane can be used for water mineralization of agricultural purposes.

It should be noted that the lowest percentage of recycling is provided by agricultural water, as about 50% of this water are irretrievably lost. The construction of the desalination system requires optimization taking into account all the technological processes taking place inside the system. There is a variant of the technological scheme in which water of agricultural purpose is obtained with the help of a nanofiltration unit. In this case, the share of electricity consumed for desalination process increases, and the share of thermal energy decreases.

The unit for sludge drying and compaction is designed for harvesting and subsequent dehydration of coagulated suspensions and concentrates. Pollutant concentrates enter the sludge treatment unit for dehydration, after this they can be used in the manufacture of concrete and asphalt in order to increase the dispersion of the product. The leach obtained after evaporation in the electrodialysis unit can be used in the production of fertilizers, as well as a source for the microelements production.

3. Conclusions

The concept of environmentally friendly and energy-efficient water supply systems development on the basis of a desalination complex is considered to solve the problem of water deficiency, reduce the ecological load on the environment and provide fresh water to regions in which a lack of fresh water is found out. In the paper, a brand-new hybrid scheme of the water supply system is proposed, with the production of water containing reduced deuterium, based on desalination technology and recirculated water supply (recycling) with the help of membrane and distillation methods, as well as improved pretreatment and water purification technology. The problem of concentrating brines and sediments to a level ensuring their utilization is considered. For example, providing a small city with a population up to one hundred thousand people with the fresh water, the possibility of using a low-power nuclear reactor or gaseous turbine as an energy source, is considered.

Thereby, developed scheme of the water supply system is economically-viable, environmentally-friendly, can successfully solve the problem of water deficit and provide high-quality water to regions with water deficiency, including Crimea.

Acknowledgements
The work was financially supported by the Ministry of Education and Science, agreement № 14.575.21.0159. Unique project identifier RFMEFI57517X0159.

References

[1] Kostin V I, Panov Yu K, Polunichev V I and Shamanin I.E. 2007 *J. Atomic Energy* **102** № 1 31
[2] Alexandrov R A, Laguntsov N I and Kurchatov I M 2015 *J. Phys. Proc.* **72** 103
[3] Laguntsov N, Kim V, Krasnov P, Neshchimenko Yu, Borisenko M, Odintsov A and Fartunin A 2006 *J. Ars Sep. Acta* **4** 57

The issues of creating rational water supply schemes using desalination and recycling systems

I A Nechaev[1], P I Nechaev, E I Pleskach[1], R A Alexandrov[2]

[1] Joint Stock Company «Scientific-research institute of chemical technology», Kashirskoye sh., 33, Moscow, 115409, Russia
[2] National Research Nuclear University MEPhI (Moscow Engineering Physics Institute), Kashirskoye sh., 31, Moscow, 115409, Russia

E-mail: inechaev@yandex.ru

Abstract. The topic of this article is a discussion of the prospects for using methods of recycling and desalination of waters of various composition that allow the implementation of schemes for the reuse of water in various production processes, while reducing both the total water consumption and the value of the resources received. Recycling itself is seen as an integral part - the addition of a single scheme of water supply, including, including the desalination stage.

The authors of numerous publications refer to the problem of water shortage as one of the global challenges of the present, pointing to this problem as a reason of the decline in the growth of economic development or even conflicts, both between different groups of water users and between individual countries. What can be called water deficit - is the shortage of water of the appropriate quality in the required amount. Indeed, the overwhelming number of water users use fresh water with a minimum amount of pollution of natural or anthropogenic origin, and therefore, first of all water shortage is a shortage of fresh water, the total amount of which is very limited and amounts to about 1% of the total water volume on the planet. In this regard, many developers of water treatment equipment and water treatment see the solution of the problem of water shortage by desalting salt or brackish water, which should lead to increasing the volume of fresh water and thereby eliminate water shortage.

However, this problem is becoming urgent today, both for countries traditionally experiencing a shortage of fresh water resources and for possessing seemingly significant reserves of natural fresh water. The reason for such an urgent problem is, first of all in the shortage of water, suitable for further use. It concerns the lack of fresh water and anthropogenic pollution of water sources, still used yesterday for water supply systems without any serious water treatment. According to the reports about the state of the environment in the Russian Federation, there is a steady tendency to maintain unsatisfactory state of a significant number of water sources [1]. Despite the fact that for the period from 1998 to 2015 in the country water consumption for various needs has decreased almost 2 times, the volume of pollution entering the environment with the discharging of insufficiently treated or untreated sewage remains at the same level. As a result, today the actual water shortage in a number of regions of the Russian Federation is observed first of all in Kalmykia, in a number of areas of the Astrakhan, Volgograd regions, the Stavropol Territory and the Crimea - where there is simply no natural fresh water in sufficient quantity, and a hidden water shortage is a lack of water of the appropriate quality in view of

IOP Conf. Series: Journal of Physics: Conf. Series **1099** (2018) 012029 doi:10.1088/1742-6596/1099/1/012029

the anthropogenic pollution of water sources, which is typical for the Central Federal District, where the main part of the population of the Russian Federation lives. Thus, in fact, a situation has been created where the shortage of water resources is exacerbated by the pollution of water sources by discharges of insufficiently treated sewage.

It is quite obvious that a number of specialists are offered a variety of technologies, technical and design solutions, including desalination and / or transfer / transportation of water over significant distances to solve the water supply problem. Most of the technologies listed are found and will be used in the field of water supply, however, before discussing the prospects of certain technologies or technical solutions, it is necessary to determine the feasibility of water treatment to certain standards, assess the possibilities of using water of various qualities, and introduce certain limitations in the cost of the product received, intended for various groups of consumers. Today, there is an unspoken, but well-established cost of water resources on the international market. This cost is not used for direct water trade between countries or individual consumers, it is a certain maximal cost of a water resource put into the cost of a product. So, the industry is ready to pay the highest price for water - within the limits of 1-3 US dollars / m3, the municipal services provide water supply services at the price of 0,4-0,7 US dollars / m3 (which generally corresponds to the average tariff for water supply in the Russian Federation - about 25-30 rubles per cubic meter). The largest consumer of water in the world is agriculture, ready to buy water at a price not more than 0.1 USD /m^3. These prices are a certain guide for estimating the cost of water within the commodity output produced / consumed by different groups of water use and can vary within very significant limits in each specific case. However, for a general assessment and to determine the perspective for different methods of water treatment / treatment, these prices are very interesting.

Turning to the issue of water supply, it is necessary to avoid the misconception typical for the most population that tap water - initially is spring water taken from protected sources or thawed water of high mountain glaciers, today the realities are somewhat different. So, the population of Western Europe: Switzerland, some lands in France, Holland and Germany consistently use the same water, which according to experts four times succeeds in becoming both a source of water supply and the result of the vital activity of the population, living in certain areas. This is the basin of the River Rhine, one of the most densely populated and industrialized region of the planet. At the same time, the establishment and enforcement of water legislation both at the interregional and international levels made it possible to create and maintain a more complex water management system that provides comfortable and economically efficient living of tens of millions people in very harsh conditions of scarcity of resources and the highest anthropogenic load.

What is the key to the success of creating such systems? First in an effective, scientifically based approach both to the development of water legislation, and the creation of effective economic mechanisms that ensure the implementation of modern technological and technical solutions.

If we turn to the domestic legislation in the field of water quality regulation, we can see a number of requirements for the quality of water used for drinking water supply, technical water supply, irrigated agriculture, as well as the requirements for the quality of treated sewage before discharging them into water reservoirs for various purposes. Given the lack of domestic water management as a single industry, the requirements to water quality have become the result of interdepartmental disputes and contradictions.

The strictest of the above standards are standards for the discharge of treated sewage into a water object for fish culture [2], regulating the quality of water for the largest number of indicators and requiring the deepest elimination of the vast majority of listed pollution. It should be noted that these standards are formally imposed on the majority of treatment plants operating in the territory of the Russian Federation, are mandatory, but according to statistics they correspond to approximately 10% of operating treatment facilities [3].

Assessing the standards for discharge into a water object for fish culture, two groups of pollution should be identified separately, the removal of which is most costly. This is primarily salt content - up to 1000 mg/l, with a separate regulation of the content of chlorides and sulfates; as well as the content of the nitrogen group - ammonium nitrogen, nitrites and nitrates.

The second strictest norm is the drinking water quality standard [4], for which the principal difference from the norm for discharging into a water object for fish culture is an easing in salt content - up to 1500 mg/l and less strictest requirements for most other indicators. Thus, it becomes obvious that water of drinking quality cannot be discharged into the environment without paying penalties for abnormal discharge, though from a formal point of view. At the same time water of water sources is rationed less rigidly [5], while the most "expensive" pollution - salt content, first of all, in water of water sources is not regulated, the norm only indicates the need. "For each particular source, the water purification scheme and required reagents are established on the basis of technological research or experience in the operation of structures under similar conditions ... ".

The third group of standards normalizes the water quality applied to water used for Irrigated agriculture [6]. These standards are in their form most closely approximated to the needs of consumers - farmers and are based on the calculation method of setting standards in each specific case, depending on the composition of the soil and the method of irrigation. At the same time, it is allowed to use pre-treated wastewater for irrigation, with the exception of certain types of industrial wastewater and observance of water-salt balance, which prevents saltification of soils.

The fourth group of standards, which does not currently have a definitive formalized form, refers to the requirements for industrial water used in industry. Here, in general, departmental and even corporate requirements predominate, historically formed on the basis of the so-called SEV's standards developed in the USSR for most industries. In general, these standards are limited to requirements for salt content (up to 1500-3000 mg/l), suspended solids content (15-50 mg/l), organic contaminants content (COD 50-100 mg/l) and some others.

Thus, it should be noted that the current Russian water legislation is characterized by a very serious imbalance in the standards for water for various purposes. At the same time, the most strictest requirements are imposed on water directed to discharge into natural water objects for fish culture, which is not a commodity product for the producer (water user).

Next, we should consider some basic schemes for cleaning domestic and related sewage waters, ensuring the achievement of the above standards. The current versions of number of normative documents [7] describe quite accurately and in detail the main stages of mechanical, complete biological and deep post-treatment of sewage. At the same time, for most types of effluents, various manufacturers offer a wide range of technical solutions and ready-made equipment ensuring water treatment of the required quality. In addition, a wealth of experience in the operation of such systems has been accumulated, which makes it possible to conduct, including a cost evaluation of the various stages of wastewater treatment.

The first stage of purification is a stage of mechanical cleaning that ensures the removal of the largest contaminants (waste), as well as sand and a part of the settling substances. The basic composition of the treatment facilities of the mechanical cleaning stage includes: a coarse grating with clearances of 6-8 mm; sand trap; in some cases - a filter with clearances up to 1 mm; primary settler.

The implementation of this stage of purification allows to ensure the quality of cleaning of average domestic sewage to the level of suspended solids - up to 100 mg/l, BOD - about 150 mg/l, COD - up to 200-220 mg/l. The use of the mechanical cleaning stage does not allow for an effective reduction of the total concentration of the nitrogen group, and also to ensure the removal of dissolved contaminants, especially salts. However, the addition of a stage of mechanical cleaning with a block of disinfection (elimination of parasitic contamination of water and sediments) makes it possible to obtain purified

IOP Conf. Series: Journal of Physics: Conf. Series **1099** (2018) 012029 doi:10.1088/1742-6596/1099/1/012029

water, suitable for use in irrigation. At the same time, the cost of mechanical treatment of domestic wastewater averages 3-5 rubles / cubic meter.

The next stage of treatment of domestic and close to them in the composition of sewage is the stage of biological treatment. Today, the most common methods of biological purification are the so-called nitride- denitrification schemes that ensure the parallel removal of organic contaminants, as well as nitrogen and phosphorus compounds. The use of aerotanks of nitride-dentifiers allows obtaining the quality of purification by BOD and suspended solids in the range of 10-15 mg/l, which is close to the water composition of water sources of 2 and 3 classes - i.e. sources of water supply of small and medium-sized settlements, and also used in technical water supply systems of industrial enterprises. At the same time, the cost of implementing the scheme using mechanical and biological treatment steps averages 25-30 rubles per cubic meter. The development of biological treatment technology using membrane bioreactors can significantly improve the quality of cleaning and the reliability of biological treatment facilities. The use of the membranes for separation of silt mixture and purified water, both submersible and remote - allows to obtain quality of cleaning suspended solids and BOD in the range of 1.0-3.0 mg/l, which potentially allows using purified water to replenish sources of water supply up to class 1, especially this refers to the possible recharge of groundwater through various drainage and filtration systems. The cost of using MBR technology is 30-35 rubles/cubic meter.

The third stage - deep sewage treatment to the requirements for discharge into the water objects for fish culture is the desalination stage. The implementation of such technologies on an industrial scale, with reference to sewage waters of settlements in the Russian Federation is not. The reason for this situation is the need to isolate and utilize salts in the form of solid waste, which makes the implementation of technology almost impossible. Nevertheless, by analogy with the desalination of saline and brackish waters, the cost of demineralizing domestic sewage, without taking into account the utilization of brines, will be 40-55 rubles/m^3. It should be noted that even the achievement of standards for discharges into a basin does not release the water user from environmental payments for the use of the basin as a receiver of treated sewage. The cost of such a payment today is about 0.2-0.5 rubles/cubic meter, depending on the region.

Thus, the following table can be presented.

Table 1. The specific cost of water, depending on the various stages of purification.

Treatment stage	Enlarged indicators of water quality	The enlarged cost of treatment, rubles/m^3	Possible customers
Mechanical treatment	suspended solids - up to 100 mg/l, BOD - up to 150 mg/l, COD - up to 200-220 mg/l	3,0-5,0	Irrigated agriculture
Complete biological treatment with nitride-denitrification	suspended solids - up to 10-15 mg/l, BOD - 10-15 mg/l, COD - up to 35-50 mg/l	25,0-30,0	Irrigated agriculture, replenishment of water supply sources of 2-3 classes, technical water supply
Deep biological treatment with MBR reactors	suspended solids - up to 1.0-1.5 mg/l, BOD - 2-3 mg/l, COD - up to 30-40 mg/l	30,0-35,0	Irrigated agriculture, replenishment of water supply sources of 1-2-3 classes, technical water supply

IOP Conf. Series: Journal of Physics: Conf. Series **1099** (2018) 012029 doi:10.1088/1742-6596/1099/1/012029

Desalting	suspended solids - up to 1.0-1.5 mg/l, BOD 2-3 mg/l, COD - up to 15 mg/l, salt content - up to 1000 mg/l	40,0-55,0	Irrigated agriculture, replenishment of water supply sources of 1-2-3 classes, technical water supply

Analyzing the presented table, it can be concluded that, according to the current legislation, the cost of standard cleaning of domestic and related wastewater will amount to 40-55 rubles/m^3 on average. At the same time, treated effluents can potentially be used in technical water supply systems, for replenishment sources of water supply, primarily underground sources, as well as for irrigated agriculture. Such decisions will reduce the need for water resources at least. Potentially, such solutions will make it possible to make deep treated water as a commodity that will be demanded by a lot of consumers. The transformation of the process of achieving formal standards in the process of commodity production - production of industrial water, will potentially change the attitude of operating organizations – vodokanals to the result of their labor. However, the potential economic efficiency will remain only as long as the cost of wastewater treatment is not transferred to the cost of re-use of treated water, which is impossible because the requirements for treatment quality are normative - established by law and not by the requirements of specific consumers. Thus, today's compliance with legal requirements is a potential opportunity to receive a free secondary water resource.

The potential need for water, the quality of which differs from the standards for discharging into water object for fish culture, creates the possibility of creating reuse schemes for run-off with a certain economic efficiency, both in comparison with the need for deep treatment of run-off and their discharge into the water object, and the use of desalination systems as the main stage of water treatment.

Table 2. Specific efficiency of water reuse after different treatment steps

Treatment step	The enlarged cost of treatment, rubles/m^3	Basic technology of seawater desalination	Possible consumers, specific efficiency, rubles/m^3
Mechanical treatment	3,0-5,0	40,0	Irrigated agriculture – 35,0
Complete biological treatment with nitride-denitrification	25,0-30,0	40,0	Irrigated agriculture, replenishment of water supply sources of 2-3 classes, technical water supply – 10,0
Deep biological treatment with MBR reactors	30,0-35,0	40,0	Irrigated agriculture, replenishment of water supply sources of 1-2-3 classes, technical water supply – 5,0

In General, the basic principles of waste water recycling systems can be as follows:

1. Use for the needs of drinking and domestic water supply exclusively natural water, fresh or salt/brackish.

2. Primarily for small and medium-sized rural settlements, the use of biological treatment facilities will allow the use of treated wastewater for the needs of irrigated agriculture, as well as replenishment of water supply sources-groundwater, taking into account the seasonal nature of these consumers. Such a solution will potentially receive up to 2.0 billion secondary water resources, and the economic effect of implementing this solution on the territory of the Russian Federation may amount to 20.0 billion rubles / year.

3. Primarily for medium and large settlements, the use of biological treatment facilities based on MBR reactors will allow the use of treated wastewater for the needs of technical industrial water supply, as well as replenishment of water supply sources. The potential market for the implementation of such a solution on the territory of the Russian Federation will amount to 20.0 billion secondary water resources, and the economic effect from the implementation of this decision may amount to 200.0 billion rubles / year.

Implementation of such schemes is possible within the framework of the current legislation, which provides for a significant increase in fees for the use of fresh natural water as one of the measures to reduce water consumption [8].

Acknowledgements
The work was financially supported by the Ministry of Education and Science, agreement № 14.575.21.0159. Unique project identifier RFMEFI57517X0159

References
[1] Report *"On the state of the environment in the city of Moscow in 2016"* / Ed. Kulbachevsky A O. - M.: DPiOOS; NIIPI IGSP, 2017. - 363 p.
[2] SanPiN 2.1.5.980-00 *"Hygienic requirements for the protection of surface waters"*. Research Institute of Human Ecology and Environmental Health. Sysina A N, RAMS (Corresponding Member of the Russian Academy of Medical Sciences, Professor Krasovsky G, Professor, MD Zholdakova Z I), Moscow Medical Academy Sechenov I M (professor, MD Bogdanov M V), Russian Medical Academy of Postgraduate Education (MD Plitman S I, Ph.D. Bespalko L E), The Federal Center for State Sanitary and Epidemiological Supervision of the Russian Ministry of Health (Chiburayev V I, Kudryavtseva B M, Nogogibchenko M K), the Department of State Sanitary and Epidemiological Supervision of the Ministry of Health of Russia (Rogovets A I).
[3] State report *"On the state and use of water resources of the Russian Federation in 2015"*, Rybalsky N G, Omelyanenko V A, Dumnov A D, Demin A P and others. Moscow: NIA-Nature, 2016.
[4] SanPiN 2.1.7.573-96 *"Hygienic requirements for the use of wastewater and their precipitation for irrigation andfertilization"*. Institute of Human Ecology and Environmental Health. Sysina A N RAMS (Corresponding Member of the Russian Academy of Medical Sciences, MD Georgiy Krasovskii, MD Rusakov N V, Candidate of Medical Sciences Kryatov I A, Ph.D. Tonkopiy N I, Ph.D. Pavlov V N, Bashara L A); Russian Medical Academy of Postgraduate Education of the Ministry of Health of Russia (Bespalko L E); Institute of Medical Parasitology and Tropical Medicine. E I Marcinovsky GCSEEN of the Russian Federation (Corresponding Member of the Russian Academy of Medical Sciences, MD Romanenko N A, MD Khizhnyak N I, Candidate of Medical Science Khromenkova E P, Cand. Skripova L V, MD Gafurova Z M, Piven V I, Evdokimov V V, Evdokimov V I); Scientific Research Institute for Agricultural Use of Wastewater NIISSV "Progress" (Ph.D. L A Muzychenko, Ph.D. Kovaleva N A, Ph.D. Kutepov L E, Mishin S I, Sytin V Z, Perepelkin S N, Ph.D. Zhirkov E I, Candidate of Biological Sciences Tereshina A N, Tararin Y I, Kaminskaya N I); All-Russian Institute of Fertilizers and Agricultural Science. Pryanishnikova D N (doctor of medical sciences Merzlaya G E, doctor of physical and mathematical sciences Afanasyev R A);

Engineering-geological and geoecological research center of the IGTs of the Russian Academy of Sciences (Prof. Goldberg V M, VSKhIZO (Ph.D. Sayapin V P)

[5] GOST 2761-84. *"Sources of centralized household and drinking water supply. Hygienic, technical requirements and selection rules"*. Interstate standard. Item 2.3.

[6] SanPiN 2.1.4.1074-01 *"Drinking water. Hygienic requirements for water quality of centralized drinking water supply systems. Quality control. Hygienic requirements for ensuring the safety of hot water systems"*.

[7] SNiP 2.04.03-85 *"Sewerage. External network construction"*. Soyuzvodokanalproekt (Mironchik G M - the head of the topic, Berdichevsky D A, Vysota A E, Yaroslavsky L V) with the participation of VNII VODGEO, Donetsk PromstroyNIIproekt and NIIOSP named after Gersevanov N M Gosstroy USSR, Scientific Research Institute of Communal Water Supply and Water Purification Academy of Public Utilities named after Panfilov K D and Giprokommunvodokanal of the Ministry of Housing and Communal Services of the RSFSR, TsNIIEP of engineering equipment of the State Civil Engineering Agency, MosvodokanalNIIproekt and Mosinzhproekt of the Moscow City Executive Committee, the Research and Design and Technological Institute of Municipal Economy and the UkrkommunNIIproekt of the Ministry of Housing and Utilities of the Ukrainian SSR, Institute of Mechanics and Earthquake Resistance named after Urazbaev M T Academy of Sciences of the Uzbek SSR, Moscow Engineering and Construction Institute named after Kuibyshev V V USSR Ministry of Higher Education, Leningrad Engineering and Construction Institute of the Ministry of Higher Education of the RSFSR.

[8] Decree of the Government of the Russian Federation № 1509 of December 26, 2014 *"On Rates of Fees for the Use of Water Objects in Federal Ownership and Amendments to Section I Rates of Fees for the Use of Water Objects in Federal Ownership"*.

IOP Conf. Series: Journal of Physics: Conf. Series **1099** (2018) 012030 doi:10.1088/1742-6596/1099/1/012030

Investigation of the three-adsorber scheme used in the hybrid membrane-sorption system

A A Tishin[1,2], V N Gurkin[1,2], I E Skuridin[1,2], I M Kurchatov[2]

[1]National Research Nuclear University MEPhI, NESPI, Kashirskoe sh. 31, Moscow 115409 Russia,
[2] Public joint-stock company "Aquaservice", Kashirskoe sh. 3/2 st 4, Moscow 115230 Russia

E-mail: aquaserv@mail.ru

Abstract. Investigation of the three-adsorber scheme used in the hybrid membrane-sorption system was carried out. The three-adsorber scheme allows to obtain enriched with oxygen system's product flow up to 70% without using additional devices and methods of sorbent regeneration on the PSA stage. It should be noticed, that it is impossible to reach similar oxygen concentration in the product flow without using proposed organization of the system's flows. These conditions allow system to enter stationary mode much faster. The ways of system improvement for obtaining oxygen concentration higher than 90% are proposed in this work. Such systems will find a use in industrial air separation to the components, in development of mobile and compact energy-efficient gas separation systems based on hybrid membrane-sorption technologies, oxygen generators which are suitable for feed of artificial lung ventilation devices, oxygen concentrators for rehabilitation during post-surgery period and also for the prevention of various diseases including cardiovascular diseases. Furthermore, such hybrid systems have a potential of application in industrial gas drying as well as natural gas drying.

1. Introduction

Nowadays the main methods of oxygen and nitrogen industrial obtaining are: rectification, membrane and sorption. The choice of the method is usually determined by the required product purity and the required productivity. Operation principle of the low-temperature rectification method is in the difference between liquefaction temperatures of the air components. The method of low-temperature rectification provides a production of tons of pure air components: nitrogen, oxygen, argon, helium. The advantages of rectification are: the possibility of obtaining high-purity products, the reliability of equipment, the possibility of obtaining simultaneously several products of separation in the liquid and gas phases. However, the method has a number of significant drawbacks: high capital and operating costs, low profitability at small and medium production volumes, the mass and dimensions parameters of installations, the complexity of hardware and the long start-up period. The principle of the membrane separation method consists in the selective penetration of air mixture components through the membrane surface. Advantages of the membrane method are: small dimensions and time to start the operating mode, simplicity of design and maintenance, high selectivity, quick start and stop. The shortcomings include: the complexity of obtaining high purity oxygen (the maximum achievable concentration of oxygen in the product gas mixture is 50% of volume), low productivity, high energy costs when using recirculation schemes. A short-cycle, non-heated adsorption with variable pressure

(PSA) is most often used as a the sorption method for air separation. In this method the separation is carried out due to the difference in the sorption capacities of the sorbent at various pressures and the alternation of the stages of adsorption and desorption in special tanks - adsorbers. The main advantages of the PSA method are: low capital costs, a long period of non-stop operation, ease of operation, quick start-up, high productivity and enrichment level, and the main disadvantage is low extraction degree and also there is the need to remove sorbent abrasion products from the product flow. Developed membrane and sorption systems for the air separation are now widely used in the oil and gas industry, in the chemical, metallurgical and other industries, as well as in medicine.

Recently, special interest has been given to hybrid air separation systems. Systems possessing the advantages of several methods, as: membrane and adsorption, membrane and cryogenic, adsorption and cryogenic at the same time, due to mutual compensation of the shortcomings of each method separately, have a rather high potential in industry and medicine. With the example of membrane adsorption systems, hybridization allows one to get rid of the drawbacks of both membrane and adsorption methods due to the optimal combination of the advantages of each separation method: ease of implementation and high recovery for membrane and high enrichment for adsorption.

The use of hybrid membrane-adsorption gas separation systems allows to reduce significantly the energy consumption of the plants [1], and also to level off such disadvantages as contamination of the product flow by the products of abrasion of sorbents, through the use of a high-selective polymeric membrane, and the restriction on the oxygen concentration using a single membrane stage after the adsorption stage. Thus, the hybrid membrane-adsorption system is capable of solving effectively the problem of air separation into components.

In addition, it is possible to create mobile and compact gas separation systems on the basis of hybrid membrane-sorption systems, including oxygen generators suitable for feeding artificial respiration apparatus and oxygen concentrators for use during the post-surgery period, as well as for the prevention of various diseases as well as cardiovascular diseases. Hybrid systems also have the potential to be used in industrial gases drying, including natural gas.

2. Statement of the problem

Nowadays, the practical interest is in the creation of portable oxygen generator suitable for feeding of respiratory medical devices based on hybrid membrane-adsorption gas separation systems.

However, at the present time, the most complete and general model and methodology for calculating and constructing of hybrid membrane-adsorption systems does not exist. First of all, it's connected to the complexity of the adsorption and membrane stages matching, as well as to the choice of sorbent and the method of operation of the adsorption stage. Requirements for sorbent are: high selectivity for pair nitrogen / oxygen , high durability and nitrogen sorption capacity. Researches [2, 3] have shown that zeolites of A and X groups, as well as their derivatives, are the most suitable sorbents for these parameters.

A scheme (Fig. 1) of a hybrid membrane-sorption system for effective air separation into components was proposed in [4]. The present work is devoted to experimental research of the separation characteristics of the developed hybrid scheme, its effective work area is shown, and experimental studies of air enrichment with oxygen are carried out. Experimental studies were carried out on a sample of a hybrid system, the pneumatic scheme of the system is shown in Fig. 1.

The work of the hybrid system is organized in a certain way. Compressor 1 supplies the compressed gas to one of the adsorbers 3-5, for example, to the adsorber A, through the filter system 7 and the mixing ejector 2. The solenoid valves 10 and 11 are closed, the valve 9 is open. The pressure in the adsorber 3 is kept constant, the gas injected by the compressor 1 through the valve 9 is pumped through the adsorber where the gas mixture is enriched with oxygen and through the valve 18 enters the membrane module 6 and filling adsorber B. The separation of the flow pumped through the adsorber A is realized by regulating valves 22 and 23, the valve 22 passes through itself a flow directed to fill the adsorber B through the open valve 16, while the valves 15 and 17 are closed, and the valve 23 serves to supply the oxygen-enriched gas in the membrane module 6, through the flowing surge vessel 25. Simultaneously, regeneration of the sorbent at atmospheric pressure occurs in the adsorber C, the valve 14 is opened and the gas is discharged into the atmosphere, while nitrogen desorption takes place on the sorbent. This operation of the PSA unit continues until the pressure in the adsorption unit B reaches the operating pressure of the system, then the adsorbers change in their functions: A - regeneration, B - pumping, C - filling. The gas which supplied to the membrane module 6 is also divided into the product gas that has passed through the membrane, which bypasses the second flowing surge vessel 26 through the control valve 8 and then supplied to the consumer, and into the waste gas low-enriched with oxygen which is fed through the valve 21 to the ejector where it mixes with the PSA flow of the stage. Thus, the system provides a stable product flow with a constant concentration at high pressure values up to 6-8 atm. The operating pressure of the PSA stage for switching of the adsorber functions is equal to 8 atm.

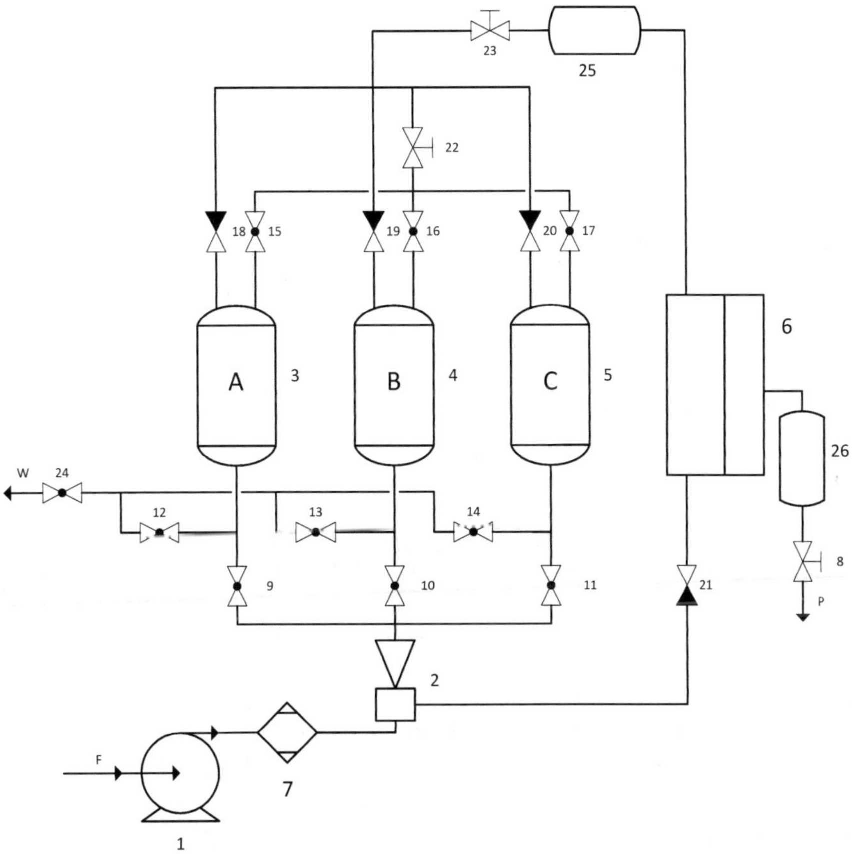

Figure 1. Principal scheme of hybrid membrane-adsorption device.

Features of the developed hybrid system:

- A mixing ejector is used for the waste flow recycling of the membrane module, which allows mixing the oxygen-depleted flow on the membrane stage with the feed flow of the PSA stage;
- The filling of the adsorber up to the operating pressure is organized by a part of the product flow of the adsorber passing the displacement stage;
- The membrane module is fed by a part of the adsorbers product flow of the PSA stage.

Above mentioned features make it possible to increase the energy efficiency of the hybrid membrane-sorption system in comparison with hybrid analogs [3] and individual methods in common. The proposed hybrid system consists of two drying stages. At the first stage of the system, the three-

IOP Conf. Series: Journal of Physics: Conf. Series **1099** (2018) 012030 doi:10.1088/1742-6596/1099/1/012030

adsorption PSA system is used as the main drying block. The second stage used a closed-type membrane steam separator. The main feature of the proposed hybrid system is the use of the flow the dried by membrane steam separator to regenerate the sorbent on the PSA stage of the system. This scheme allows solving the problem of the return of the product flow of the PSA system for the regeneration of the sorbent.

3. Results and discussion

The main feature of the proposed hybrid membrane-sorption system is in the particular approach to the organization of the PSA system stage operation. The operation scheme of the stage is shown in Fig. 2.

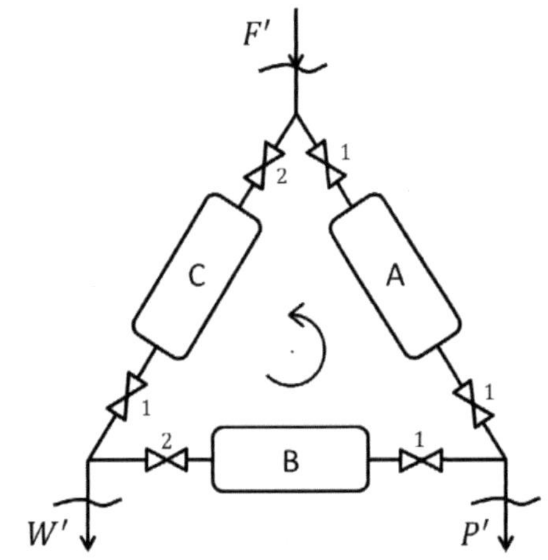

Figure 2. Operation scheme of the PSA stage of the hybrid membrane-sorption system.

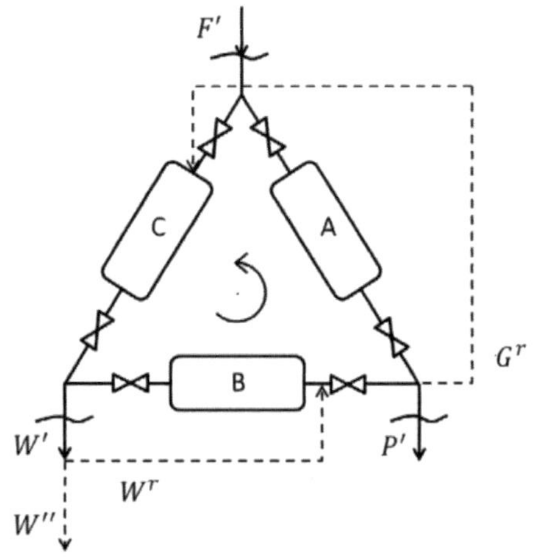

Figure 3. Operation scheme of the PSA stage of the system with countercurrent expulsion and compensation stage.

The adsorbers and the valve system are presented on the scheme, the adsorber in position A always passes the expulsion stage, the adsorber in position B passes the filling stage, and the adsorber C – the stage of discharge. The valve marked 1 are opened, 2 – closed. Adsorbers imaginary move in a counterclockwise direction, changing its functions, while the valve opening scheme is maintained. The adsorber B is filled with a part of the product flow from adsorber A, the rest of the product flow is withdrawn from the PSA stage. The regeneration of the adsorber C is carried out at atmospheric pressure by venting of gas into the atmosphere. The balance equation for this PSA stage is:

$$F' = P' + W', \quad (1)$$

where F' - feed flow, supplied by compressor to the PSA stage, P' is the product flow of the PSA stage, supplied to the membrane stage of the system, W' - the waste flow of the PSA stage, which excluded from the system in a whole.

On the model sample of a hybrid membrane-sorption system developed according to the scheme (Figure 1), the PSA stage of which operates according to the scheme presented in Fig. 2, experimental researches of the hybrid system were carried out.

During the experiments the following parameters were maintained in the system: working pressure of the PSA stage of the system was 8 atm., pressure of sorbent regeneration was 1 atm., full cycle duration – 60 seconds, cycle time of one adsorber work (N) – 20 seconds, which corresponds to a third of the cycle, the pressure of the product flow of the PSA stage – 8 atm., pressure of the membrane stage and hybrid system as a whole – 6 atm. System performance was maintained at 10 normal liters per minute. This performance was chosen, cause it corresponds to the average capacity level of the oxygen concentrators used for feeding the respiratory medical devices, introduced to the market.

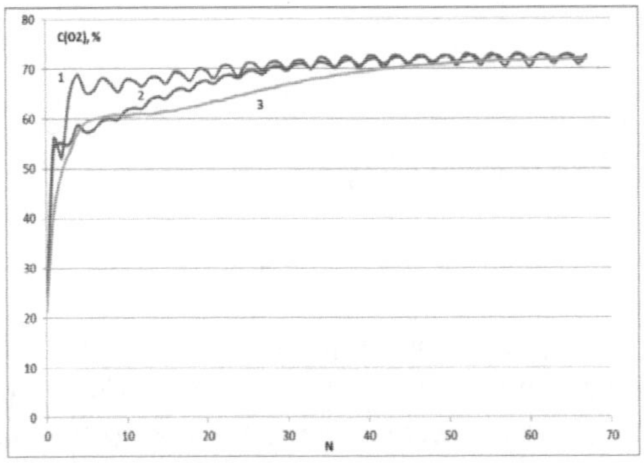

Figure 4. Experimental dependence between the oxygen concentration in the product flow and the number of operating cycles of adsorbers of the PSA stage: the PSA stage of the hybrid system (1), the hybrid membrane-sorption system (2) and the hybrid system with the flowing surge vessel (3).

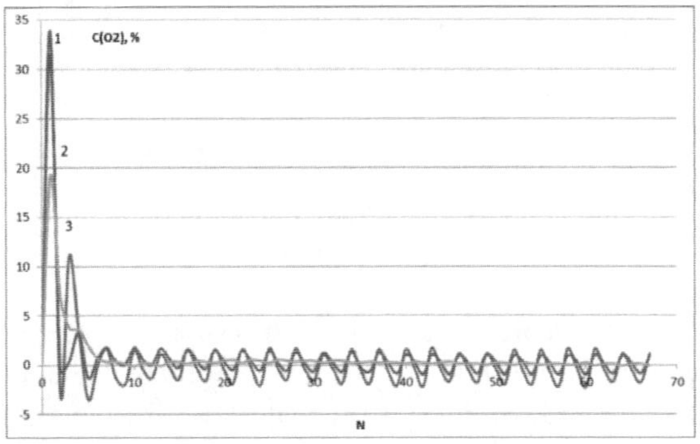

Figure 5. Experimental dependence between oscillation amplitude of oxygen concentration in the product flow and the number of operating cycles of adsorbers of the PSA stage: the PSA stage of the hybrid system (1), the hybrid membrane-sorption system (2) and the hybrid system with the flowing surge vessel (3).

Experimental research of the hybrid membrane-sorption system have shown its rather high efficiency. On the figure 4. the chart ratio of the oxygen concentration in the product flow of the PSA stage of the hybrid system (curve 1), the hybrid membrane-sorption system with the regeneration of the sorbent at atmospheric pressure (curve 2) and the hybrid membrane-sorption system with the flowing surge vessel (curve 3) in dependence to the number of operating cycles of adsorbers of the PSA stage was presented. It is shown on the plot, that the curve 1 has maximum oscillation amplitude of oxygen concentration while entering the stationary mode of the PSA stage operation, concentration is equal to 6 % of oxygen. The number of cycles required to enter the operating mode of the PSA stage is minimal and amounts to 10-20 cycles of adsorption operation or 3-6 cycles of the PSA. At the same time, the concentration of product oxygen in the stationary mode of the PSA stage is about 70-72%. A membrane unit addition to the system, as seen on curve 2, allows to reduce the oscillation amplitude of oxygen concentration, when entering the stationary mode, to 2%, however, the number of cycles which is necessary to reach the stationary mode increases to 8-10 cycles. Compared to the PSA stage, the oxygen concentration in the product flow of the hybrid system increases slowly, due to the fact that the pressure fall on the membrane module during the operation of the hybrid system decreases rapidly and the membrane begins to function as a filter. Adding flowing surge vessels to the system, as seen in curve 3, virtually eliminated fluctuations in the product oxygen concentration, which now amount to 0.1%. but the output to the stationary mode of such a system amounted to about 13 cycles. The experimental dependences of the oscillation amplitude of the oxygen concentration as a function of the number of operating cycles of the PSA adsorbers of the step are shown in Fig. 5. The data on the amplitudes of the oscillation of the oxygen concentration and the number of cycles to enter the regime during the experimental studies are presented in Table 1. An addition to the system of flowing surge vessels allows to decrease the oscillations of product oxygen concentration up to 0.1%, but the output to the stationary mode of such a system amounted to about 13 cycles. The experimental dependences of the oscillation amplitude of the oxygen concentration as a function of the number of operating cycles of adsorbers of the PSA stage are shown in Fig. 5. The data on the oscillation amplitudes of the oxygen concentration and the number of cycles for entering the operation mode during the experimental research are presented in Table 1.

Table 1. The characteristics of the hybrid system operation modes.

	A_{max}, O_2%	A, O_2%	N_{eq}	N_{op}
PSA	34	6	10-20	1
HMS	32	2	25-30	1
HMSR	19	0.1	40	3

In the table 1 value A_{max} corresponds to the maximum fluctuation of concentration in each of the processes, value A corresponds to average deviation of concentration in the stationary mode, N_{eq} – the number of operating cycles of the adsorbers of the PSA stage before reaching the stationary mode, and N_{op} is the number of operating cycles of the adsorbers of the PSA stage before entering the operation mode, when oxygen concentration in the product flow reaches the value, higher than 50%.

To compare the obtained experimental data with the obtained data of previous works, similar in design and in purpose of the systems, the data on the operation of the PSA of the system with two adsorbers and regeneration at atmospheric pressure [3] - curve 1, the PSA of the system using a countercurrent expulsion and compensation stage [5] - curve 2 and the investigated three-absorber system - curve 3, as well as a single membrane module [6] - curve 4 is presented on the Figure 6.

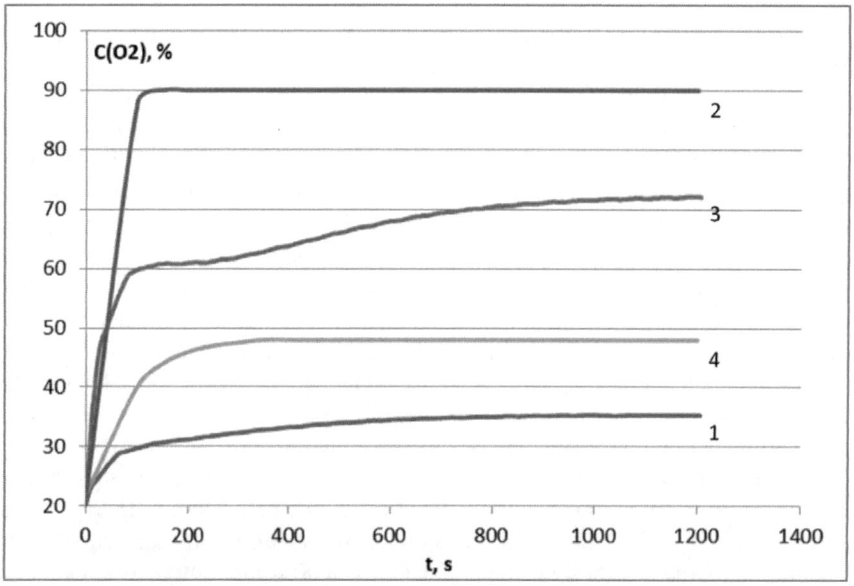

Figure 6. Oxygen concentration in the waste flow: the PSA of the system with two adsorbers and regeneration at atmospheric pressure (1); the PSA of the system with the use of countercurrent expulsion and compensation stage (2); proposed hybrid membrane-sorption (3); single membrane module (4).

It can be seen from the figure that the maximum concentration of product oxygen – 90% in the PSA process can be obtained by using both countercurrent expulsion and vacuum regeneration, but due to the fact that a large part of the product flow is actually discharged during expulsion, such a system loses in the efficiency to the system offered in this work. The membrane module, which allows to obtain an oxygen concentration of not more than 50%, is in advance of the PSA scheme with regeneration of the sorbent at atmospheric pressure without countercurrent expulsion, which allows obtaining only 30-35% of oxygen, while the proposed method gives 70% oxygen under the same conditions, which in 2 times greater than the PSA method without expulsion and is more effective than the method of the PSA with the expulsion, by saving an expulsion flow in the system.

Taking into account the positive effect of the introduction of the countercurrent expulsion stage and the compensation stage shown in Figure 6, it can be assumed that the application of similar stages into the proposed hybrid membrane-sorption system will increase the concentration of the product flow of the system up to 90-93%, and also increase system performance in whole. On the figure 3 a scheme of organization of the flows with the stage of countercurrent expulsion and the compensation stage is proposed. The scheme of flows, experimentally tested in this work, is taken as the basis of the scheme. The scheme assumes a part of the product flow from the adsorber A, which was previously sent to the adsorber B, to be redirected to the countercurrent expulsion of the adsorber C, and the flow discharged from the adsorber C at this time is redirected to the adsorber B to shorten the filling stage. With such a flow organization, the general balance equation for flows (1) will be preserved and will look like:

$$F' = P' + W'', \tag{2}$$

where W'' will be determined by the following relations:

$$W'' = W' - W^r, \tag{3}$$

$$W^r = G^r, \tag{4}$$

where W' - total waste flow of the PSA stage of the hybrid system, W^r – the part of the waste flow, directed to fill the adsorber B during the compensation stage, and G^r – the part of product flow of the adsorber A, directed to the countercurrent expulsion of the adsorber C.

According to this scheme, the organization of the flows in proposed hybrid membrane-sorption system would allow to obtain the concentration of the produced oxygen equals to the level of 90-93%, with increased energy efficiency in comparison with the system [5].

4. Conclusions

Investigation of the three-adsorber scheme used in the hybrid membrane-sorption system was carried out. System consists of sorption and membrane units, connected into the recirculating contour. The three-adsorber scheme allows to obtain enriched with oxygen system's product flow up to 70% without using additional devices and methods of sorbent regeneration on the PSA stage, that is impossible to reach similar oxygen concentration in the product flow without using proposed organization of the flows in the system. Moreover, there is a need to conduct less than 40 work cycles of adsorbers to achieve similar concentration, which allows to reach the stationary mode of the system much faster. The ways of system's improvement for obtaining oxygen concentration higher than 90% are proposed in this work.

Such systems will find a use in industrial air separation to the components, in development of mobile and compact energy-efficient gas separation systems based on hybrid membrane-sorption technologies, oxygen generators which are suitable for feeding of artificial lung ventilation devices, oxygen concentrators for rehabilitation during post-surgery period and also for the prevention of various diseases including cardiovascular diseases. Furthermore, such hybrid systems have a potential of application in industrial gas drying as well as natural gas drying.

References

[1] Laguntsov N. I., Kurchatov I. M., Korolev M. V. 2015 *Petrol. Chem.* **9** 738-745.
[2] Watanabe K., Austin N., Stapleton M. R. 1995 *Molecular Simulation*. **4** 197-221.
[3] Tishin A.A., Gurkin V.N., Laguntsov N.I., Kurchatov I.M. 2018 *Petrol. Chem.* **2** 55–60.
[4] Tishin A.A. et all. 2018 *Petrol. Chem.* **4** 338–345.
[5] Akulinin E. I. et al. 2012 *Transaction of Tambov state technical university*. **1** 157.
[6] Coombe H. S., Nieh S. 2007 *Energy Conversion and Management* **5** 1499-1505.

IOP Conf. Series: Journal of Physics: Conf. Series **1099** (2018) 012031 doi:10.1088/1742-6596/1099/1/012031

PAN filtration membranes with extended solvent stability

A A Yushkin[1], M N Efimov[1], A A Vasilev[1], G P Karpacheva[1], A V Volkov[1]

[1] A.V.Topchiev Institute of Petrochemical Synthesis, RAS, Moscow, Russia
Halex@ips.ac.ru

Abstract. This work focuses on preparation of PAN membranes stable in aprotic solvents. Laboratory samples of PAN with different molecular weight were synthesized by varying the polymerization conditions. Membranes were prepared by non-solvent induced phase inversion method. Membranes prepared with higher molecular weight PAN had stable separation characteristics at pressures up to 30 bar and higher retention but lower permeability than for membranes prepared from PAN with lower molecular weight. At the same time membranes from PAN with lower molecular weight had higher but unstable in time permeabilities. The retention of Bovine serum albumin (M_w=69 kDa) from water solutions reached the value up to 98%. After IR irradiation treatment membrane permeability for water were the same as for initial membranes and varied from 5 to 1500 kg/m^2·h·bar. At the same time PAN membranes was stable toward aprotic solvents NMP, DMSO, DMFA, DMAA although initial polymer was soluble in such solvents.

1. Introduction

Polyacrylonitrile (PAN) has received wide acceptance as a material for the development of chemically resistant membranes used in the filtration of liquids [1,2]. PAN has high strength, good resistance to solvents, thermal and chemical stability, and low cost [2,3]. In comparison to other membrane materials like polyvinylidene fluoride, polysulfone and polyethersulfone, membranes from PAN are less prone to fouling during operation [4-8]. However, the solubility of this polymer in polar aprotic solvents and in aqueous electrolyte solutions with high ionic strength limits an application of such membranes.

Usually porous membranes from PAN are prepared by phase separation induced by non-solvent (for example, water) [2, 9-12], although other non-solvents, for example, supercritical CO_2, can be used [13]. Important parameters for the membrane preparation are the composition of the casting solution, its temperature, additives, solvent and non-solvent used. In addition, the structure of the resulting membranes depends significantly on the viscosity of the polymer solution and non-solvent.

Membrane characteristics can be improved by addition special poreforming additives, such as calcium chloride, poly(ethylene glycol), or polyvinylpyrrolidone [14, 15]. In [16], the addition of carbon nanotubes to PAN-based membranes was used for improve their filtration properties.

Another way to improve membrane characteristics is plasma treatment, grafting of different functional groups, or hydrolysis [16–23]. Modification of PAN based membranes with calcium carbonate help to obtain membranes with retention of Congo Red (Mw~700) up to 98% and water permeability 25.5 L/(m^2·h·bar) [24]. It was shown that hydrolysis of PAN membranes also can increase membrane permeability [25].

Modification often used for increase membrane chemical stability for solvents. For example, cross linking of PAN by solutions of hydrazine hydrate at elevated temperature is often used for obtain membranes stable in different solvents [26-27]. Temperature treatment also can be used for cross

IOP Conf. Series: Journal of Physics: Conf. Series **1099** (2018) 012031 doi:10.1088/1742-6596/1099/1/012031

linking of PAN by cyclization of the nitrile groups and dehydrogenation [28]. Recently IR treatment was used to improve resistance of PAN membranes for aprotic solvents [11,12]. In the case of IR radiation treatment of PAN, the mechanism of chemical transformation does not differ from thermal modification but takes place much faster. The application of IR radiation makes it possible reduce time of processing to several minutes and also to decrease processing temperature [11,12].

The synthesis of PAN is carried out by radical polymerization of acrylonitrile in a solvent, aqueous emulsion or suspension. As initiators, azo and diazo compounds, peroxides, organoelement compounds, and oxidation-reduction systems are used. The synthesis time, temperature and gas medium in which the polymerization is carried out also have a significant effect on the molecular weight characteristics of the polymer. By varying the conditions for polymer synthesis, a wide range of molecular weight characteristics can be obtained.

This work focuses on preparation of PAN membranes stable in aprotic solvents using laboratory synthesized PAN with different molecular weight. The stability of membranes was achieved by the modification of the polymer with IR irradiation treatment after membrane preparation. Membranes were prepared by non-solvent induced phase separation method.

2. Membrane preparation and modification

Laboratory samples of PAN were synthesized by varying the polymerization conditions. The synthesis was carried out in an aqueous medium in the presence of an oxidation-reduction system from ammonium peroxodisulfate ($(NH_4)_2S_2O_8$) and sodium dithionite ($Na_2S_2O_4$) as initiators. The polymerization was carried out under the following conditions: $[(NH_4)_2S_2O_8]=1.47\times10^{-3}$ mol/l, $[Na_2S_2O_4]=6.29\times10^{-4}$ mol/l, $[H_2SO_4]=1.9\times10^{-2}$ mol/l, and [acrylonitrile]=1.45 mol/l. The initial concentrations of ammonium peroxodisulfate and sodium dithionite varied by a factor of 0.5; 1; 1.5; 2. Water, monomer and sulfuric acid were charged into the reactor in two stages. The initiator, which is a redox system of ammonium peroxodisulfate and sodium dithionite, was charged to a reactor containing 3/4 of the required volume of water, monomer and acid, followed by a reaction for 20 or 40 minutes at 60°C. Then, the remaining 1/4 solution was loaded, after which the reaction was continued for 4 hours. The polymer was then filtered, washed sequentially in a solution of potassium carbonate in water and methanol to remove sulfuric acid, and then dried in a vacuum to constant weight. The polymer samples were named after the synthesis conditions of PANab, where «a» is the coefficient of change in initiator concentration, and «b» is the duration of the first polymerization step in minutes. The molecular mass characteristics of the obtained polymer were studied by GPC. The study was carried out on a GPC-120 chromatograph from PolymerLabs. The analysis was carried out at 50°C in DMF.

When varying the conditions for polymer synthesis, it was found that an increase of the concentration of ammonium peroxodisulfate and sodium dithionite leads to a decrease in the molecular weight of PAN (Table 1). This may be due to the fact that a large number of radicals formed as a result of the introduction of a higher concentration of ($(NH_4)_2S_2O_8$) and ($Na_2S_2O_4$) (and vice versa) leads to the initiation of chain growth on a larger number of monomer molecules. As a consequence, the probability of recombination of two radicals at the ends of short polymer chains increases, on which the growth of the polymer chain ceases. As a result of varying the concentration of initiators, samples of a polymer with different molecular weights were obtained (Table 1).

Table 1. Molecular-mass characteristics of laboratory synthesized PAN samples.

Sample	M_n, $\times10^{-3}$	M_w, $\times10^{-3}$	M_w/M_n
$PAN_1$20	57.7	267.2	4.63
$PAN_1$40	79.1	280.0	3.54
$PAN_{1.5}$40	47.9	183.0	3.82
$PAN_2$40	41.9	146.5	3.49

Solutions of synthesized PAN samples in DMSO were prepared with concentration 9%. For this purpose, the PAN samples in the required proportions were dissolved in DMSO. The solution was stirred for 72 hours, after which an additional 50 minutes was treated in an ultrasonic bath. The resulting polymer solution with a thin layer (200 μm) was applied to glass and precipitated in distilled water. The time between application of the solution on the glass and immersion in water did not exceed 10 seconds. The rate of application of the polymer solution was 2.5 m/min. The resulting flat membrane was sequentially washed with distilled water, ethanol, and hexane (each time for an hour) and dried at room temperature in air.

Thermal treatment of membranes was carried out in a laboratory setup of pulsed IR radiation at temperature 130°C under a nitrogen atmosphere for 5 minutes. Halogen lamps KG-220 were used as a source of incoherent IR radiation. The radiation maximum of which falls within the range 0.9-1.2 μm. The lamps were arranged in such a way as to ensure a uniform heating of the entire sample. The membrane was fixed in a graphite cassette, which was placed in a cylindrical quartz reactor. The intensity of the IR radiation was controlled by the sample heating temperature, since sample temperature increased only due to IR radiation and not from separate heat source. Temperature was measured with a chromel-alumel thermocouple placed directly below the sample. The temperature regulation was accurate to within 0.25°C.

3. Membrane characteristics

A study of filtration characteristics of membranes was carried out in dead-end cells at a transmembrane pressure of 5-30 bar equipped with a magnetic stirrer to reduce the effect of concentration polarization. The active area of the membranes in the cells was 8 cm^2. The pressure in the system was created with the aid of an inert gas (helium). The permeate flux (*J*) was determined gravimetrically as follows:

$$J = \frac{m}{S \cdot \Delta t}, \qquad (1)$$

where *m* is the mass of permeate (kg) that passed through the membrane with area *S* (m^2) per time Δt (h). The membrane permeability (*P*) was defined as the ratio of flux (*J*) to the applied transmembrane pressure Δp (bar):

$$P = \frac{J}{\Delta p} = \frac{m}{S \cdot \Delta t \cdot \Delta p} \qquad (2)$$

To characterize the separation properties of the membrane, the retention *R* (%) was used:

$$R = \left(1 - \frac{c_P}{c_0}\right) \cdot 100\%, \qquad (3)$$

where c_0 and c_P are the concentration of solute in the feed and permeate, respectively. Bovine serum albumin (BSA, MM = 69 kDa), Ovalbumin (45 kDa), Pepsin (34.5 kDa), Lysozyme (14.3 kDa) were used as model solutes. The concentration of dissolved substances in the solution was determined spectrophotometrically at a wavelength of 280 nm.

It is established that the molecular weight of the polymer has a significant effect on the characteristics of the membranes obtained. Membranes were characterized by water permeability and solutes retention at transmembrane pressure 5 bar. The highest permeability was observed for membranes obtained from the PAN_240 sample which has lowest molecular weight (Figure 1). At the same time studies of the dependence of membrane retention on the molecular weight of the polymer shows that membranes prepared from polymers with a lower molecular weight have generally lower retention values (Figure 2), but difference was much lower than for permeability. Although membranes prepared from PAN_240 and $PAN_{1.5}40$ have permeability much higher than from other samples at elevated pressures 10-30 bar their permeability constantly decreased with time up to 20% per hour. From the other hand permeabilities of PAN_140 and PAN_120 were stable during whole time of experiment even at higher pressure 30 bars.

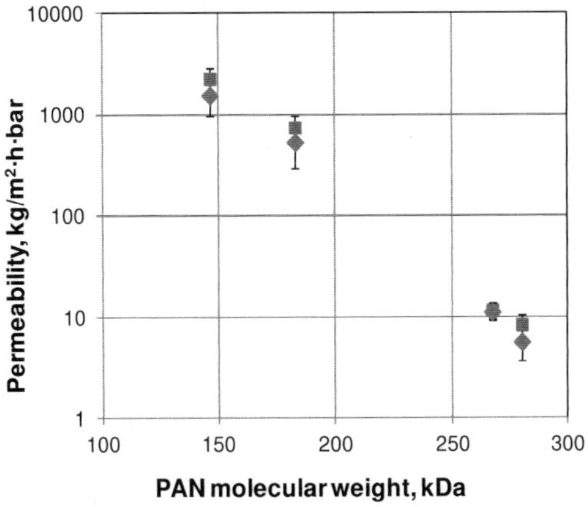

Figure 1. Water permeability of membranes prepared from PAN with different molecular weight at 5 bar; ■ – before IR radiation treatment; ♦ – after IR radiation treatment.

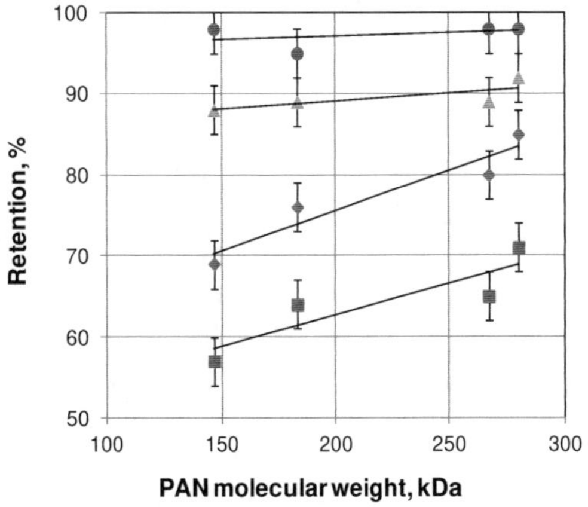

Figure 2. Retention of solutes by membranes prepared from PAN with different molecular weight; ■ – Lysozyme (14.3 kDa); ♦ – Pepsin (34.5 kDa); ▲ – Ovalbumin (45 kDa); ● – BSA (69 kDa).

After IR irradiation treatment membrane permeability for water were slightly lower than for initial samples but difference did not exceed experimental errors. At the same time such membranes prepared from PAN_140 and PAN_120 have stable permeabilities toward aprotic solvents NMP (3-5 kg/m^2·h·bar), DMSO (3-5 kg/m^2·h·bar), DMFA (7-12 kg/m^2·h·bar) and DMAA (6-11 kg/m^2·h·bar) although initial polymer was soluble in it. For PAN_240 and $PAN_{1.5}40$ stable permeabilities of aprotic solvents did not observed as well as for water with these polymers but membranes still not dissolved in these solvents. This means that membrane permeability decreases due to membrane compaction instead of polymer dissolution.

4. Conclusion

In this work laboratory samples of PAN with different molecular weight were synthesized by varying the polymerization conditions. Membranes were prepared by non-solvent induced phase inversion method. Membranes prepared with higher molecular weight had stable separation characteristics at pressures up to 30 bar and higher retention but lower permeability than for membranes prepared from PAN with lower molecular weight. The retention of Bovine serum albumin (M_w=69 kDa) from water solutions reached the value up to 98%. After IR irradiation treatment membrane permeability for water were the same as for initial membranes and varied from 5 to 1500 kg/m^2·h·bar. At the same time PAN membranes was stable toward aprotic solvents NMP, DMSO, DMFA, DMAA although initial polymer was soluble in such solvents.

Acknowledgments
Alexey Yushkin acknowledges Russian Foundation of Basic Research (RFBR project no. 16-38-60152) for financial support.

References

[1] Kumar M, Shevate R, Hilke R, Peinemann K-V 2016 Novel adsorptive ultrafiltration membranes derived from polyvinyltetrazole-co-polyacrylonitrile for Cu(II) ions removal *Chem Eng J* **301** 306
[2] Scharnagl N and Buschatz H 2001 Polyacrylonitrile (PAN) membranes for ultra-and

microfiltration *Desalination* **139** (1-3) 191
[3] Tran T D, Mori S, Suzuki M 2007 Plasma modification of polyacrylonitrile ultrafiltration membrane *Thin Solid Films* **515** (9) 4148
[4] Marbelia L, Mulier M, Vandamme D, Muylaert K, Szymczyk A, Vankelecom I F 2016 Polyacrylonitrile membranes for microalgae filtration: Influence of porosity, surface charge and microalgae species on membrane fouling *Algal Research* **19** 128
[5] Guillen G R, Pan Y, Li M, Hoek E M 2011 Preparation and characterization of membranes formed by nonsolvent induced phase separation: a review *Industrial & Engineering Chemistry Research* **50** (7) 3798
[6] Cornelissen E R, Van den Boomgaard T, Strathmann H 1998 Physicochemical aspects of polymer selection for ultrafiltration and microfiltration membranes *Colloids and surfaces A: Physicochemical and engineering aspects* **138** (2-3) 283
[7] Lohokare H, Bhole Y, Taralkar S, Kharul U 2011 Poly (acrylonitrile) based ultrafiltration membranes: Optimization of preparation parameters *Desalination* **282** 46
[8] Klaysom C, Hermans S, Gahlaut A, Van Craenenbroeck S, Vankelecom I F 2013 Polyamide/Polyacrylonitrile (PA/PAN) thin film composite osmosis membranes: Film optimization, characterization and performance evaluation *Journal of membrane science* **445** 25
[9] Kim I C, Yun H G, Lee K H 2002 Preparation of asymmetric polyacrylonitrile membrane with small pore size by phase inversion and post-treatment process *Journal of Membrane Science* **199** (1-2) 75
[10] Yang S and Liu Z 2003 Preparation and characterization of polyacrylonitrile ultrafiltration membranes *Journal of membrane science* **222** (1-2) 87
[11] Yushkin A A, Efimov M N, Vasil'ev A A, Ivanov V I, Bogdanova Y G, Dolzhikova V D, Karpacheva G P, Bondarenko G N, Volkov A V 2017 Effect of IR Radiation on the Properties of Polyacrylonitrile and Membranes on Its Basis *Polymer Science, Series A* **59** (6) 880
[12] Yushkin A A, Efimov M N, Vasilev A A, Bogdanova Y G, Dolzhikova V D, Karpacheva G P, Volkov A V 2017 Modification of polyacrylonitrile membranes by incoherent IR radiation *Petroleum Chemistry* **57** (4) 341
[13] Wang H, Yu J, Bai H, Li L 2016 Preparation of PAN nanofiltration membranes by supercritical-CO_2-induced phase separation *The Journal of Supercritical Fluids* **118** 89
[14] Linkov V M, Sanderson R D, Jacobs E P 1994 Carbon membranes from precursors containing low-carbon residual polymers *Polymer international* **35** (3) 239
[15] Kamińska G, Bohdziewicz J, Palacio L, Hernández A, Prádanos P 2016 Polyacrylonitrile membranes modified with carbon nanotubes: characterization and micropollutants removal analysis *Desalination and Water Treatment* **57** (3) 1344
[16] Fei Z D, Wan L S, Wang W M, Zhong M Q, Xu Z K 2013 Thermo-responsive polyacrylonitrile membranes prepared with poly (acrylonitrile-g-isopropylacrylamide) as an additive *Journal of membrane science* **432** 42
[17] Khulbe K C, Feng C, Matsuura T 2010 The art of surface modification of synthetic polymeric membranes *Journal of Applied Polymer Science* **115** (2) 855
[18] Zhao Z P, Li J, Zhang D X, Chen C X 2004 Nanofiltration membrane prepared from polyacrylonitrile ultrafiltration membrane by low-temperature plasma: I. Graft of acrylic acid in gas *Journal of Membrane Science* **232** (1-2) 1
[19] Zhao Z P, Li J, Wang D, Chen C X 2005 Nanofiltration membrane prepared from polyacrylonitrile ultrafiltration membrane by low-temperature plasma: 4. Grafting of N-vinylpyrrolidone in aqueous solution *Desalination* **184** (1-3) 37
[20] Chen J, Li J, Zhao Z P, Wang D, Chen C X 2007 Nanofiltration membrane prepared from polyacrylonitrile ultrafiltration membrane by low-temperature plasma: 5. Grafting of styrene in vapor phase and its application *Surface and Coatings Technology* **201** (15) 6789

[21] Ulbricht M, Oechel A, Lehmann C, Tomaschewski G, Hicke H G 1995 Gas-phase photoinduced graft polymerization of acrylic acid onto polyacrylonitrile ultrafiltration membranes *Journal of applied polymer science* **55** (13) 1707

[22] Zhang G, Song X, Li J, Ji S, Liu Z 2010 Single-side hydrolysis of hollow fiber polyacrylonitrile membrane by an interfacial hydrolysis of a solvent-impregnated membrane *Journal of Membrane Science* **350** (1-2) 211

[23] Wang J, Yue Z, Ince J S, Economy J 2006 Preparation of nanofiltration membranes from polyacrylonitrile ultrafiltration membranes *Journal of membrane science* **286** (1-2) 333

[24] Chen X N, Wan L S, Wu Q Y, Zhi S H, Xu Z K 2013 Mineralized polyacrylonitrile-based ultrafiltration membranes with improved water flux and rejection towards dye *Journal of membrane science* **441** 112

[25] Bao W, Xu Z, Yang H 2009 Electrokinetic and permeation characterization of hydrolyzed polyacrylonitrile (PAN) hollow fiber ultrafiltration membrane *Science in China Series B: Chemistry* **52** (5) 683

[26] Kasperchik V P, Yaskevich A L, Bil'dyukevich A V 2005 *Membranes, Series Kriticheskie Tekhnologii* **28** (4) 35

[27] Pérez-Manríquez L, Aburabi'e J, Neelakanda P, Peinemann K V 2015 Cross-linked PAN-based thin-film composite membranes for non-aqueous nanofiltration *Reactive and Functional Polymers* **86** 243

[28] Tsai H A, Chen Y L, Lee K R, Lai J Y 2012 Preparation of heat-treated PAN hollow fiber membranes for pervaporation of NMP/H_2O mixtures *Separation and purification technology* **100** 97

Reagent purification and pre-treatment of water in water supply systems with desalination complex

R A Alexandrov, N I Laguntsov, V A Salomasov, E A Ananyeva and S N Tikhonov

National Research Nuclear University MEPhI (Moscow Engineering Physics Institute), Kashirskoye sh., 31, Moscow, 115409, Russia

E-mail: plazma-06@bk.ru

Abstract. The work is devoted to the investigation of the process of reagent purification of seawater, using the example of the Black Sea water, purification of recycled water, as well as the development of a technological scheme for the purification and preparation of sea and recycled water system, taking into account the composition of pollutants, salt content and the productivity of the water treatment unit for the water supply system with desalination complex.

1. Introduction

The marine area is continuously connected with its external environment. It is influenced by various external and internal conditions such as: a variety of natural phenomena, industry, industrial and communal construction, household activities and much more. The consequence of these influences is the introduction into the aquatic environment of new, uncharacteristic substances - pollutants, which impair the quality of water. In addition, natural marine and oceanic waters in themselves are characterized by a fairly high hardness and a high content of such toxic ions, having in some countries maximum permissible concentrations (MPCs), like bromine, fluorine, iodine, boron.

For long-term operation of water supply systems with desalination complexes [1], the quality of the original desalinated and recycled water must meet certain stringent requirements. This entails the need to develop a water purification system, including its preliminary preparation before desalination. At the same time, the cost of this system can be two or three times higher than the cost of the desalination plant itself.

When water is desalinated by distillation in evaporators, serious difficulties are caused by rapid overgrowth by the scum of heat exchange surfaces. This is due to the high rigidity inherent in the seawaters. The formation of scale on the heat transfer surfaces of the evaporators leads to a sharp decrease in their efficiency; there is a need for frequent stopping and cleaning of evaporators, the use of antiscale reagent, which is associated with high operating costs. When membrane methods of desalination are used, the high sensitivity of membranes to various kinds of impurities of organic and inorganic character requires the use of advanced pre-treatment schemes in reverse osmosis plants. Both for reverse osmosis and for electrodialysis, hardness salts, especially calcium stiffness, are the most dangerous. In this connection, the problem of pre-treatment and softening of seawater before its desalination is highly topical.

After the use of water by consumers obtained in the water supply system with desalination complex, the water from the household and industrial water supply system of the settlement gets into the sewage system, from which the waste water enters the additional pre-treatment unit [1], whose task should be to bring the quality of this recycled water to the level of MPC.

2. Subject and method of research

2.1. Subject of research

Researches of the process of reagent purification of seawater were carried out using the example of the Black Sea water. A water sample was taken in the vicinity of Nizhneimeretinskaya Bay in Sochi (Nizhneimeretinskaya Bay, Sochi, Krasnodar Territory, Russia 43.397218, 39.966488).

Researches of the process of reagent purification of recycled water are carried out on the example of model and real sewage (recycled) waters of a non-ferrous metallurgy plant containing such types of pollutants as: suspended substances, petroleum products, heavy metals, hardness salts.

2.2. Methodology of research

The main technological problems of preparation and purification of marine and industrial wastewater include the need to select the most efficient coagulant-flocculant and to create an optimal technological regime for its use. The hybrid aluminum-silicon reagent FCS-9 [2] was chosen as the base reagent in the work, because of its high relative coagulating and flocculating ability. Reagent treatment of water was carried out with FCS-9 reagent with activating additives, which used a soda solution and sodium orthophosphate. The reagent concentration (in terms of Al_2O_3) was varied from 20 to 70 mg/l.

Methods of atomic spectrometry, ion chromatography and titrimetric analysis were used to determine the concentration of various cations and anions in seawater. Determination of the content of the main cations in seawater was carried out by the method of atomic emission spectrometry with inductively coupled plasma. The measurements were carried out using an emission spectrometer with inductively coupled plasma Profile Plus. The method of ion chromatography was used to determine the concentration of the main anions in seawater. The measurements were carried out on a chromatograph "Stayer". Determination of the total mineralization (dry residue) of seawater was carried out by evaporation. The determination of the total rigidity was carried out by the titrimetric method.

Modeling of recycled water composition was carried out according to the rules of model solutions preparation. Model waters with a given concentration of iron and copper were prepared by adding to the distilled water a known amount of their salts $FeCl_3{\cdot}6H_2O$ and $CuSO_4{\cdot}5H_2O$ respectively. The prepared water was settled until a fixed concentration of pollutants was established, after which it was filtered through a filter with a pore size of 100-200 μm. Model waters containing petroleum products were prepared by dispersing transformer oil in water. The initial concentration of all pollutants (iron, copper and petroleum products) in model solutions was 10 mg/l, and the pH of solutions was in the range of 6.8 – 7.5.

Model waters with a given concentration of suspended solids were prepared by adding kaolin (clay) suspension to distilled water. To study the kinetics of sedimentation of suspended solids during spontaneous settling (without reagents), and also during sedimentation using FCS-9 reagent, model waters with an initial clay content of 150 mg/l were used.

Methods of electrochemical, colorimetric and photocolorimetric analysis were used to determine the main characteristics and physico-chemical composition of model and real recycled waters. The concentrations of iron, copper and aluminum were determined by photometric method using a Spekol 1300 spectrophotometer. The essence of the method for determining petroleum products is associated with the determination of the content of an oil-soluble dye (sudan, azobenzene), previously dissolved in a model hydrocarbon (diesel fuel, gasoline, toluene and others).

3. Research results

3.1. Results of reagent treatment of the Black Sea water

3.1.1. Results of preliminary softening of seawater. As a result of preliminary studies, it has been established that in connection with the fact that seawater is characterized by a rather high rigidity (≈ 1350 ppm for the Black Sea), the capacity of the hybrid reagent is not sufficient to reduce it and effectively coagulate pollutants, including toxic ions. To eliminate this problem, it is proposed in the work to first reduce the rigidity of seawater using reagent methods.

Reagent methods of water softening are based on water treatment by substances that bind calcium and magnesium ions to practically insoluble compounds [3].

For the most effective softening of seawater, the method of successive water treatment with Na_2CO_3 solution for pre-softening and then sodium orthophosphate solution $Na_3PO_4{\cdot}12H_2O$ together with FCS-9 reagent for softening and removal of "residual" phosphates is proposed.

The process of reducing carbonate and non-carbonate hardness using a soda solution and sodium orthophosphate is described by the following reactions (for magnesium, similarly):

$$Ca(HCO_3)_2 + Na_2CO_3 \rightarrow Ca(CO_3)\downarrow + 2NaHCO_3;$$

$$CaCl_2 + Na_2CO_3 \rightarrow Ca(CO_3)\downarrow + 2NaCl;$$

$$CaSO_4 + Na_2CO_3 \rightarrow Ca(CO_3)\downarrow + Na_2SO_4;$$

$$3Ca(HCO_3)_2 + 2Na_3PO_4 \rightarrow Ca_3(PO_4)_2\downarrow + 6NaHCO_3;$$

$$3CaCl_2 + 2Na_3PO_4 \rightarrow Ca_3(PO_4)_2\downarrow + 6NaCl;$$

$$3CaSO_4 + 2Na_3PO_4 \rightarrow Ca_3(PO_4)_2\downarrow + 3Na_2SO_4$$

The use of FCS-9 reagent allows to significantly increase the rate of these reactions, intensify the purification process, and also reduce the phosphate ions and sodium introduced into the water by adding a soda solution and sodium orthophosphate.

In the course of experimental researches it was established that the degree of purification from the hardness salts strongly depends on the kind of activating additives introduced into the water during the purification process. To soften very hard water (600 ppm and more), it is necessary to use a solution of soda together with a solution of sodium orthophosphate and reagent FCS-9. In this method, the degree of purification reaches values of about 95%. When softening water of medium hardness (up to 600 ppm) it is sufficient to use only a soda solution together with FCS-9 to intensify the process. It is possible to obtain drinking water, the total hardness of which corresponds to the norm [4].

Table 1 presents the results of studies on the softening of the highly rigid seawater of the Black Sea with an initial rigidity of 1350 ppm.

Table 1. Results of softening of seawater.

Parameter	Initial	Method		
		1	2	3
Hardness, ppm	1350	600	225	50
Phosphates, mg/l	0,0	0,00	15,00	3,50
Sodium, g/l	0,12	0,40	4,00	0,20
pH	7,61	7,75	8,00	7,50

Softening was carried out in three different ways:

1. Adding to the water 10 ml/l 10% soda solution.
2. Adding to the water 10 ml/l 10% soda solution and 5 ml/l 10% sodium orthophosphate solution.
3. Adding to the water 10 ml/l 10% soda solution, 5 ml/l 10% sodium orthophosphate solution and 30 mg/l FCS-9 (in terms of Al_2O_3).

When 10 ml/l of 10% soda solution is added to the seawater, followed by sedimentation and filtration, the water hardness decreases from 1350 to 600 ppm (Method 1). When added together with a soda solution 5 ml/l 10% sodium orthophosphate solution, the intensity of clarification rises sharply, while the water hardness decreases to 225 ppm (Method 2). However, in this case there is an increased concentration of phosphate ions in the treated water, which is 15 mg/l. With the simultaneous introduction of a solution of soda, sodium orthophosphate and 30 mg/l FCS-9 (in terms of Al_2O_3) it is possible to reduce the concentration of hardness salts to 50 ppm (Method 3). In this case, the content of "residual" phosphate ions decreases to a value corresponding to the norm [4]. This fact confirms the effectiveness of the use of a hybrid aluminum-silicon reagent to intensify the process of purification and removal of residual phosphorus ions. It should be noted that, as mentioned above, the use of the reagent FCS-9 without activating additives does not lead to a noticeable decrease in water hardness.

3.1.2 Results of re-treatment of seawater. After reducing the overall hardness of seawater, a hybrid aluminum silicon reagent FCS-9 can be reused to reduce the concentration of various impurities, including toxic ones. Based on the results of preliminary studies conducted with samples of seawater, an optimal dosage was chosen to be 60 mg/l (in terms of aluminum oxide) FCS-9. At this dosage, the pH value was 6.0. The content of residual aluminum and iron in the treated seawater did not exceed the level of maximum permissible concentration. With the addition of large doses of the reagent (> 60 mg/l), the residual aluminum and iron content in the treated water significantly increased and exceeded the MPC level. Table 2 presents the results of reagent treatment of the softened seawater of the Black Sea using FCS-9 reagent.

Table 2. Results of reagent treatment of the Black Sea water.

Parameter	Initial water	Water after reagent treatment	MPC
Total mineralization, g/l	17,0	15,0	-
pH	6,42	6,00	6,5-9,5
Boron, mg/l	0,2	0,15	1,0
Bromine, mg/l	0,25	0,2	-
Fluoride, mg/l	1,0	0,5	1,5
Iodine, mg/l	0,07	0,05	-

It is established that during reagent treatment the total mineralization of water decreases by 2 g/l, which may be associated with the removal of a part of the insoluble salts. The content of toxic ions in the test sample of the Black Sea water does not exceed the MPC level. However, reagent treatment using FCS-9 reagent tends to reduce the concentration of toxic ions in seawater.

3.2. Results of reagent treatment of model waters simulating the composition of recycled waters

When using the reagent method of recycled water treatment, it is necessary to calculate the amount of reagent required for water treatment of different composition. For accurate prediction of the use of these or other reagents, the calculated doses are established on the basis of data obtained as a result of trial water treatment in laboratory conditions (trial coagulation), taking into account the permissible residual doses of reagents in water, established by the standards [4].

The results of experiments on primary treatment with hybrid aluminum-silicon reagent FCS-9 model waters with total iron, copper and petroleum products are shown in figure 1.

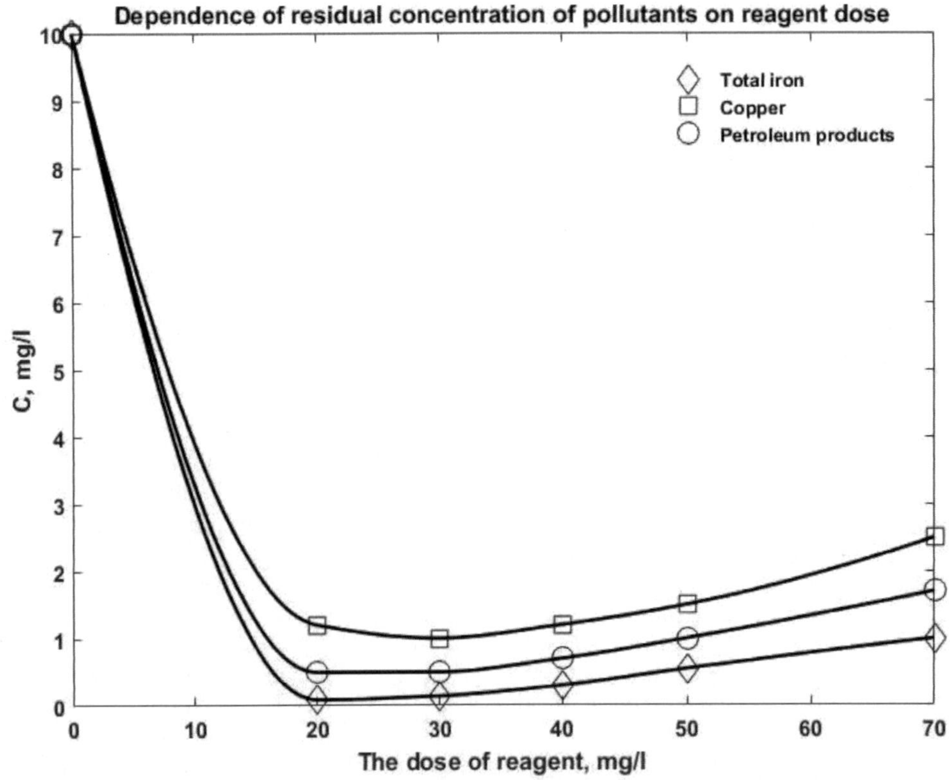

Figure 1. Dependence of the residual concentration of model pollutants on the reagent dose

It was found that the optimal dose of FCS-9 for effective water purification from total iron is a dose of 20-30 mg/l (figure 1). Wherein the degree of purification reaches values over 99%. The optimal dose for water purification from copper and petroleum products is also 20-30 mg/l. Wherein the concentration of copper is reduced by 90%, and the concentration of oil products by 95%.

The minimum value of residual aluminum in treated water is observed at doses of FCS-9 20-40 mg/l. In this case, the amount of residual aluminum in the treated water is 5-10 times lower than the MPC norm, equal to 0.2 mg/l. Corresponding to these doses, pH values are in the range of 6.0-8.0.

Figure 2 shows a graph of the time dependence of the suspended solids concentration. It can be seen that the highest degree of clarification is achieved in the case of the reagent FCS-9. Already 10 minutes after the introduction of the reagent, the suspended solids concentration was less than 40 mg/l, while without the use of the reagent, the concentration was about 60 mg/l. This fact testifies to the high efficiency of the use of the FCS-9 reagent as an intensifier of the clarification process of clay-containing waters.

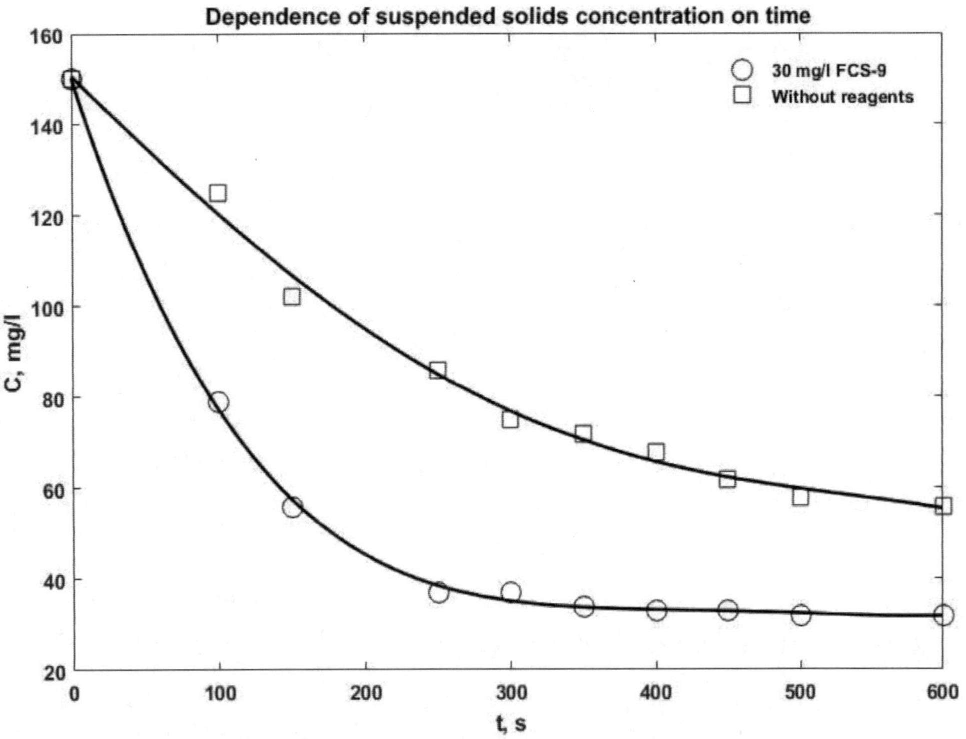

Figure 2. Dependence of suspended solids concentration on time

It should be noted that after preliminary reduction of rigidity, re-treatment of seawater with reagent FCS-9 and its purification from such pollutants as iron, copper, petroleum products, suspended solids, does not differ from the treatment of recycled water containing these pollutants. In this regard, the FCS-9 reagent can also be used to purify seawater from these pollutants, after preliminary softening.

4. Development of the technological scheme of the purification and pre-treatment system

The basic technological scheme of the purification system and the pre-treatment of sea and recycled water (Figure 3) includes the following main nodes:

- softening module;
- reagent processing module [5];
- a flotation phase separation module, equipped with a blowing and collecting device for foam, zones of formation, sedimentation and removal of sediment, a unit for the collection of purified water, with a total residence time of water in the flotator of at least 10 minutes;
- fine cleaning module on SHS filters;
- drainage and sedimentation module.

The main parts of the purification and pre-treatment system are the reagent processing module and the flotation phase separation module (flotator). Wherein there may be several flotators. Flotators are located directly in the water and are divided into different groups. In each group of flotators it is possible to obtain waters of different quality (depending on requirements). This method of location of the flotators was introduced during the development of technology for surface water treatment facilities at the Moscow International Business Centre (MIBC), also known as "Moscow City", with a capacity of 5000 m^3/h.

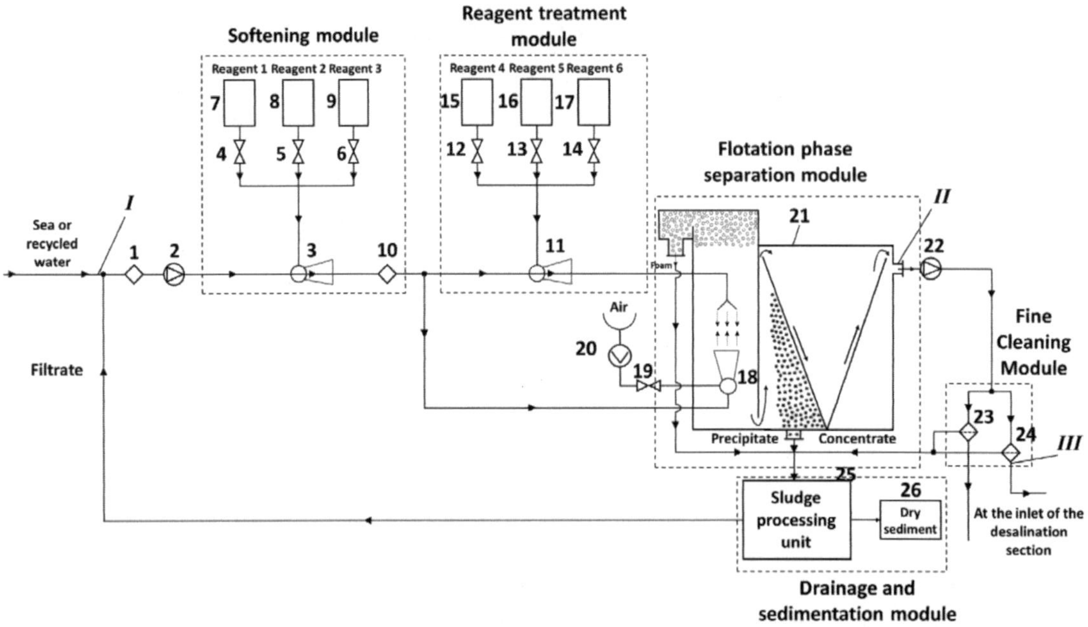

Figure 3. Technological scheme of the purification and the pre-treatment system

The initial seawater is fed through the pre-coarse filter 1 by means of the pump 2 to the softening module, where it is softened using three reagents 7, 8, 9 (soda solution, sodium orthophosphate and FCS-9 solution). Reagents are fed by means of regulating valves 4, 5, 6 and ejector 3. After softening, water through the additional filter 10 enters the module of reagent treatment and flotation phase separation module in order to release fine and dissolved impurities, including toxic ones. In the reagent treatment module, water is treated with FCS-9 reagent and, if necessary, by activating additives (containers 15, 16, 17). The supply of reagents and additives is carried out by means of regulating valves 12, 13, 14 and an ejector 11. In the flotation phase separation module, the flow of reagents with water is sent to the reaction chamber of the flotation unit 21, where it freely flows and collides with the oppositely directed counterflow of water enriched, by means of the water-air ejector 18, with air. The air flow is controlled by the regulating valve 19 and the rotameter 20. Thus, effective mixing of streams takes place in the reaction chamber of the flotator 21, favoring the rapid distribution of reagents throughout the entire volume of treated water and the formation of coagulated flakes and sediments, some of which are removed at the top and bottom of the flotation cell, and clarified water enters further filtration. In the fine cleaning module, water is supplied by pump 22 and is filtered on SHS filters 23 and 24, operating on the principle of filtration from the tangential flow. Here, the water is divided into two streams: water that does not pass through the filter elements and contains the bulk of the impurities (concentrate) retained by the filter element and the filtered water. Water flows that do not pass through the filter elements are sent to the sludge processing unit 25, and the filtered water flows enters the inlet of the desalination section.

The flotation sludge (foam) of the flotation removal zone of the flotation reactor's chamber and the flocculent sediment from the sludge removal zone also enter to the sludge processing unit 25. In the sludge processing unit 25, the sludge is dewatered and compacted into the dry sediment 26, and the liberated water is sent to the head of the system before the coarse filter (filtrate).

Recycled waters formed as a result of water users activities enter the purification stage, where they are treated in a similar way and can be further used for various agricultural and other purposes.

IOP Conf. Series: Journal of Physics: Conf. Series **1099** (2018) 012032 doi:10.1088/1742-6596/1099/1/012032

4.1. Results of reagent treatment of recycled (waste) waters

A study of the efficiency of treatment of recycled water containing all types of mentioned model pollutants according to the technological scheme of the purification and pre-treatment system (without a softening module) was carried out on an experimental stand using the example of real waste (recycled) waters of a non-ferrous metallurgy plant. The injection inlet of the mixing ejector was fed a dose of 30 mg/l of the reagent FCS-9 in terms of aluminum oxide. Sampling of processed water was provided for water analysis at different stages of treatment. Sampling zones are indicated in figure 3:

I – initial contaminated water sample;
II – water sample after reagent treatment;
III – filtrate sample after filtration module

Table 3 shows the test results.

Table 3. Results of reagent treatment of recycled (waste) water.

Parameter	Unit of measurement	Maximum permissible concentration [4]	Water sampling zone		
			I	***II***	***III***
			Initial contaminated water	Water after reagent treatment	Water after the filtration module
pH	pH units	**6,5-9,5**	6,91	6,63	7,30
Total mineralization	mg/l	-	2536	975	349
Total hardness	ppm	-	600	390	225
Petroleum products	mg/l	-	>20	0,8	0,05
Total iron	mg/l	**0,2**	6,0	0,1	0,01
Manganese	mg/l	**0,05**	5,0	0,1	0,01
Copper	mg/l	**2,0**	28,0	5,4	0,2
Aluminum	mg/l	**0,2**	0,2	0,05	0,01

Based on the data provided in table 3, it can be concluded that the water parameters obtained after treatment at the experimental stand meet the standards [4]. On the basis of this, it can be concluded that the water obtained after purification can be used for various agricultural, domestic and technological needs.

5. Conclusion

The process of reagent treatment of seawater in the Black Sea has been studied, the concentrations of the main cations and anions contained in the Black Sea water have been determined. It has been established that after a preliminary softening and re-treatment using the FCS-9 hybrid aluminum-silicon reagent, the total mineralization of water decreases by 2 g/l, which may be due to the removal of sparingly soluble salts. The content of toxic ions in the waters of the Black Sea practically does not exceed the MPC level. However, in the reagent treatment using the FCS-9 hybrid reagent, the concentration of toxic ions in seawater tends to decrease.

It was found that the optimal dose of FCS-9 for effective purification of model water with total iron is 20-30 mg/l. At the same time, the purification degree reaches values over 99%. The optimal dose for water purification from copper and petroleum products is also 20-30 mg/l. The concentration of copper is reduced by 90%, and the concentration of petroleum products is reduced by 95%. In addition, an example of the high efficiency of the FCS-9 reagent as an intensifier of the clarification of clay-containing waters is shown.

The technological scheme and the experimental stand of the purification system and the pre-treatment of sea and recycled water have been developed. It has been established that with the reagent treatment of real recycled waters, the concentration of heavy metals and petroleum products is reduced by 99%, and the residual aluminum content in purified water is an order of magnitude lower than in the original one.

Acknowledgements
The work was financially supported by the Ministry of Education and Science, agreement № 14.575.21.0159. Unique project identifier RFMEFI57517X0159.

References

[1] Feklistov D Yu, Alexandrov R A, Laguntsov N I and Kurchatov I M 2017 *XXXIII Int. Conf. Research and development* vol 1 № 11 (33) (Moscow: Eurasian Scientific Association) pp 60-62
[2] Laguntsov N, Kim V, Krasnov P, Neshchimenko Yu, Borisenko M, Odintsov A and Fartunin A 2006 *J. Ars Sep. Acta* **4** 57
[3] Alekssev L S and Gladkov V A 1994 *Improvement of soft water quality* (Moscow: Stroyizdat)
[4] Council directive 98/83/EC of 3 November 1998 on the quality of water intended for human consumption. Official Journal L 330 , 05/12/1998 P. 0032 – 0054.
[5] Alexandrov R A, Laguntsov N I and Kurchatov I M 2015 *J. Phys. Proc.* **72** 103

Hybrid solutions of compressed gas drying

I M Kurchatov[1], A A Tishin[1,2], M D Karaseva[1,2], V N Gurkin[1,2]

[1]Public joint-stock company "Aquaservice", Kashirskoe sh. 3/2 st 4, Moscow 115230 Russia
[2]National Research Nuclear University MEPhI, NESPI, Kashirskoe sh. 31, Moscow 115409 Russia

E-mail: aquaserv@mail.ru

Abstract. In this paper, the schemes of membrane-refrigeration drying and vapor removal from compressed gas are considered for use in industry. The efficiency analysis of the hybrid membrane-sorption technologies usage for gas drying was carried out. The scheme of the hybrid steam separator with two recirculation circuits is proposed, and this scheme consists from three stages. The first stage of the scheme is represented by PSA block, consisted from three adsorbers, which are filled with a solid zeolite. These adsorbers are good in water vapor absorption. The second stage is a membrane module, which is fed by waste flow of the PSA stage. Herein membrane material has a significant value of the permeability for water vapor. The third stage – refrigerator is needed for condensing of water vapor from a gas flow, permeated through membrane. Water vapor is condensed and removed from the system while dry gas is returned to the system. According to proposed scheme of hybrid system organization, deep drying can be performed without pressure loss and without change in the components of the flow.

1. Introduction

Dry air has found its application in many branches of industry and agriculture, in exact mechanical engineering, pharmaceuticals and medicine, aviation and shipbuilding. Currently on an industrial scale, the drying air is carried out by two main methods: adsorption and refrigeration. The nature of the refrigeration method is to cool the air to the required dew point. However, this method is quite energy-intensive and does not allow reaching low dew points. The most popular drying method is the adsorption method based on cyclic processes. The essence of the method lies in the alternation of adsorption and desorption processes carried out at different temperatures or pressures. The most widely used method of PSA (Pressure Swing Adsorption), proposed by C. W. Skarstrom [1]. A distinctive feature of this method is that adsorption and desorption are carried out at a constant temperature, and the partial pressures of the adsorbed components with adsorption are greater than when desorbed. However, the PSA method has a significant drawback, for the regeneration of the sorbent, it is necessary to use either additional pumps or a significant part of the product flow that is returned to the system to restore the properties of the sorbent. A lot of papers have been devoted to the problem of returning the flow to the regeneration of the sorbent [2, 3].

In recent years, special attention has been paid to the membrane technology of gas drying. One of the advantages of using membrane technology in the drying of compressed gases is the ability to extract readily penetrating impurities simultaneously with the removing of water vapor, as well as the separation of gas mixtures into components. An example is the task of preparing natural gas for

IOP Conf. Series: Journal of Physics: Conf. Series **1099** (2018) 012033 doi:10.1088/1742-6596/1099/1/012033

liquefied [4]. In addition, such schemes are used to purify natural gas from CO2 proposed in the patent [5].

2. Statement of the problem

Practical interest in the field of compressed gas drying in the coming years will be the development and creation of drying systems based on hybrid membrane-adsorption systems, which in the future should reduce the energy consumption of compressed gas drying systems. One of the variants of membranes that are supposed to be used in such systems of drying is PVTMS (vinyltrimethylsilane) membrane having a rather high permeability for water vapor.

As examples of a hybrid drying system using a membrane vapor separator, one can consider the schemes of open and closed single-module steam separator and a closed two-module system.

Figure 2 shows the schematic diagram of a single-module membrane steam separator of open type. The scheme works as follows. The input flow F by compressor 1 is compressed and fed to the membrane module 2. A pressure drop is created by pump 3 connected to the permeate of the membrane module. Freely penetrating water vapor permeates through the membrane material and enters the refrigerator 4, where condensation and moisture removal take place - flow C. Dry air that doesn't pass through the membrane (flow P) is supplied to the consumer.

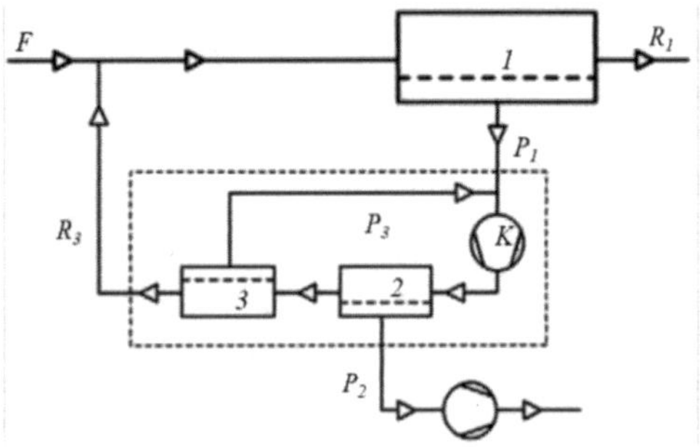

Figure 1. Two-stage two-circuit counter flow cascade [4].

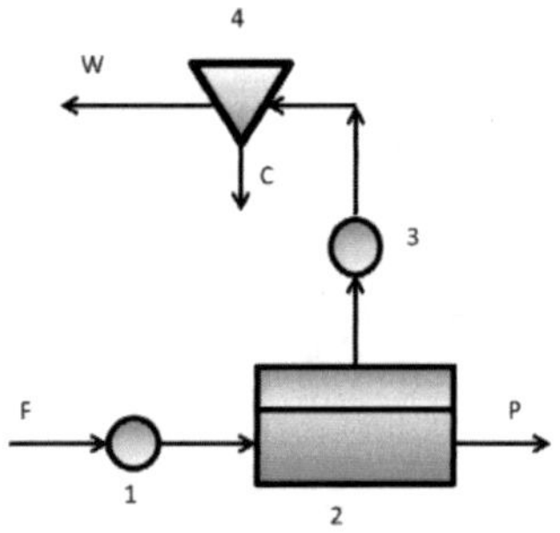

Figure 2. Schematic diagram of a single-module membrane steam separator of open type.

The main disadvantage of open-type systems is the presence of the flow W coming out of the dryer, it is often thrown out, which reduces the energy efficiency of the system as a whole. However, the concentration of water vapor in such a flow after passing through the refrigerator can be significantly lower than in the flow F. Therefore, to preserve the flow W in the system and increase the overall efficiency of the system, closed circuits are more often used. The schematic diagram of the closed-type single-module membrane steam separator is shown in figure 3. Its peculiarity is that the flow W is not output from the system, but is returned to the input of compressor 1. This solution significantly improves the efficiency of the drying process with a single membrane steam separator. Additionally, in a system organized such method in the future, it is possible to get rid of the pump 3 using a compressor 1 to create a differential pressure.

To increase the efficiency and depth of drying, it is also possible to use the scheme of a two-module closed-type steam separator with two condensers, shown in Fig. 4. In such a drying system, the dehydrated gas in the first membrane module enters the second membrane module 4 where deeper gas drying takes place. Thus, using two consecutive membrane separators it is possible to achieve deeper degrees of drying. In addition, the system is organized in such a way that the permeates of both membrane elements are connected to the input of the compressor, which ensures a differential pressure them. However, due to the fact that the permeate of the membrane separator 3 is more moist with respect to the permeate of the membrane separator 4, it is necessary to install a second capacitor

IOP Conf. Series: Journal of Physics: Conf. Series **1099** (2018) 012033 doi:10.1088/1742-6596/1099/1/012033

thereon. Condenser 1 allows pre-drying of compressed air and prevents condensed moisture from entering the membrane module.

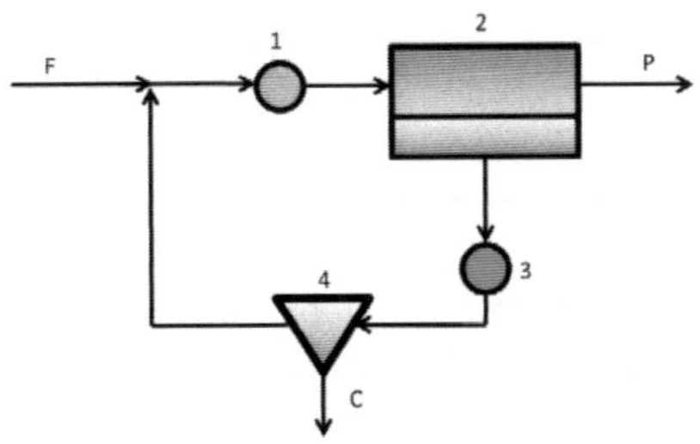

Figure 3. Schematic diagram of a single-module closed-type membrane steam separator

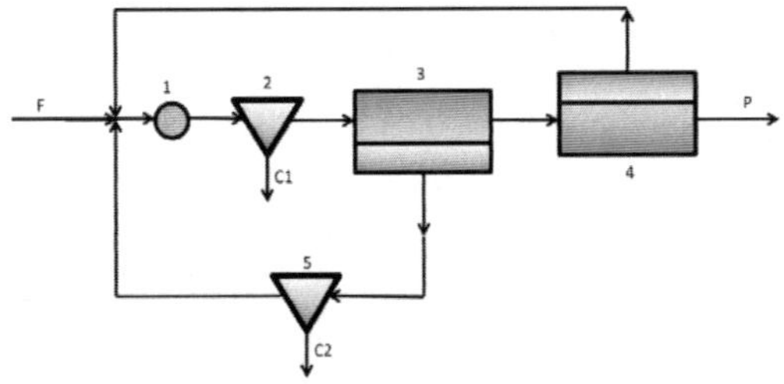

Figure 4. Schematic diagram of a two-module closed steam trap with two condensers.

However, the membrane direction is not the only promising direction in the field of gas drying. In this paper propose a scheme of a hybrid system for drying. The essence of hybrid methods is the simultaneous use of several methods of drying to improve the energy efficiency of the process, its productivity and purity of the product. In this paper, we consider a hybrid system consisting of a refrigerator, adsorption and membrane drying blocks.

The proposed hybrid system consists of two drying stages. At the first stage of the system, the three-adsorption PSA system is used as the main drying block. The second stage used a closed-type membrane steam separator. The main feature of the proposed hybrid system is the use of the flow the dried by membrane steam separator to regenerate the sorbent on the PSA stage of the system. This scheme allows solving the problem of the return of the product flow of the PSA system for the regeneration of the sorbent.

3. Results and discussion

The first stage of the scheme is represented by a PSA block consisting of three adsorbers filled with a solid zeolite, which absorbs water vapor well. The second stage of the scheme is a closed-type membrane steam separator, to which the wet flow of the PSA stage is fed. The membrane material of which the membrane module is composed, has a high water vapor permeability. A schematic diagram of a hybrid membrane-adsorption drying system is shown in Figure 5.

The scheme is designed for drying compressed gases by means of sorption and membrane technologies. The main working element of the scheme is the sorption block, which is represented by three adsorbers filled with a granular adsorbent. The work on gas drying in adsorbers is performed alternately, thereby it is possible to achieve a constant in time and pressure of the flow. During the drying of the gas with one adsorber, the others are on the step of regeneration and raising the pressure.

Dehumidification is carried out under high pressure, so the pressure rises in the adsorber before passing through it moist air. Adsorption of moisture from the gas stream is carried out during its transmission through the layer of regenerated sorbent. After the sorbent consumes its sorption resource, the gas on the drying starts to be fed to a pre-prepared adsorber, in which the operating pressure is set, and the adsorbent has passed the stage of regeneration.

The sorption stage of the system, consisting of three adsorption columns, operates in a cyclic mode. In one cycle of adsorber passes three steps: filling, production (displacement) and discharge (regeneration). A similar organization of work for the PSA system is described in detail in [6] with respect to the separation of air into components.

The step of regeneration of the spent sorbent is carried out in several stages. The first stage of regeneration is to reduce the pressure in the adsorber by discharging gas into the membrane vapor separator. In the second stage, the adsorber is purged with the dried gas supplied from the membrane module. During purging of the spent sorbent with dry air, moisture is desorbed from its surface and is carried away with a dry stream making it wet.

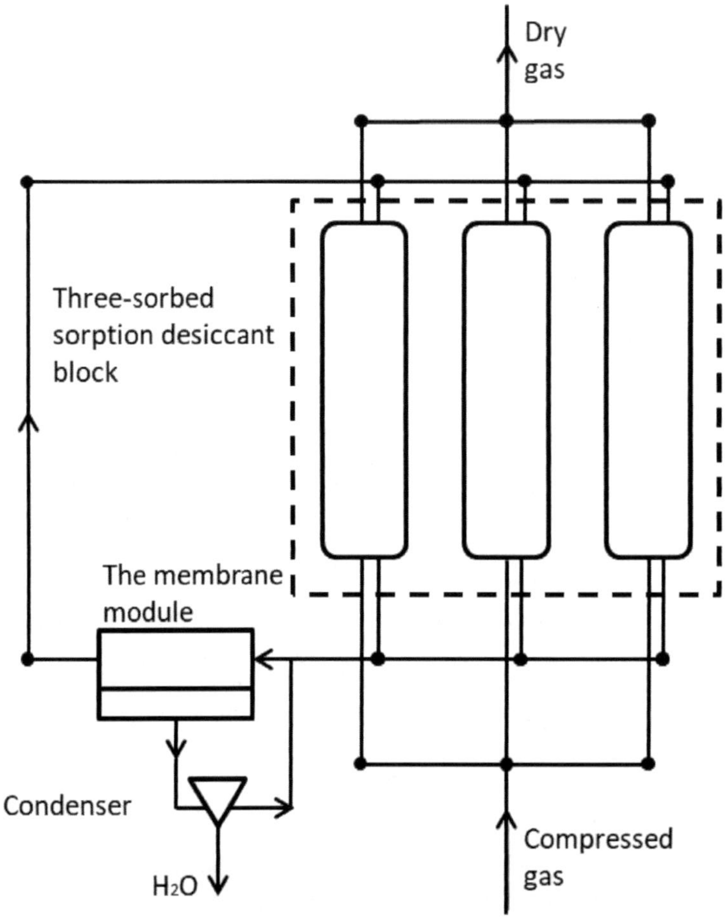

Figure 5. Schematic diagram of a hybrid membrane-adsorption system for drying compressed gas.

The wet flow after the regeneration step of the sorbent is sent to the membrane module where it is dried by the membrane method. The gas and moisture that has passed through the membrane enters the condenser where the gas and moisture are separated. Moisture is removed from the system, and again the dehumidified gas returns to the feed stream of the membrane module. Thus, a closed regeneration loop is organized in the system.

Through the organization of the two-circuit system, most of the product gas is supplied to the consumer. There is practically no waste part of the gas in the system.

Compressed gas is supplied to the block of sorption gas drying. The work of the sorption block is based on the work of the cycle with variable pressure. At the first step of the cyclorama operation, the adsorber is filled up to the working pressure, at the second step the dry gas production stage passes, and in the third step the sorbent properties in the adsorber are restored. Simultaneously, all three adsorbers pass through different stages of the cycloramas, thus achieving a constant product flow of the dried gas of the PSA stage.

The first recycling circuit combines the sorption (PSA) and membrane stage as follows. The wet flow of PSA stage is fed to the membrane module. The non-permeated air returns to the PSA stage to regenerate the zeolite. The second circuit combines a membrane module and a condenser in which the flow of moist air is cooled. Water vapor condenses and leaves the system, and dry gas returns to the system.

4. Conclusions

In the paper, schemes of membrane-refrigeration drying and removal of water vapor from compressed gas for use in industry are considered. A scheme of a hybrid steam separator with two recirculation circuits is proposed, consisting of three stages.

The first stage of the scheme is represented by a PSA block consisting of three adsorbers filled with a solid zeolite, which absorbs water vapor well. The second stage of the circuit is the membrane module, to which the wet flow of the PSA stage is fed, the membrane material of which the membrane module is composed, has a high water vapor permeability value. The third stage - the condenser serves to condense water vapor from the gas flow penetrated through the membrane.

The first recirculation loop combines the sorption (PSA) and membrane stage as follows. The wet flow of the PSA stage is fed to the membrane module. The non-permeated air returns to the PSA stage to regenerate the zeolite. The second circuit combines a membrane module and a condenser in which the flow of moist air is cooled. Water vapor condenses and leaves the system, and dry gas returns to the system.

Through to the proposed scheme for organizing a hybrid system, it is possible to carry out deep dehydration of the compressed gas without loss of pressure in the flow and loss of a part of the product stream that is usually returned to regenerate the sorbent in the PSA drying method.

Acknowledgements

The work is financially supported by the Ministry of Education and Science, agreement №14.576.21.0097. Unique project identifier RFMEFI57617X0097.

References

[1] Skarstrom C. W. 1960 *US Pat.* 2944627.
[2] Smith A. R., Klosek J. 2001 *Fuel processing technology* **70** 2 115-134.
[3] Franklin E. et al. *AIChE Journal* **64** 2 406-415.
[4] Laguntsov N. I. et al. 2017 *Petrol. Chem.* **57** 2 132-138.
[5] T. Cnop, S.-J. Doong, D. Farr, and G. K., Xomeritakis 2012 *WO Pat.* 2012050816.
[6] Tishin A.A., Gurkin V.N., Laguntsov N.I., Kurchatov I.M. 2018 *Petrol. Chem.* **58** 4 338-345.

THE EFFICIENCY OF THE COUNTERCURRENT BLOWDOWN IN THE MEMBRANE-SORPTION SYSTEM WITH THREE ADSORBERS

V N Gurkin[1,2], N I Laguntsov[1,2], I E Skuridin[1,2], A A Tishin[1,2]

[1]National Research Nuclear University MEPhI, NESPI, Kashirskoe sh. 31, Moscow 115409 Russia
[2]Public joint-stock company "Aquaservice", Kashirskoe sh. 3/2 st 4, Moscow, 115230 Russia

E -mail:Wladimir0594@yandex.ru, aquaserv@mail.ru, NILaguntsov@mephi.ru

Abstract. Hybrid membrane-sorption systems are used to solve the problems of the gas mixtures separation, as air enrichment with oxygen or compressed air drying. Investigation of the waste flow of the PSA stage showed that the flow contains a significant amount of the target component. Thus, it is possible to increase the efficiency of the plant without additional energy costs by implementing a scheme, in which waste flow is used for expulsion. Considered scheme with the sorbent regeneration by the method of blowdown is applicable for various gas mixtures separation, in which it is necessary to separate out the low-sorption component. A demanded application of this scheme can be the solution of the air drying problem.

1. Introduction

The gas drying is one of the actual gas separation problems. In many technological procedures operating on the compressed gas, for example, wherein air is divided into nitrogen and oxygen, moisture adversely affects the processes occurring in them and can disserve the system. To solve these problems, membrane and adsorption technologies are often used. These technologies allow obtaining the necessary gases directly at the point of use. Due to the continuous development of the membrane and sorption technologies, the devices created on their basis become more mobile and compact, which expands the field of their application.

The combinations of different methods have been applied for better results. The investigations of the hybrid systems [1-3] has shown that their use is more effective than the use of individual methods. The combination of several methods that operate on different physical principles and in different technological modes has difficulties, which, as a rule, are connected with the work coordination of the individual modules and the choice of the optimal operating mode.

A scheme of a membrane-adsorption plant for air enrichment with oxygen is described in [4]. The system is designed for the air production enriched with oxygen. Such an installation can be used as a feed supply for medical oxygen equipment. However, the system can be improved to solve other problems, for example when choosing the necessary sorbent, it can be used for the drying of wet compressed gas.

The use of a adsorption unit with three adsorbers allows to provide the membrane module with a constant feed flow. This is achieved due to the coincidence of the number of stages, the duration of which is approximately the same, in the cyclogram and adsorption columns. At any one time, one of the columns is at the stage of producing a stream enriched in oxygen. The other two columns are in the stages of filling and regeneration.

2. Main body

Each adsorption column operates in a specific cycle, which includes several stages. In the simplest version, the cycle consists of two stages: the production of the target component and regeneration. The effectiveness of the production stage largely depends on how effectively the regeneration stage has passed. Several methods for the adsorbent regenerating are described in [5]. Each of these methods has its own peculiarities of application. So, regeneration due to the heating of the adsorbent is very effective, but significantly increases the cycle times, since the process of heating and then cooling of the sorbent takes a long time. Regeneration by pressure reducing can be conducted much faster.

The regeneration by pressure change underlies plants operating by the method of pressure swing adsorption (PSA) [6]. The essence of this method is in the rapidly changing stages of adsorption-desorption. The adsorption stage proceeds at elevated pressure, and the desorption stage under reduced pressure. The faster the adsorption stage passes, the higher the concentration of the desired component can be obtained. However, at short adsorption cycles and to the end of the production step the residual concentration of the target component is still high in a gas phase of adsorber. During the regeneration stage, this gas enriched with the target component is dropped into the atmosphere.

Figure 1 shows the graphs of the concentration dependencies in the vented flow on the pressure in the adsorber. Each curve corresponds to a different preliminary saturation of the sorbent with oxygen. In order to achieve different initial conditions, the sorbent layer was blown with air with different concentrations of oxygen until equilibrium state was established, i.e. up to the time when the flow rate and the composition of the gas supplied to the adsorber will be equal to the flow and to the gas composition leaving the adsorber. After that, the work of the adsorption column in the PSA plant was simulated. For some time, air was blown into the adsorber under operating pressure for oxygen production.

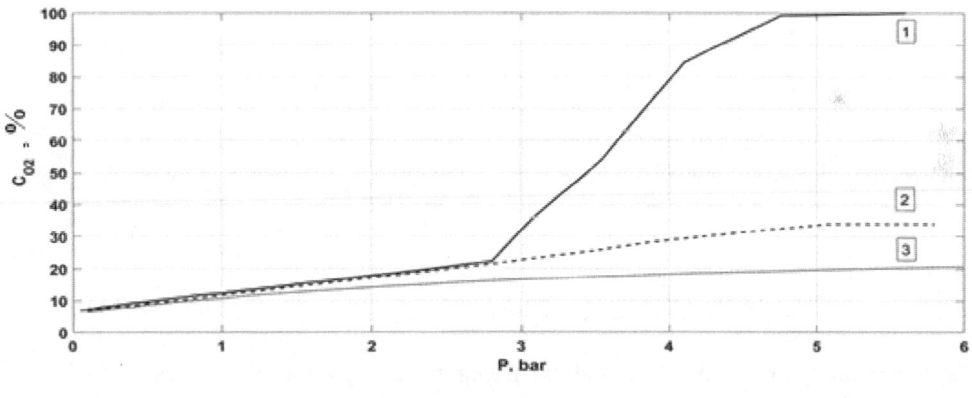

Figure 1. The graph of the concentration dependencies in the vented flow on the pressure in the adsorber

From the graphs in Figure 1, it can be seen that desorption has several distinctive stages. Thus, in the first stage, the concentration of the weakly-sorbed component (oxygen) remains constant and equals to the equilibrium concentration established before the beginning of the displacement stage. In the second stage, the concentration of the weakly-sorbed component is sharply reduced. This happens due to the active desorption process of the highly-sorbed component (nitrogen). The desorbed nitrogen is carried away with the gas flow created by the gradient of concentration. In the third section, the concentration of the weakly sorbed component also decreases, but this section has the peculiarity. The desorption curves are close to each other, independently of the initial equilibrium concentration.

To improve the efficiency of the system, various methods are used. So often the stage of the pressure equalizing between the adsorber, which has passed the regeneration stage by the pressure drop, and the adsorber, in which the production stage has been already completed, is used.

Equalization of the pressure allows reducing energy costs associated with the pressure growth in the adsorber [7]. In addition, as can be seen from Fig. 1, the adsorber will be filled with a gas enriched with the target component.

It is known that in order to achieve high concentrations of the target component, the stage of countercurrent expulsion has the great importance, which allows to regenerate the adsorbent deeper. The expulsion of the sorbent is carried out by a gas enriched with the target component, in the direction opposite to the adsorption flow. In this case, the regeneration takes place due to a decrease in the partial pressure of the highly-sorbed gas.

The objectives of the countercurrent expulsion are:

- The desorption of the highly-sorbed component and its removal from the adsorber with the gas flow;
- Creation of the most profitable for adsorption an initial concentration distribution between phases.

For investigation of the expulsion process, 100% oxygen was supplied to the input of the adsorber at a different flow rate. The output of the adsorber was connected with the atmosphere, while the main parameters of the output flow were recorded (concentration and the flow rate).

Figure 2 shows the time variation of the oxygen concentration in the output gas at different expulsion velocity.

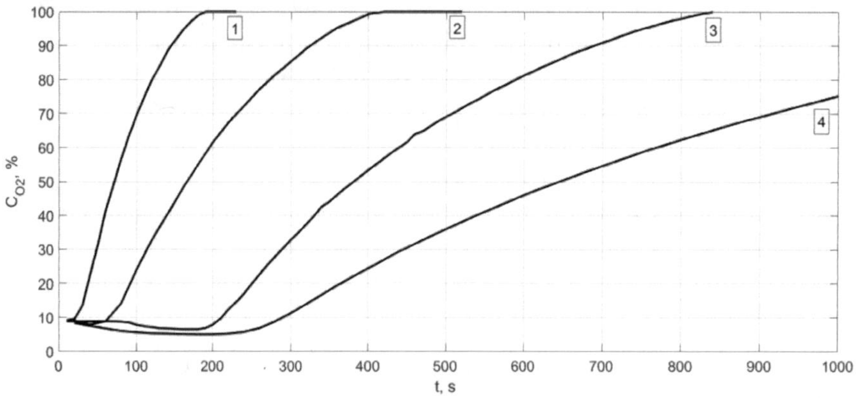

Figure 2. Oxygen concentration dependence in the outlet gas on the time at the different expulsion velocity:
1 – 2.5 l/min; 2 – 2 l/min; 3 – 1.7 l/min; 4 – 1.2 l/min;

The presaturation of the adsorbent with the nitrogen was the same. It is corresponded to the value at which the adsorbed nitrogen was in equilibrium state with the gas phase at a pressure of 1 atm. and an oxygen concentration of 8%. This composition of the gas phase is achieved during desorption by reducing the pressure to atmospheric (Figure 1).

On the resulting graphs, two characteristic areas can be distinguished. The first section, where the concentration remains constant, corresponds to the concentration of oxygen in the preceding stage. The presence of this area indicates that the state between the gas phase and certain sorbent layers, including the last in the direction of the gas flow, is close to equilibrium. The second section is characterized by a regular fall in the nitrogen content in the output gas flow.

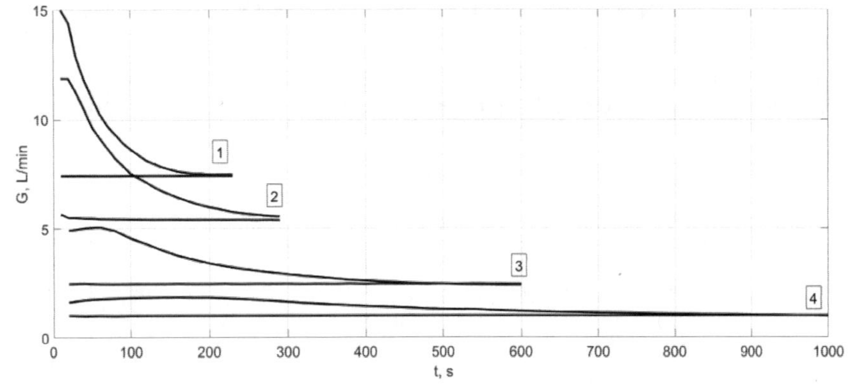

Figure 3. The graphs of the waste and feed flow dependence on time at the different expulsion velocity:
1 – 7.4 l/min; 2 – 5.4 l/min; 3 – 3.5 l/min; 4 – 1 l/min.

As it can be seen from the obtained results, that at the initial time the waste flow rate is significantly higher than the feed flow rate. This occurs due to the fact that the incoming oxygen disturbs the equilibrium state between the solid and gaseous nitrogen phase, as a result the process of nitrogen desorption from the surface of the sorbent starts actively. As the desorption front moves along the sorbent layer, the flow rate of the waste flow approaches to the value of the feed flow. When the desorption front reaches the outlet, the flows become equal, and at the outlet the oxygen concentration reaches the level of 100%.

Experiments, had been carried out at different flow rates, showed that the desorption degree at a constant temperature is approximately independent of the flow velocity and depends only on the amount of the expulsion flow.

3. Conclusions

The characteristics investigation of the sorption unit of the hybrid membrane-sorption system with three adsorbers was carried out. This characteristics are connected with sorbents regeneration.

In the framework of the task, experiments were carried out to investigate the waste flow of the sorption unit at pressure dump in the adsorption-desorption cycle as well as at the expulsion of used sorbent by oxygen flow.

As a result, it was shown that the gas dropped during the desorption stage into the atmosphere is enriched with the target component, so, it is used to fill another adsorber in order to increase the efficiency of the system. However, experiments have shown that the sorbents expulsion is an effective method of sorbents regenerating. Thus, instead of filling the adsorber, the gas flow can be sent to the adsorber expulsion and the part of the nitrogen will be removed from the sorbent layer and the gas phase. This method of regeneration can be used in schemes with three or more adsorbers. Since two adsorbers are involved in the expulsion process, and therefore a two-adsorber scheme will not allow to obtain the constant flow.

Acknowledgements: The work is financially supported by the Ministry of Education and Science, agreement №14.576.21.0097. Unique project identifier RFMEFI57617X0097.

References

[1] Laguntsov N I, Kurchatov I M, Korolev M V 2015 *Petroleum Chemistry* **55** 738-745
[2] Kudinov A N et al 2014 *Theoretical Foundations of Chemical Engineering* **48** 832-836
[3] Kudinov A N, Kurchatov I M, Laguntsov N I 2014 *Theoretical Foundations of Chemical Engineering* **48** 352-359

IOP Conf. Series: Journal of Physics: Conf. Series **1099** (2018) 012034 doi:10.1088/1742-6596/1099/1/012034

[4] Tishin A A, Gurkin V N, Laguntsov N I, Kurchatov I M 2018 *Petroleum Chemistry* **58** 338–345
[5] Zahra M, Jafar T, Masoud M 2008 *World Academy of Sci. Eng. and Technology* **47** 99-103
[6] Sircar S 1988 "Air Fractionation by Adsorption, Separation Science Technology" **23** 2379-2396
[7] Cruz P, Santos J C, Magalhaes F D and Mendes A 2003 *Chem. Eng. Sci.* **58** 3143 – 3158

IOP Conf. Series: Journal of Physics: Conf. Series **1099** (2018) 012035 doi:10.1088/1742-6596/1099/1/012035

Production of Water with Reduced Content of Deuterium for Water Supply System with Desalination Installation

A Yu Smirnov, G A Sulaberidze

National Research Nuclear University MEPhI
31 Kashirskoe Shosse, Moscow 115409, Russian Federation

E-mail: a.y.smirnoff@rambler.ru

Abstract. Industrial methods for obtaining light water have been analyzed in order to find an optimal variant of a separation plant for reducing deuterium content in desalinated sea water for its further use in various applications (domestic needs, medicine, nuclear energy). It is assumed that the installation should ensure the reduction of deuterium content to the level of the cleanest freshwater ponds, working as an integral part of the water supply system based on the nuclear desalination complex.

It was shown that the most promising option for a deuterium content reduction in desalinated water is an installation based on the distillation method. Analysis of the results of the study allows us to formulate proposals and recommendations for the implementation of a subsequent scientific research and feasibility study to assess the key characteristics of nuclear desalination plants and make decisions about their implementation.

1. Introduction

Over the last years the mankind has faced an ever-growing problem involving the deficiency of freshwater suitable for drinking as well as domestic and hygienic applications. Many experts note that clean water becomes more and more significant strategic reserve for the development and safety of particular countries and the mankind in whole. This fact makes them regard the world ocean as a potential source of water. However, the high content of salts in sea water makes it unsuitable for water supply and drinking needs. The solution to this problem is possible with desalination techniques [1, 2]. Under the project being implemented in the RF on creation of a nuclear desalination complex a scientific basis is assumed to be developed for water desalination technique environmentally-friendly and involving the development of a water-supply hybrid system based on the up-to-date low-power nuclear reactors in order to solve the water deficiency problem, mitigate the ecological impact on the environment and provide the water-deficient regions with fresh water. The plant being developed, as compared with the ones already implemented, involves the combination of the desalination complex with a mobile nuclear power plant that may be developed based on the Russian KLT-40S and RITM-200 reactors (designer: JSC AKBM Africantov). The said power plants are designed based on the naval reactors and may be installed on an offshore platform, which ensures their mobility. They can also operate in self-contained mode for a long time (up to 7 years) on single fueling. Thus, the complex being developed is simultaneously both a mobile electric power station capable to feed relatively small communities (up to 100 thousand people) and a desalination plant using not only the electric power generated in the reactor, but also its "exhaust" heat (waste steam heat).

The complex being developed has one more specific feature, i.e. a plant available therein for reducing deuterium content in water. As a matter of fact, the desalinated water thus obtained may be unrestrictedly used for only technical use. It is a problem to use it for drinking due to a higher total deuterium content therein as compared with freshwater ponds. For example, according to the ocean water standard (VSMOW) the deuterium content in ocean water amounts to approximately 155 ppm which is in about 10% excess over the deuterium average content in freshwater. In some experts' opinion, using such water for drinking is not advisable since the results of studies concerning the effect of the deuterium contained in water on vital processes of some animals and insects show that an increase of its content has a negative impact on metabolic processes and immune activity of organisms [3-5].

Thus, to obtain drinking water in the nuclear desalination complex, it is necessary to correct its isotopic composition in order to reduce the deuterium content therein to the average level for freshwater (120-130 ppm) and less.

To reduce the deuterium concentration in desalinated water means to separate the hydrogen isotopes. The solution to this problem demands appropriate separation techniques to be applied. To separate isotopes of the biogenic elements including hydrogen, the so-called two-phase separation methods have gained widespread use [6-10] including distillation, chemical isotope exchange, et al [10]. It should be stressed that the problem to obtain water with a reduced deuterium content (or "light" water) is not typical for the separation of hydrogen isotopes since the major studies in this respect are aimed at heavy water (D_2O) [6-10] to be obtained or cleared from protium and tritium isotopes [11-14]. The said factors impose particular conditions on the process of protium and deuterium separation since, unlike the case of deuterium enrichment and heavy water production, this task implies that the above process runs mainly in the H_2O-HDO, but not in the H_2O-D_2O system, which affects the values of separation coefficients. In addition, in view of the low content of deuterium in water, it is actually about further depletion of water over a given isotope.

To create the nuclear desalination complex, it is necessary to consider the following criteria for choosing the method of water isotopic composition correction for deuterium:

- Using water as working agent
- Possibility to utilize energy of reactor coolant
- Relatively easy implementation and availability of techniques
- Fire and explosion safety
- Possibility to integrate the plant into the water desalination scheme with minimum of additional components.

In this article the industrial methods for separation of hydrogen isotopes have been analyzed in order to find an optimal variant of a separation plant for reducing deuterium content in desalinated sea water It is shown that the water rectification allowing utilizing the exhaust heat of nuclear power plant directly is the most promising method to implement the deuterium separation plant. Specific energy consumption to obtain the product of required quality by this method has been assessed in terms of drinking water supply to a community with a population of 100 thousand people. Aspects of further efforts to create the deuterium separation plant within the nuclear desalination complex have been formulated.

2. Choice of Hydrogen Isotope Separation Method

The problem of the hydrogen isotope separation occurred to appear at different times and for different purposes. For instance, at the rise of nuclear power industry and military programs the main interest was to obtain heavy water, but several decades later the major problem became to clear hydrogen isotope mixtures from tritium [11-14]. The versatility of the problems gave rise to various techniques for their solutions. To separate the hydrogen isotopes at industrial level, such methods have been applied as water electrolysis [11], rectification of various hydrogen-containing compounds (ammonia, water, molecular hydrogen) [6-11], chemical isotope exchange [11]. Hereinafter find a brief critical analysis of the hydrogen isotope separation methods in

IOP Conf. Series: Journal of Physics: Conf. Series **1099** (2018) 012035 doi:10.1088/1742-6596/1099/1/012035

terms of their applicability to implement the deuterium separation plant within the nuclear desalination complex.

2.1 Electrolytic Method

The method means the following: at water decomposition into hydrogen and oxygen under electric current the isotope separation effect appears. To multiply the single separation effect, the electrolyzers are connected in series thereby implementing the cascading principle.

The main advantage of electrolysis, as compared with other methods, are rather high (1.5-10) separation coefficients. However, this gain is negated several times (or even by orders) due to high energy consumption of the traditional electrolysis relative to other methods of hydrogen isotope separation, e.g. when heavy water is obtained [11]. One more major shortcoming of the electrolysis is its fire and explosion hazard due to the generation of hydrogen and oxygen in one apparatus, which imposes strict demands on electrolyzer design.

It should also be stressed that the electrolytic method can be implemented subject to direct use of electric power and thus makes it difficult to utilize the waste steam energy in turbine of the mobile NPP. This means that such installation will be capable of full power operation during only power plant "off-peak" hours. However, this may prove to be insufficient to ensure the required water volumes with reduced deuterium content.

Summarizing the above, it should be noted that this method is not advisable to be considered as a key one to obtain water with reduced deuterium content within the nuclear desalination complex for the following major reasons:

- High energy consumption
- Failure to directly use nuclear reactor exhaust heat.

2.2 Chemical Isotope Exchange Method

Chemical isotope exchange (CIE) method [10, 11] is another method applicable for the hydrogen isotope separation. The method is based on the fact that in a system of two chemically not interacting substances an isotope exchange reaction may run with equilibrium constant differing from the value corresponding to distribution of isotopes with equal probability between the molecules involved in the reaction [10]. The process of separation (isotope exchange) may run in the water-hydrogen system:

$$H_2O + HD \leftrightarrow HDO + H_2. \qquad (1)$$

The characteristic values of separation coefficients for the chemical isotope exchange reactions in such system are approximately 3-3.5 [11], therefore the CIE processes have become dominant in the up-to-date techniques of heavy water production.

In practice the CIE may be implemented in column-type separation device. Reactions of chemical isotope exchange between molecules incessantly run in the volume of the column. Flow conversion nodes are the major parts of the vessel (column): upper flow node and lower flow node. The function of the lower flow node is to quantitatively convert the total liquid flow leaving the column into a gaseous working agent of the chemical exchange system (water into hydrogen for the above reaction). In the upper flow node this process needs to be conducted in the opposite direction (hydrogen into water).

For the water-hydrogen system the flow conversion nodes can be implemented as follows: the lower flow node is meant for water electrolysis, while the upper one – for burning or hydrogen catalytic oxidation [11]. This system has a "weak point" which is the necessity to use a hydrogen-activating catalyst capable of working effectively in a system with liquid water [11].

Thus, despite the proved effectiveness of the CIE method for hydrogen isotope separation, this method is not the best choice to implement the deuterium separation plant in the desalination complex. The major reasons: failure to directly use nuclear reactor exhaust heat as well as additional catalysts to be used for the separation in the water-hydrogen system.

2.3 Rectification Method

The rectification method was widely used for the hydrogen isotope separation at different times [6-11]. In terms of this method the separation occurs in the volume of a column-type separation device at redistribution of isotopes between liquid and gaseous phases of the working substance, which may be hydrogen, water or ammonia [6-11]. In the nuclear desalination complex water is the only acceptable option. In this case the following reaction is running [11]:

$$H_2O_{liquid} + HDO_{\text{steam}} \leftrightarrow HDO_{\text{liquid}} + H_2O_{\text{steam}}. \quad (2)$$

At water rectification protium and deuterium concentrate in gaseous and liquid phases, respectively. Thus, there is a counterflow of two phases (liquid-steam) occurring in the column with the incessant isotope exchange between them.

Among all the above methods of hydrogen isotope separation the water rectification is least effective as for separation coefficients. Thus, depending on separation process conditions (operating pressures and temperatures) the characteristic values of protium-deuterium separation coefficients may fall in a range of 1.022-1.069 [10]. However, unlike the previous methods, the water rectification allows direct utilization of the turbine waste steam energy to heat up the water evaporators.

For contact interface buildup and, hence, separation process intensification there are various contact devices used in columns. At water rectification the so-called tray-type and packing-type contact devices are used [6-11].

Packing that fill the column volume is the most effective contact device. The packing is known to be preformed, i.e. forming regular-shaped channels for steam passage, and random (or dumped), when the packing elements being positioned in a random way form curved channels of arbitrary shape [11]. The random packing has, as a rule, better mass-exchange characteristics (height of equivalent theoretical separation stage, etc.) and ensures the best separation conditions. However, its cost is in excess to the preformed option, which can impede its use in the large-scale industrial production of the isotope product required.

The preformed packing is inferior to the random one for mass-exchange characteristics, but demonstrates higher throughput and lower cost. The preformed roll-type helical band packing offered by D.I. Mendeleev RHTU [15] is one of the promising domestic options. This packing may be made of stainless steel gauze by winding either corrugated bands only or, in turn, corrugated and flat bands at an angle to each other. The advantages of this packing are as follows: (a) relatively easy production, (b) the roll can be made in size to column shell. This makes this packing accessible, effective and easily adaptable to columns of various diameters According to the data in [15] such packing is competitive with known foreign analogs for a number of characteristics and, hence, may be used to implement the deuterium separation plant with rectification.

Summarizing the above, it should be noted that the water rectification is one of the most suitable methods to obtain water with reduced deuterium content based on the nuclear desalination complex. The major reasons of this are as follows:

- Using water as working agent
- Developed and proved at industrial level the RF-located manufacturing processes for columns themselves and packing to them
- Relative safety of the method.

However, its major shortcoming is tangible energy consumption due to heating and evaporation of water. This problem can be solved by utilizing the nuclear reactor exhaust heat as well as applying the effective schemes of energy recuperation allowing the energy of steam condensation in the condenser to be returned to heat up the evaporator. To implement such conditions, it is necessary to subsequently choose an optimal heat balance scheme for such installation. To find a heat balance scheme for the installation considering its "coupling" with the reactor on one side and desalination plant on other side is the key stage of the work being done and its main feature as compared with the already implemented desalination plants or rectification cascades for hydrogen isotope separation. When choosing the optimal heat balance scheme, it is necessary to consider, as inlet conditions, the parameters of the turbine waste heat (mass flow, temperature, etc.) and the temperature of the desalinated water inlet flow from the desalination plant. Based on the heat balance scheme found, it will become possible to work out proposals on design features of the flow conversion nodes, which will be conditioned by heat balance between the amount of heat necessary for functioning of the evaporators providing the specified steam flow in the columns and the energy of the steam having passed through the turbine.

3. Specific Energy Consumption Estimation

To create the nuclear desalination complex with the deuterium separation plant, an integrated installation is assumed to be developed to obtain water with deuterium content reduced relative to its level in the desalinated sea water. Such plant is mainly intended to obtain water with the deuterium content at the level of freshwater ponds (~120-130 ppm) to provide the communities connected to the desalinated plant with water suitable for drinking. According to the requirements imposed such plant should cover the needs of a community with a population of up to 100 thousand citizens. At a standard consumption rate of drinking water averaged to 3 l/day per person the plant throughput should be ~12.5 t/h. It should be observed that the water must enter the deuterium separation plant as having been preliminarily cleared from salts, but being mixed up at outlet with a salted solution to provide the final product. Assuming the initial water as containing deuterium at a level of ~155 ppm, the approximate mixing proportion should be 1:1. Proceeding from this fact, the further development of the deuterium separation plant should be conducted for the following external conditions specified:

- Deuterium content in the feeding mixture is 155 ppm
- Deuterium content in the product flow is 100 ppm (so that the water to be obtained after mixing with a deuterium content of ~120-125 ppm)
- Separation plant does not have the exhausting part and the water at the waste end must be repeatedly directed into the evaporator. In this case the deuterium content in the inlet section of the column will only slightly (not more than by 5 ppm) exceed the corresponding content in the steam flow entering the column
- Product flow rate is 6.25 t/h.

Proceeding from the provided analysis of hydrogen isotope separation methods, it follows that the water rectification is the most promising method to be used for the development of the deuterium separation plant. It should be stressed, however, that its choice is dictated to a great extent by the fact that the problem set forth demands a relatively minor change in the deuterium concentration, as well as the fact that the water rectification will allow involving the waste heat of the nuclear reactor.

To proceed with the design of the deuterium separation plant, its technological analysis and feasibility study, it is necessary to receive estimates on specific energy consumption for production and material flow rates.

Let us estimate the material flow rates required to implement the above external conditions for a supposed binary separation (protium-deuterium). According to [10] the minimal liquid flow through the section of the column/columns at a specified throughput of the separation plant can be calculated as:

$$L_{\min} = \frac{P\left(\alpha C_P - \varepsilon C_F C_P - C_F\right)}{\varepsilon C_F\left(1 - C_F\right)} \tag{3}$$

where C_P and C_F – concentrations of the component being enriched in product flow and feed flow, α – separation coefficient, ε – enrichment coefficient, P –product flow.

Let us consider the most easily implemented case of the rectification at atmospheric pressure, which runs at a temperature of 373 K and with a H_2O-HDO separation coefficient of α=1.026 [10]. Subject to such assumptions and with P=6.25 t/h, C_P=99.99 atomic %, C_F=99.845 atomic % as above specified the calculated Lmin amounts to 91.6 t/h. The liquid flow rate in the real installation may be estimated as L=1.5·Lmin=137.4 t/h. In this case the steam flow rate at column inlet (G) will be G=L+P=143.65 t/h. Based on the G value thus calculated, it is an easy task to estimate the power for heating and evaporation of the water mass required. Assuming the temperature of the feed water to the evaporator as 293 K, the specific energy for water heating and evaporation will be 18.74 kW·h/kg. The value of specific energy consumption thus received should be construed as "overestimate" since this calculation does not consider such factor as return of the heated liquid from the column to the evaporator. Besides, the received value may be reduced subject to application of effective manufacturing process schemes that allow utilization of the heat releasing at condensation of the working agent in the condenser to heat up the evaporator. According to the data in [14] such schemes allow reducing the energy consumption by at least 2-3 times. In addition to the above, the separation equipment as itself demands a proper heat insulation, which can allow further reduction of the specific energy consumption.

In the context with the above the said value of the specific energy consumption for production undoubtedly demands its verification in the course of the further project implementation. Nevertheless, based on the data received, it is necessary to assess the possibility to satisfy the energy consumption as specified for the installation by utilizing the turbine waste steam of the mobile NPP providing the needs of the desalination plant itself are taken into consideration as well.

The schematic diagram of the deuterium separation plant and options of its implementation will demand further development and analysis in order to find the final solution most effective. The steps to conduct the work are as follows:

- Verify the process parameters (initial temperature of the feedwater, characteristics of the turbine waste steam, etc.)
- Choose the options for implementation of separation equipment and the materials used (packing type and size, etc.)
- Define the design parameters of the separation equipment and its operating characteristics (diameters and number of the columns, height of equivalent theoretical stage and other mass-exchange characteristics, options of columns commutation, etc.)
- Make a detail thermo-fluid calculation for parameters of the evaporators and condensers, piping, as well as heat pumps suitable for energy recuperation
- Work out a schematic diagram for commutation of the plant components and its provision with automatic control systems.

4. Conclusion

Critical analysis on industrial methods of hydrogen isotope separation has been conducted. As a basic method to create the deuterium separation plant the water rectification method has been preliminarily chosen. This choice is conditioned by using water as working substance, relative easiness of its implementation, possibility to utilize the NPP turbine waste steam energy in order to heat up the evaporator.

Specific energy consumption to obtain the required product by water rectification at atmospheric pressure has been estimated as 18.74 kW·h/kg.

Further efforts will be directed toward elaboration and analysis of the plant schematic diagram, options of its implementation and "interfacing" with the mobile nuclear power plant so that to find the heat balance and manufacturing process schemes most effective.

The accomplishment of the above efforts will enable a feasibility study to be conducted allowing estimates on the specific consumption for production.

Acknowledgments

The work was financially supported by the Ministry of Education and Science, agreement № 14.575.21.0159. Unique project identifier RFMEFI57517X0159.

References

[1] Obraztsov S V 2011 *Complex processing of sea, reservoir, surface and waste waters: textbook* (Tomsk Polytechnic University. - Tomsk: Publishing house of Tomsk Polytechnic University) p. 83 (in Russian)

[2] Mosin O V 2012 *Consciousness and physical reality* **1** 19 (in Russian)

[3] Barishev M G, Bolotin S N, Frolov V Yu, Dzhimak S S, Pikula A A, Dolgov M A, Shashkov D I, Petriev I S 2013 *Ecol. Bull. of Res. Cent. Black Sea Econom. Cooper.* **1** 13 (in Russian)

[4] Samkov A A , Dzhimak S S, Barishev M G, Volchenko N N, Khudokormov A A, Samkova S M, Karaseva E V 2015 *Biophysics* **60**(1) 107

[5] Dzhimak S S, Barishev M G, Basov A A, Timakov A A 2014 *Biophysics* **59**(4) 614

[6] Andreev B M, Zelvensky Ya D, Katalnikov S G 1982 *Separation of stable isotopes by physicochemical methods* (Moscow: Energoatomizdat) 208 p. (in Russian)

[7] Rosen A M 1960 *The theory of isotope separation in columns* (Moscow: Atomizdat) (in Russian)

[8] Andreev B M, Zelvensky Ya D, Katalnikov S G *Heavy isotopes of hydrogen in nuclear engineering* (Moscow: Energoatomizdat) 456 p. (in Russian)

[9] *Isotopes: Properties. Receiving. Application* 2005 ed. V Yu Baranov (Moscow: Fizmatlit) (in Russian)

[10] Andreev B M, Magomedbekov E P, Reitman A A et al. *Separation of isotopes of biogenic elements in two-phase systems* (Moscow: Izdat) 376 p.

[11] Magomedbekov E P, Rastunova I L, Rozenkevich M B 2014 *Issues At. Sci. Technol. Ser.: Mater. Sci. New Mater.* **3(78)** 70 (in Russian)

[12] Magomedbekov E P, Belkin D Yu, Rastunova I L, Sazonov A B, Selivanenko I L, Kulov N N 2017 *Theor. Found. Chem Engng.* **51(4)** 384

[13] Magomedbekov E P, Belkin D Yu, Rastunova I L, Sazonov A B, Selivanenko I L, Kulov N N 2017 *Theor. Found. Chem. Engng.* **51(2)** 133

[14] Niculescu A, Constantin T, Ana G, Draghia M. 2017 *Fus. Engng. and Des.* **124** 752

[15] Magomedbekov E P, Belkin D Yu, Selivanenko I L, Rastunova I L 2016 *Theor. Found. Chem. Engng.* **50(4)** 398

Effect of carbon nanotubes on a gases permeability of polymer PMMA

Yu S Eremin[1], D Bakhtin[2], S V Pavlov[1] and A M Grekhov[1,2]

[1]National Research Nuclear University MEPhI (Moscow Engineering Physics Institute) Molecular Physics Department, 115409, Russia, Moscow, Kashirskoe highway, 31
[2]Topchiev Institute of Petrochemical Synthesis, Russian Academy of Sciences, 119991, Russia, Moscow,Leninsky prospect, 29

E-mail: yseremin@yandex.ru

Abstract. Mixed matrix membranes (MMMs) with unique transport characteristics can be prepared by the addition of the minor amounts of carbon nanotubes. Changes in the membrane performance are shown to be provided by the formation of a percolation cluster composed of nanotubes. For MMMs based on Poly(methyl methacrylate) (PMMA) containing carbon nanotubes (CNT), due to the formation of the CNT percolation cluster, gas permeability increases by a factor of 2.5-10.

1. Introduction

In the recent years research is being actively carried out in the field of the development of hybrid membrane materials (mixed matrix membranes) [1,2] and hybrid membrane materials by incorporation of these polymers with nanoparticles[3]. The interest in such polymer modification is connected with the improvement of the membrane properties: the increase in permeability coefficients, selectivity, mechanical properties, ect. [4]. These effects are connected with the change of the polymer chain mobility, the degree of crystallinity and the structure of polymer free volume which interacts with nanoparticles [5]. In most cases inorganic nanoparticles (TiO_2, ZnO, $A1_2O_3$, SiO_2, ect.), nanoporous particles with high selectivity (e.g. zeolites), nanoparticles with high aspect ratio (carbon nanotubes, carbon fibers) are used for the development of hybrid membranes [6,7]. The non-linear change of different characteristics of hybrid membranes with CNT is described in the published works: mechanical properties, electroconductivity, magnetic properties, increase in gas and liquid permeability, etc. [8,9]. The change of permeation connected with the formation of the percolation cluster in the nanostructured domains is confirmed by the decrease of the critical concentration during the orientation of MCNT in the polymer. Nevertheless, despite the well-understood connection of the change in the permeability of the polymers and the percolation structure which is formed with the incorporation of nanoparticles, the mechanism of the selective transport in such materials is still ambiguous. In this work the research was carried out on the permeability gases through the hybrid membranes based on the polymer Poly(methyl methacrylate comprising CNT (less than 3% wt.) and the critical concentration of the percolation cluster formation from the domains containing CNT was determined. And confirmed what addition of CNTs above the critical concentration leads to the formation in PMMA percolation cluster, the permeability of which is several times greater than polymer permeability.

IOP Conf. Series: Journal of Physics: Conf. Series **1099** (2018) 012036 doi:10.1088/1742-6596/1099/1/012036

2. Methods and materials

In this work, we used the multiwalled carbon nanotubes (CNT) (TAUNIT™), NanoTekhCenter LLC, Russia) prepared by chemical precipitation from a gaseous phase. According to technical specifications, TAUNIT™ carbon nanomaterial is composed of fibrils based on polycrystalline graphite with a length below 2 microns; their outer diameter varied from 20 to 70 nm, specific surface was 120-130 m^2/g, and the bulk density was 0.4-0.6 g/cm^3. After the treatment with nitric acid according to the standard procedure [10], density of the CNT samples was 2.26±0.03 g/cm^3, specific volume was V = 0.20±0.02 cm^3/g, and the surface area was S = 150±5 m^2/g.

We used the commercial polymer PMMA sample with a density of 1.18 g/cm^3: the permeability coefficients were P(He) = 12, P(O_2) = 0.23, P(N_2) = 0.039 [11]. In this experiment were obtained permeability coefficients in PMMA: P(He)=11.3, P(O_2)=0.27, P(N_2)=0.06 barrer. To create hybrid PMMA / CNT, PMMA membranes were dissolved in chloroform, then a CNT / chloroform slurry was added to the PMMA / CNT solution, an ultrasonic disperser was used to disperse the CNT agglomerates to separate CNTs. In order to select the optimal concentration of CNTs in chloroform for casting membranes, to determine the optimum processing time for ultrasonic scanning and the distribution function of CNTs by dimensions, measurements were made on a laser particle size analyzer "nanotrac 150/250" and on a spectrophotometer "Hach dr 5000". In the experiments, we used the solutions containing 0.01 wt % of CNTs. According to the DSL method, the dimensions of agglomerates in the solution fit the dimensions of individual nanotubes with a diameter varying from 20 to 70 nm and with a length a length below 2 microns.

The chloroform solution of PMMA (3 wt %) was prepared under normal conditions then and filtered. Concentrations of PMMA and CNTs in the chloroform solutions was controlled by their optical density. The membranes were prepared by casting the PMMA and CNT solutions onto a cellophane support. Chloroform was evaporated under normal conditions (atmospheric pressure, T = 25°C) until the constant weight was attained. Thickness of the prepared membranes was 25±2 mkm. We prepared the PMMA /CNT membranes containing 0.25, 0.50, 0.75, 1.00, 1.50, 2.00, 2.50 and 3.00 wt. % of CNTs.

Gas permeability of nitrogen, oxygen, methane, and propane through the PMMA/CNT membranes were measured by the Dyness-Barrer method, the pressure gradient was 1 bar. and the temperature was 30°C.

3. Results

Figure. 1 presents the gas permeability of the mixed matrix PMMA/CNT membranes for nitrogen, oxygen, methane, helium, and propane plotted against the CNT concentration in the membrane, and this plot also shows three well-pronounced concentration intervals. When the CNT content in the membrane is lower than 1.5 wt %, gas permeability of the matrix mixed membranes for all gases under study is equal to the permeability of the neat PMMA. When the CNT content is changed from 1.5 to 2 wt %, gas permeability increases in the following manner: by a factor of 2 for helium, by a factor of 4 for hydrogen, by a factor of 12 for nitrogen, by a factor of 10 for oxygen, 70 for methane. When the CNT content is higher than 2 wt %, the growth in the gas permeability is ceased and this value remains virtually constant or slow degrease. This behavior was also observed for other glassy polymers. [14, 15, 16].

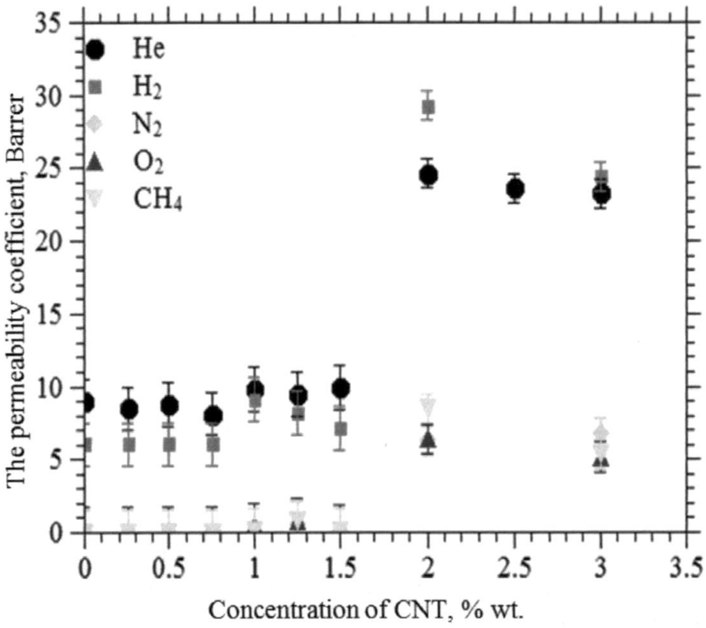

Figure1. The coefficient of gas permeability (He, H_2, N_2, O_2, CH_4) through samples of hybrid PMMA

This nonlinear change in the permeability of the mixed matrix membranes suggests that permeability of the mixed matrix membranes is controlled by the parameters of the formed percolation cluster. Traditionally, when two different materials are mixed, any additive changes in their characteristics can be described by the mixture rule. If changes in the net membrane permeability are provided by the development of highly permeable open channels in the membrane, permeability starts to increase only when CNTs are organized into the percolation cluster; in this case, permeability should be proportional to the volume of the percolation cluster. By analogy, if permeability of the mixed matrix membranes depends on the content of nanotubes, this parameter can be described by the following relationship:

$$P_{MMM} = P_{CNT} \cdot C_{cluster\ of\ CNTs} + P_{polymer} \cdot C_{polymer}$$

Where PMMA is the permeability coefficient of the hybrid polymer/CNT membrane, P_{CNT} is the permeability coefficient of the percolation cluster of the CNT formed, $C_{cluster\ of\ CNTs}$ is the volume fraction of the percolation cluster formed from CNT, $P_{Polymer}$ is the polymer permeability coefficient, $C_{Polymer}$ - volume fraction of polymer. Changes in the membrane permeability and $C_{cluster\ of\ CNTs}$ are nonlinear when concentration of nanotubes in the membrane is above the percolation threshold [16]. At low CNT concentrations, the percolation cluster does exist or, in other words, $C_{cluster\ of\ CNTs}$ is equal to zero and the presence of nanotubes has no effect on the membrane permeability. The effect of the percolation cluster on the permeability of the mixed matrix membranes was studied according to the Monte Carlo scheme which allows one to assess the conditions providing the formation of the percolation structures composed of nanotubes in the 3D space and their characteristics [17].

Figure 2 shows the permeability coefficients of the PMMA/CNT membranes calculated after the substitution of $C_{cluster\ of\ CNTs}$ to Eq. (1). In these calculations, we assume that there is no gas transfer through inner CNT channels, and the increase in gas permeability is solely provided by gas transport through the highly permeable layer of the modified polymer near the CNT surface. This assumption is valid only for the permeability coefficient P_{CNT} but has no effect on the threshold concentrations corresponding to the formation of the percolation cluster and related changes in permeability. As follows from Fig. 3, the calculated permeability coefficients well agree with the experimental values when the concentration of nanotubes in the matrix is below 1,5%. The development of the percolation structure and threshold changes in permeability are observed when the volume concentration of nanotubes varies from 1.5 to 2.0%. When the volume concentration of CNTs is above 2 wt %, the calculated permeability coefficients appear to be appreciably higher as compared with the experimental values. This disagreement can be explained by different parameters of the percolation cluster due to the agglomeration of nanotubes with increasing their content in the membrane. As follows from calculation, when the average diameter of capsules increases due to the coalescence and agglomeration of nanotubes, the volume fraction of highly permeable regions at the same concentration of the capsules in the matrix decreases.

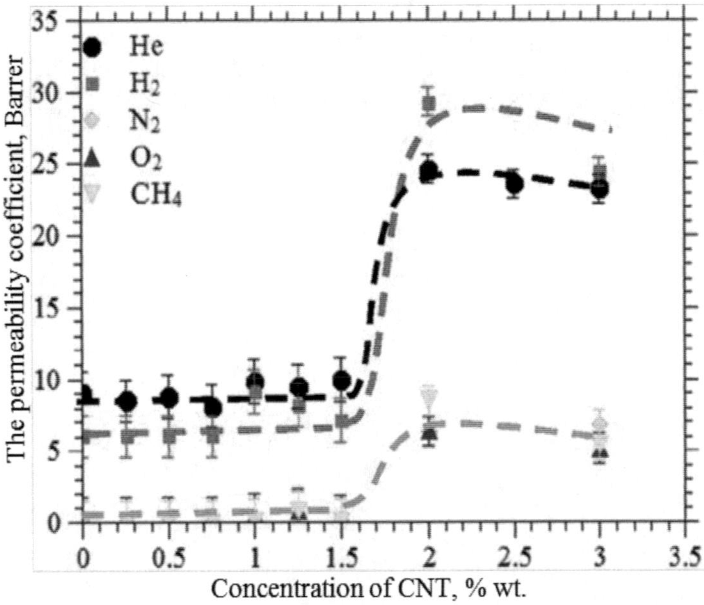

Figure 2. Gas permeability for model gases through the PMMA/CNT membranes with the account for agglomeration

4. Conclusion

This work shows that, for the mixed matrix CNT-containing membranes, there exists the concentration interval where gas permeability dramatically increases. At high (excessive) concentrations of CNTs, due to agglomeration of nanotubes and deterioration of the percolation structure, further increase in the CNT concentration does not provide any improvement in the membrane performance. According to the experimental results, as the CNT concentration varies from 1.5% to 2 wt %, permeability of the CNT-containing membranes markedly increases: by a factor of 2 for helium, by a factor of 4 for hydrogen, by a factor of 12 for nitrogen, by a factor of 10 for oxygen, 70 for methane. However, further increase in the CNT content does to slow degrease in permeability.

This behavior (effect, phenomenon) was explained by the numerical simulation of a percolation cluster composed of capsules in the 3D matrix with the dimensions equivalent to nanotubes and membranes. Experimentally observed changes in the membrane permeability at high CNT concentrations can be explained with the account for the agglomeration of nanotubes (capsules) and related changes in the dimensions of the percolation cluster.

References

[1] Chen L, Wu Y, Dong H, Meng M, et al., 2018 *Sep. Purif. Technol.* **197** 70.
[2] Sunarsoa J, Hashima S S, Linb Y S, et al., 2017 *Sep. Purif. Technol.* **176** 335.
[3] Li X, Sotto A, Li J, et al., 1998 *J. of Membr. Sci.* **524** 502.
[4] Ji Y, Qian W, Yu Y, et al., 2017 *Chi. J. of Chem. Engin.* **25** 1639.
[5] Sotto A, Boromand A, Balta S, et al., 2011 *J. of Mat. Chem.* **21** 10311.
[6] Li J F, Xu Z L, Yang H, et al., 2009 *J. Appl. Polym. Sci.* **255** 4725.
[7] Weidong L, Fusheng P, Yimeng S, et al., 2017 *Chi. J. of Chem. Eng.* **25** 1563.

IOP Conf. Series: Journal of Physics: Conf. Series **1099** (2018) 012036 doi:10.1088/1742-6596/1099/1/012036

[8] Xiong J, Zheng Z, Qin X, et al., 2006 *Carbon* **44** 2701.
[9] Kodgire P V, Bhattacharyya A R, Bose S, et al., 2006 *Chem. Phys. Lett.* **432** 480.
[10] Hung N T, Anoshkin I V, Dementjev A P, et al., 2008 *Inorg. Mater.* **44** 270.
[11] Yusuke N, Hiroaki Y,Yoshiharu T. 2006 *Polym. J.* **38** 376.
[12] Grekhov A M, Eremin Yu S. 2015 *Nanotech. in Rus.* **4** 1.
[13] Grekhov A M, Belogorlov A A, Yushkin A A, et al., 2012 *J. of Membr. Sci.* **390** 160.
[14] Grekhov A M, Eremin Yu S, Dibrov G, et al., 2013, *Petrol. Chem.* **53** 549.
[15] Aani S A, Gomez V, Wright C J, et al., 2017 *Chem. Eng. J.* **326** 721.
[16] Ahmad A L, Jawad Z A, Low S C, et al., 2012 *J. of Membr. Sci.* **451** 55.
[17] Grekhov A M, Eremin Yu S, 2015 *J. of Membr. Sci.* **485** 42.

IOP Conf. Series: Journal of Physics: Conf. Series **1099** (2018) 012037 doi:10.1088/1742-6596/1099/1/012037

Fabrication of composite hollow fiber membranes with thin film selective layers from highly permeable polymers

A Ovcharova[1], S Bazhenov[1], V Volkov[1,2], T Plisko[3] and A Bildyukevich[3]

[1] A.V. Topchiev Institute of Petrochemical Synthesis, RAS, Leninsky prospect, 29, Moscow, 119991, Russia
[2] National Research Nuclear University MEPhI (Moscow Engineering Physics Institute), Kashirskoe shosse 31, Moscow, 115409, Russia
[3] Institute of Physical Organic Chemistry National Academy of Sciences, Surganov street 13, Minsk, 220072, Belarus

E-mail: ovcharoff@ips.ac.ru

Abstract. Fabrication of highly efficient composite gas separation membranes involves selection and pretreatment of the support material. In the present work, porous support pretreatment was for the first time studied for deposition of thin selective layers made from highly permeable organosilicon polymers. Asymmetric polysulfone hollow fiber membranes with mesoporous skin layer structure were employed as supports. It was shown that preliminary impregnation of the support pores is essential for composite membranes fabrication. If poly[1-(trimethylsilyl)-1-propyne] layers from casting solution are deposited onto the polysulfone support without impregnation, the casting solution penetrates the mesoporous layer to a depth of > 1 μm, thus predominantly (> 90%) contributing the overall mass transfer resistance, as estimated by the resistance-in-series model. Such membranes provide relatively poor transport properties: carbon dioxide permeance is 0.2-0.3 $m^3/(m^2 \cdot h \cdot bar)$. Preliminary impregnation of porous support by distilled water or aqueous glycerol solution prevents casting solution penetration into porous support and results in defect-free thin polymer layer. According to SEM, the deposited selective layer thickness is less than 1 μm. Composite membranes fabricated via this technique provide permeance increase more than order of magnitude. The composite membranes with poly[1-(trimethylsilyl)-1-propyne] and polydimethylsiloxane selective layers show high transport properties in gas-liquid membrane contactors.

1. Introduction

In recent years, membranes have been frequently proposed for fixing gas-liquid interfacial areas [1]. Such types of non-dispersive gas-liquid contactors are generally termed as gas-liquid membrane contactors. In contrast to other membrane processes, the membrane in a contactor acts as a barrier and as a means of bringing two immiscible fluid phases, such as gas and liquid, into contact with each other without dispersion [2]. The membrane contactor acts as a mass-transfer device with continuous contact, like, i.e., packed columns. However, in case of contactor, there is no need to disperse physically one phase in another or to separate the phases at the end of the process. One of the key benefits of this scheme is high specific surface area (phase contact area), making the membrane contactor different from other membrane processes. Whereas conventional packed columns have specific surface area ~ 20–500 m^2/m^3, the hollow fiber membrane contactors provide 500–5000 m^2/m^3 [3].

The main disadvantage of separation process in a gas-liquid membrane contactor is the fact that the membrane itself contributes significantly in the overall mass transfer resistance value. If liquid phase penetrates into the membrane (i.e., wetting phenomenon occurs), the membrane resistance increases [4,5]. For this reason, different researches proposed various approaches to prevent membrane wetting in contactors: employment of hydrophobic membranes or composite membranes with dense layer on the side of phases contact area, modification of membranes [6-9], selection of absorption liquids with suitable surface tension [10], and optimization of the process parameters. One of the solutions for gas-liquid membrane contactor configuration is employment of composite membranes constituted by a thin dense layer of polymer (selective layer) on a porous support. The dense layer is exposed to liquid phase side, thus stabilizing the membrane properties and preventing penetration of absorbent into the porous substructure. When choosing the selective layer material for composite membrane, one should consider the ratio between gas and liquid permeabilities – the higher this value, the more suitable is the polymer [11]. Furthermore, it is preferable to use the polymers possessing good hydrophobic properties.

Polysulfone (PSf) is widely used as membrane material due to its mechanical stability as well good thermal and chemical stability. Also, it is well suited for producing hollow fiber membranes with controlled pore size [12]. In the present work, the composite hollow fiber membranes with thin selective layers based on highly permeable organosilicon polymers – poly[1-(trimethylsilyl)-1-propyne] (PTMSP) and polydimethylsiloxane (PDMS) – are fabricated on asymmetric polysulfone hollow fiber supports with mesoporous structure of the skin layer. The major problem under deposition of selective layers onto porous supports is that the solution of modifying polymer penetrates into porous matrix of the support to a depth of > 1 μm, thus predominantly (> 90%) contributing the overall mass transfer resistance, as estimated by the resistance-in-series model. Such membranes provide relatively poor transport properties: carbon dioxide permeance is 0.2-0.3 $m^3/(m^2 \cdot h \cdot bar)$ [13]. To solve this problem, it is proposed to use preliminary impregnation of porous support, which results in deposition of thin selective layer on the membrane surface. The fabricated composite membranes were characterized by various techniques to study their morphological and transport properties.

2. Experimental

Asymmetric hollow fiber supports were fabricated using the dry-wet phase inversion technique in the free spinning mode in air when bore fluid was brought into liquid polymer solution orifice, resulting in the selective layer appearance on the fibers lumen side. The spinning solution included the polymer – polysulfone (Ultrason® S 6010, BASF), the solvent – N-methylpyrrolidone (NMP, Acros Organics 99% extra pure) and the pore forming agent – polyethylene glycol with average molecular weight 400 g/mol (PEG-400, Acros Organics).

For deposition of selective layer, the solutions of PTMSP (1 wt. %) in hexane and PDMS (1 wt. %) were prepared. The impregnation liquids used for fabrication of composite membranes were water or aqueous glycerol solution (20 wt. %) in case of PTMSP and ethyl or isopropyl alcohols in case of PDMS. The impregnation liquid was injected into the fibers by a syringe. After that, the excessive liquid was removed with air flush and the polymer solution was deposited on the lumen surface of the fibers. The composite membranes fabricated were rinsed with distilled water and dried at ambient conditions to constant weight.

The membranes were characterized using scanning electron microscopy (SEM) technique and measurement of gas permeance and contact angle. To obtain SEM micrographs of the membranes cross sections, we used Hitachi Table top TM 3030 Plus microscope with proprietary highly sensitive low-vacuum secondary electron detector; accelerating voltage was 15 kV. The contact angle values were measured by via the conventional sessile drop technique using the LK-1 goniometer (Open Science, Russia). Measurement error was ± 2°. Gas permeance of the hollow fibers was determined via the volumetric technique using the laboratory setup described elsewhere [7]. Ideal selectivity value was determined as the ratio between individual gases permeability values ($\alpha = P_1/P_2$).

3. Results and discussion

Table 1 shows gas transport properties and water contact angle values for both porous supports and composite membranes based on PTMSP and PDMS.

Table 1. Nitrogen permeance and water contact angle values of porous supports and composite membranes based on PTMSP and PDMS

Membrane	Impregnation liquid	P/l (N_2), $m^3/m^2 \cdot h \cdot bar$	θ (H_2O), °
Porous support	–	270	78
Composite (PDMS)	Ethanol	2.3	147
	Isopropyl alcohol	0.4	132
Composite (PTMSP)	Distilled water	3.5	98
	Aqueous glycerol solution	1.7	92

As can be seen from Table 1, the type of impregnation liquid effects the properties of the composite membrane fabricated. For the membranes with PDMS selective layer, impregnation by isopropyl alcohol leads to poor gas permeance as well as membrane defects. In contrast, use of ethanol results in enhanced gas transport properties of the membranes. Furthermore, these membranes showed drastic increase in hydrophobicity: contact angle value increases from 78° to 147°.

The composite membranes with PTMSP layer also differ depending on the type of impregnation liquid. With minor difference in water contact angle values, the membranes treated by distilled water show nitrogen permeance 2 times higher than that of membranes treated by glycerol solution.

In order to prove that the selective layer of PTMSP composite membrane had no defects, we also measured the membrane gas permeance by carbon dioxide, P/l (CO_2), and ideal selectivity value α (CO_2/N_2), for both composite membrane specimens. The results are given in Table 2.

Table 2. Gas transport properties of PTMSP composite membranes depending on the type of impregnation liquid

Impregnation liquid	P/l (N_2), $m^3/m^2 \cdot h \cdot bar$	P/l (CO_2), $m^3/m^2 \cdot h \cdot bar$	α (CO_2/N_2)
Distilled water	3.5	4.1	1.2
Aqueous glycerol solution	1.7	9	5.3

As can be seen from Table 2, the selectivity of PTMSP-based composite membrane, fabricated with the use of distilled water, almost fully corresponds to the porous support selectivity, which indicates that the polymer coating is inhomogeneous and the selective layer has defects. At lower permeance, the composite membranes fabricated with glycerol solutions pretreatment show selectivity close to that of the dense PTMSP film (α (CO_2/N_2) = 5.7). So, it can be concluded that aqueous glycerol solution is preferable for fabrication of such membranes.

Fig. 1 shows SEM micrographs of the composite PTMSP-based hollow fiber membrane which was fabricated via preliminary impregnation by aqueous glycerol solution.

As can be seen from Fig. 1 (a), the PTMSP composite membranes possess pronounced asymmetric morphology constituted by thick macroporous support, spongy transitional layer and thin selective layer. Fig. 1 (b) shows enlarged image of the PTMSP layer. It can be seen that the coating is uniform and homogeneous. The estimated PTMSP layer selectivity is < 1 μm, which, presumably, accounts for relatively high nitrogen permeance of the membrane.

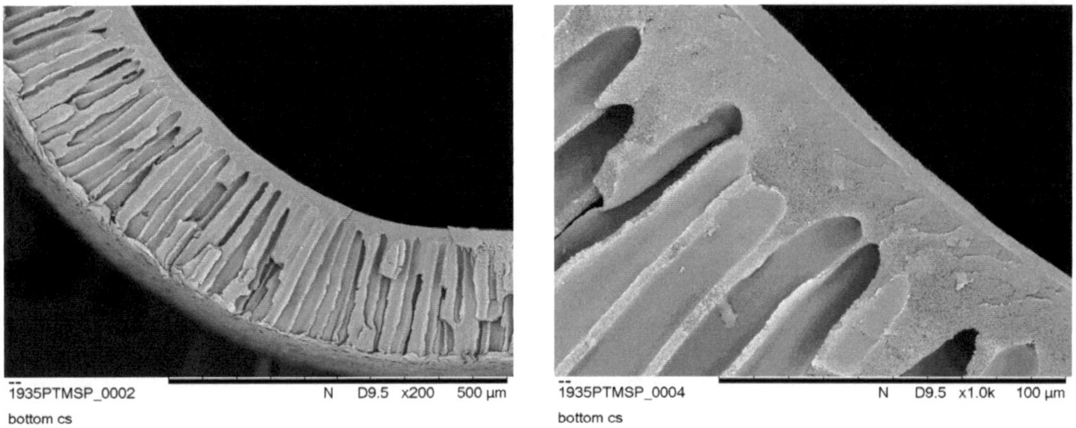

a **b**

Figure 1. SEM micrographs of the composite PTMSP membrane (cross-section): a – general view, magnification 200x; b – enlarged image of the selective layer, magnification 1000x

4. Conclusion

Fabrication of composite hollow fiber membranes based on PTMSP was studied. Mesoporous polysulfone supports were pretreated via liquid impregnation. The membranes were characterized using gas permeance and contact angle measurements; SEM micrographs are provided. The PTMSP composite membranes show selectivity corresponding to that of the dense polymer film, at reasonable permeance. The PDMS composite membranes provide higher gas permeance combined with high selective layer hydrophobicity (water contact angle value ~ 150°). Depending on the particular tasks, both types of the composite membranes fabricated are promising for employing them in gas-liquid membrane contactors.

Acknowledgments

This work was performed at the Topchiev Institute of Petrochemical Synthesis and supported by the Russian Science Foundation, project no. 14-49-00101.

References

[1] Dindore V, Brilman D and Versteeg G 2005 *Chem. Eng. Sci* **60** 467
[2] Pabby A, Sastre A 2013 *J. Membr. Sci.* **430** 263
[3] Iversen S, Bhatia V, Dam-Johansen K and Jonsson G 1997 *J. Membr. Sci.* **130** 205
[4] Mosadegh-Sedghi S, Rodrigue D, Brisson J and Iliuta M 2014 *J. Membr. Sci.* **452** 332
[5] Kirsch V, Roldugin V, Bildyukevich A and Volkov V 2016 *Sep. Purif. Technol.* **167** 63
[6] Tarboush B, Rana D, Matsuura T, Arafat H and Narbaitz R 2008 *J. Membr. Sci.* **325** 166
[7] Ovcharova A, Vasilevsky V, Borisov I, Bazhenov S, Volkov A, Bildyukevich A and Volkov V 2017 *Sep. Purif. Technol.* **183** 162
[8] Peter J, Peinemann K 2009 *J. Membr. Sci.* **340** 62
[9] Bildyukevich A, Plisko T, Liubimova A, Volkov V and Usosky V 2017 *J. Membr. Sci.* **524** 537
[10] Ortiz A, Ruiz A, Gorri D and Ortiz I 2008 *Sep. Purif. Technol.* **63** 311
[11] Nymeijer D, Folkers B, Breebaart I, Mulder M and Wessling M 2004 *J. Appl. Polym. Sci.* **92** 323
[12] Ismail A, Mansourizadeh A 2010 *J. Membr. Sci.* **365** 319
[13] Volkov V, Bildukevich A, Dibrov G, Usoskiy V, Kasperchik V, Vasilevsky V and Novitsky E 2013 *Petrol. Chem.* **53** 619

IOP Conf. Series: Journal of Physics: Conf. Series **1099** (2018) 012038 doi:10.1088/1742-6596/1099/1/012038

Pressure effect at permeability and deformation of the hollow fiber membrane gas separation module

N I Laguntsov, M D Karaseva

Public joint-stock company "Aquaservice", Kashirskoe sh. 3/2 st 4, Moscow 115230 Russia

E -mail: aquaserv@mail.ru, NILaguntsov@mephi.ru, rita09karaseva@mail.ru

Abstract. In this paper, the deformation effect in the gas separation process of the hollow fiber membrane module is considered. A technique of gas permeability measurement in a commercial gas separation hollow fiber membrane module is provided. It is determined that permeability and selectivity are nonlinearly dependent on the pressure of the feed stream, this is justified by competing factors. The results of this work will allow introducing the membrane modules into hybrid gas separation units, in particular, into the drying unit of compressed gas.

1. Introduction
Membrane technologies are widely used in industry. The hollow fiber gas separation membrane modules can be used in various installations [1, 2] for helium extraction from natural gas, oxygen enrichment, drying of hydrocarbon gases, etc. Gas separation units can be membrane (consist of one or more membrane modules) or hybrid (fox example, these units have adsorption or absorption stage in combination with a refrigerator or a membrane stage).

A bundle of tightly packed hollow fibers is installed in a cylindrical body - a membrane module. Depending on the design of the membrane module, the feed flow is supplied into the hollow fiber or to the inter-fiber space. Typically, the hollow fiber of the gas separation module consists of a porous support and a selective layer applied to the outer surface.

2. Problem statement
Companies that manufacture membranes and membrane modules, primarily deal with the complex solution of industrial problems, less often the implementation of membrane modules. The technical characteristics of the membrane modules don't contain information about the material, permeability and membrane area. Consumers of ready-made membrane modules are companies that are engaged in the creation of installations for industry and transport. The absence of parameters for a commercially available hollow fiber membrane module complicates the process of the development and the creation of gas separation units.

To determine these parameters, a technique of gas permeability measurement has been developed on the basis of experimental investigations.

We should take into account the possible deformation of the hollow fiber in the nonstationary gas flow regime. The outer and inner diameter of the hollow fiber can be increased by supplying flow at a high pressure inside of the hollow fiber or can be narrowing if the feed flow is supplied to the inter-fiber space.

In the case of supplying the feed flow into the hollow fiber, it is appropriate to talk about deformation (expansion) of the fiber (Figure.1.).

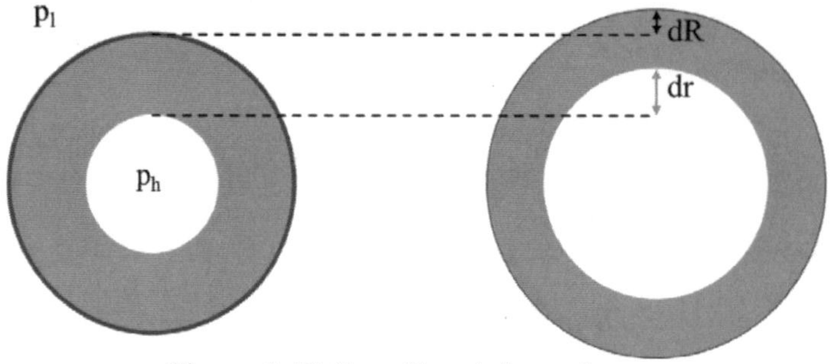

Figure.1. Hollow fiber deformation

The hollow fiber consists of a thin diffusion (selective) layer of several nanometers thick [3] and a porous support with a gradient porosity. The pore size in the hollow fiber decreases from the inner radius to the outer radius.

The deformation of a hollow fiber is approximately described by the variation equations between inner and outer radius of a thick-wall cylinder:

$$dr = \frac{r}{E}\left[P_h\left(\frac{R^2+r^2}{R^2-r^2}+\mu\right) - 2P_l\frac{R^2}{R^2-r^2}\right], \quad (1)$$

$$dR = \frac{R}{E}\left[2P_h\frac{r^2}{R^2-r^2} - P_l\left(\frac{R^2+r^2}{R^2-r^2}+\mu\right)\right], \quad (2)$$

where r – inner radius of the hollow fiber, R – outer radius, μ – Poisson ratio, E – modulus of elasticity, P_h – high pressure of the feed flow P_l – low pressure of the permeate flow. It is important that equations (1) and (2) are linearly dependent on high and low pressures.

To establish the parameters of commercial gas separation hollow fiber membrane modules, experimental investigations were carried out with a clean gas (nitrogen or oxygen) in the "locked" regime (the permeate flow is blocked). Gas from the high pressure cylinder was supplied to the input of the unit. The pressure of the feed flow was raised by a gas pressure regulator to 10 ATMG for oxygen and 14 ATMG for nitrogen at a pitch of 1 ATMG, permeate flow was measured for each feed flow pressure value. Experimental investigations were carried out on a hollow fiber membrane module in which the feed flow is supplied into the hollow fiber. As the permeability values and the membrane area values are unknown for commercially available membrane modules, reduced permeability is determined from formula:

$$\pi_i^* = \pi_i * S = \frac{P}{(P_h-P_l)}, \quad (3)$$

where π_i – the permeability of *i* component, S – membrane area, P – permeate flow.

3. Results and discussion

Experimental investigations of the nonstationary flow of oxygen and nitrogen through a selective membrane in a "locked regime" were performed in this work with a variation of the feed flow pressure. From the experimental research, the dependence of the reduced permeability (3) on the pressure of the feed flow was obtained. Fig. 2a, 2b. The data obtained from experimental investigations is approximated by the of least square method of the second order.

On the Figures 2a and 2b, curve 1 corresponds to the process of a gradual increase in the pressure of the feed flow, curve 2 shows a gradual decrease in the pressure of the feed flow. When the pressure is decreased, the values of the reduced permeability are lower than values at the process, in which the pressure is increased, so hysteresis is observed. At a feed flow pressure of 8 ATMG, the maximum of reduced oxygen permeability is observed (Figure 2a)., This maximum is not observed for the nitrogen in the measured pressure range.

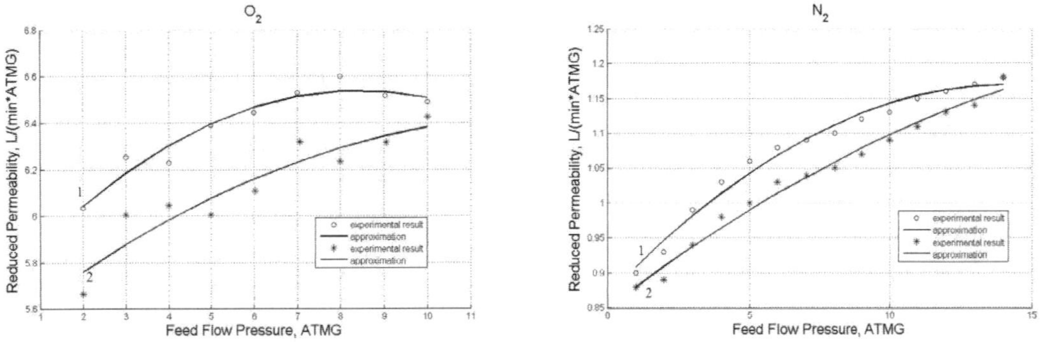

a) b)

Figure 2. The dependence of the reduced permeability on the feed flow pressure a) – oxygen, b) – nitrogen.

On the Figure.3. the nonlinear dependence of the oxygen / nitrogen average selectivity value (the average between the selectivity when the feed stream pressure increases and decreases) of the investigated membrane module from the feed flow pressure was shown. An experimental investigation has shown that permeability and selectivity are not constant values.

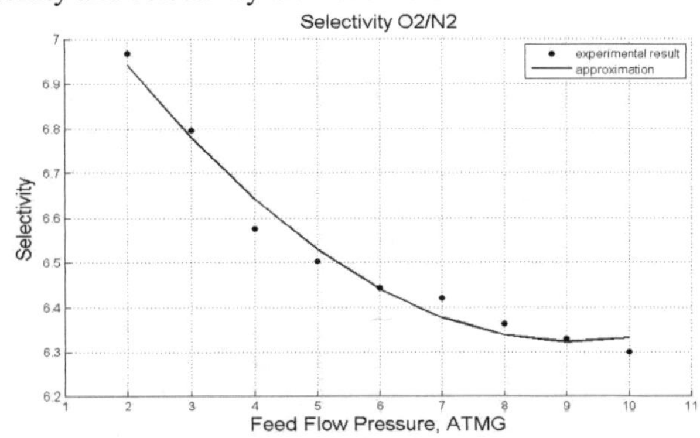

Figure 3. The dependence of O_2/N_2 selectivity on the high pressure feed flow.

Obtained results aren't described by equations (1) and (2). The reason for the nonlinear dependence of the reduced permeability and selectivity on the of the feed flow pressure can be associated with several competing factors. The pressure growth increases the permeability of the gas. As the pressure increases, the inner and outer radius of the hollow fiber increases also, and, as a consequence, the selective layer area increases, and the thickness decreases (Figure.1), this leads to the increase of permeability. The deformation of the porous support also influences on the reduced permeability, since its thickness (support packing), the area and the porosity vary over the entire thickness of the support, which leads to a decrease in the permeability of the porous support.

The presence of a maximum value of the oxygen reduced permeability at 8 ATMG (and its absence for nitrogen) is justified by the fact that the oxygen permeability is higher than the nitrogen permeability. At a critical value of the pressure of the feed flow, the effect of the porous support compacting dominates over the growth factors of permeability when the pressure of the feed flow increases.

It is also determined from experimental investigation that a change in the reduced permeability is observed with time. (Figure.4a, 4b).

The experiment was carried out as follows: the pressure of the feed flow was gradually increased to a certain pressure (3 ATMG.), then within 7 minutes the pressure of the feed stream remained constant, the permeate flow was measured. The experiment was repeated at different pressures. The experimental investigation was carried out in a similar manner at the pressure decrease for nitrogen and oxygen.

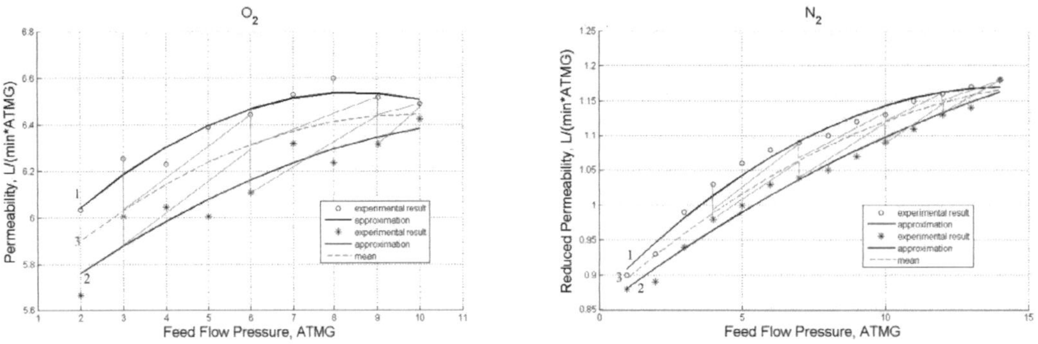

a) b)

Figure 4. The dependence of the reduced permeability on the feed flow pressure (The time was taken into account) a) – oxygen, b) – nitrogen.

The reduced permeability decreases in 3 minutes, at a constant pressure of the feed flow, and after 3 minutes the value of reduced permeability becomes constant; it was determined from experimental investigation. The flowing process of pure gas through the selective membrane becomes stationary. With a smooth pressure decrease of the feed flow (curves 2 in Figure.4.) the permeability has been transforming into a stationary value per 3-4 minutes.

The experiment was repeated after a few hours, and the values of the reduced permeability were the same. It has been established that for the values of the feed flow pressure up to 14 ATMG, the deformation of the fibers under investigation is elastic, although the relaxation time of the geometric dimensions is about 10 minutes (under the experimental conditions).

Conclusions

In the present work, the nonstationary process of nitrogen and oxygen flow in a commercially available hollow fiber membrane module has been investigated.

From the obtained experimental data, it follows that the reduced permeability increases nonlinearly with increasing pressure, and the selectivity decreases nonlinearly. Taking into account the fact that the hollow cylinder deformation model depends on the pressure linearly, this effect can be explained by a combination of competing factors: the area increase of the selective layer and the support, and the thickness decrease of the selective layer and porous support, the decrease of pores size, and a change in the porosity of the support.

It was also determined, that for the investigated membrane module, the time to enter the stationary mode is equal to 3-4 minutes. The deformation is elastic, and the relaxation time is approximately equal to10 minutes.

The results obtained in this work have the practical importance in the investigation process, development and creation of the membrane/hybrid systems for compressed gas drying, in particular, for the drying of the compressed air.

Acknowledgements: The work was financially supported by the Ministry of Education and Science, agreement № 14.576.21.0097. Unique project identifier RFMEFI57617X0097.

References

[1]. Soleimany A, Hosseini S S, Gallucci F 2017 Recent progress in developments of membrane materials and modification techniques for high performance helium separation and recovery: A review *Chemical Engineering and Processing: Process Intensification*

[2]. Sidyganov Yu N, Shamshurov D N, Kostromin D V 2015 *The International Scientific Journal* **1** 78-91

[3]. Laguntsov N I et al. 2017 *Petroleum Chemistry* **57** 132-138

IOP Conf. Series: Journal of Physics: Conf. Series **1099** (2018) 012039 doi:10.1088/1742-6596/1099/1/012039

Fabrication of cellulose-based composite membranes for organic solvent nanofiltration

T Anokhina, V Ignatenko, S Ilyin, S Antonov, and A Volkov

A.V. Topchiev Institute of Petrochemical Synthesis, Russian Academy of Sciences, 29 Leninsky prospect, 119991 Moscow, Russia

E-mail: tsanokhina@ips.ac.ru

Abstract. In this paper, the methods for obtaining delamination-resistant composite cellulose membranes were considered. Different approaches to improve the adhesion of the cellulose layer to the polyester non-woven support were analyzed. The first approach, consisting in etching the PET support with solutions of trichloroisocyanuric acid in acetone, had a negative effect on the formed membranes. The negative effect was to increase the hydrophobicity of the support, resulting in peeling off of the cellulose layer. Suppression of delamination was achieved by reduction of cellulose shrinkage: by its plasticization with glycerol or sequential treatment by liquids with gradually decrease of their affinity to cellulose. Nevertheless, treatment of the membrane with an aqueous solution of glycerol resulted in a decrease of the dye rejection. The optimal technique is sequential post-processing of composite membranes with ethanol and hexane, which makes it possible to obtain strong composite cellulose membranes with high values of the rejection of the nanosized anionic dye.

1. Introduction

Aprotic polar solvents (such as tetrahydrofuran (THF), dimethylformamide (DMF), N-methylpyrrolidone (NMP), dimethylacetamide (DMAc), dimethylsulfoxide (DMSO)) are widely applied in various fields of the chemical industry. For example, THF, NMP, and DMSO are actively used as a medium in interfacial catalysis [1]. DMF and NMP are used as extractants to remove oxidized products in the process of desulfurization of fuels [2], and also for the isolation of benzene-toluene-xylene fraction from the mixture with naphthenes, paraffins, and olefins. Due to anti-inflammatory, antiseptic, analgesic, and fibrinolytic action, DMSO is used in pharmaceutics [3, 4]. In this regard, aprotic solvents must be thoroughly cleaned and regenerated for reuse; also, a separate task is to isolate the target components from them. Traditional methods for purification of these organic solvents are energy-intensive and carry a heavy load on the environment due to the use of high temperatures and solvent evaporation. An alternative to these methods can be the process of organic solvent nanofiltration (OSN) – it is a baromembrane process of separation of liquids [5]. The main requirements for membranes used in OSN are mechanical and chemical stability in used media, high selectivity of separation, and the filtration of the target components.

Due to the sharp reduction in crude oil reserves and environmental degradation in the world, in recent years there is a trend to use biodegradable, renewable polymers, including for the production of membranes. Cellulose is the most common organic substance in nature. The content of cellulose in plants is in a wide range from 15 to 99 wt % [6, 7]. Cellulose is a linear syndiotactic homopolymer with a molecular formula $(C_6H_{10}O_5)_n$, composed of D-anhydroglucopyranose units, commonly referred to as glucose units [8]. Due to the presence of hydrogen bonds, whose energy is 17 - 34 kJ/mol, cellulose has a high chemical resistance to most organic solvents, which makes it promising as a membrane material for OSN. But at the same time, this chemical stability of the biopolymer makes it difficult to create membranes. For OSN it is possible to use cellophane [9, 10], but the traditional viscose method used to produce cellophane films is very laborious and involves a large number of reagents that are hazardous to both humans and the environment.

One of the promising and environmentally friendly solvents for cellulose is N-methylmorpholine N-oxide (MMO) [11, 12]. It is already widely employed for the production of fibers and yarns [13], as well as for the production of cellulose membranes for gas separation, pervaporation, ultrafiltration, and organic solvent nanofiltration [14-18]. The main disadvantage of the MMO, which is in the crystalline state at room temperature, is the high melting point and, correspondingly, the high temperature of the dissolution of cellulose (>100°C). Other promising solvents for cellulose can be eco-friendly ionic liquids (ILs) [19, 20]. There are works where ILs are used to make nanofiltration membranes; to reduce the viscosity of the molding solution and to reduce the cost of the process, a co-solvent (e.g., acetone [21] and DMSO [22, 23]) is added to the basic ionic solvent. In order to eliminate the deformation of the cellulose membranes due to the strong shrinkage of the natural polymer under drying, the cellulose layer is applied to a non-woven polyester support. The main problem of composite membranes is the peeling of the cellulose layer due to the poor adhesion between the support material and the cellulose layer, and also due to the strong shrinkage. Therefore, the aim of this research is to develop a technique for manufacturing delamination-stable composite cellulose membranes.

2. Experimental

Cellulose (the degree of polymerization is 600; the moisture content in cellulose is no more than 8%; alpha-cellulose content is 92%) produced at Baikal Pulp and Paper Mill (Russia) was used. 1-Ethyl-3-methylimidazolium acetate ([Emim]OAc) with a density of 1.27 g/sm^3 (Sigma-Aldrich) was used to dissolve cellulose. DMSO and DMF of reagent grade were purchased from Khimmed (Russia). Cellulose was applied on a polyethylenterephtalate (PETF) nonwoven fabric (Crane Technical Materials, United States). The nonwoven fabric had a density of 87 g/m^2, a thickness of 90–99 μm, and an air permeation of 16.0–29.6 sm^3/(sm^2·s·Pa).

All composite membranes were made from 14% cellulose solution in the mixture of [Emim]OAc with DMSO (ratio 1:1), the dissolution temperature of the cellulose was 80°C [22]. The method for preparing molding solutions, consisting in their degassing was developed, to obtain defect-free membrane samples. After completely dissolving of the cellulose in the mixture of [Emim]OAc with DMSO and obtaining of homogeneous solutions, they were transferred in the oven at the temperature 80°C for 2 hours to remove air bubbles dispersed in the solution. Further, the solutions were degassed using the vacuum pump for an hour. After complete degassing, the solutions were placed in the desiccator at 80°C. Then the hot solution of cellulose in the mixture of [Emim]OAc with DMSO was applied to the non-woven polyester support and then the membrane was formed using the laminator HLCL-1000 (ChemInstruments, USA) at the temperature 80°C. The laminated membrane was precipitated in water immediately.

The commercially available chlorinating agent Satreat® was used for etching of the support. It is the solution of trichloroisocyanuric acid (Fig. 1) in acetone. The polyester supports were held in this agent until completely wetted, and then they were dried.

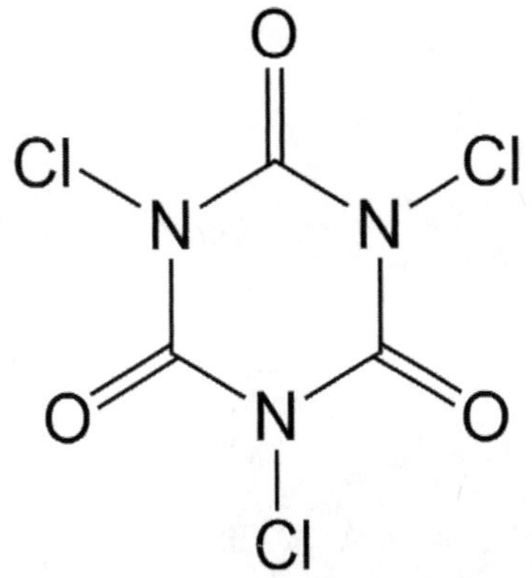

Figure 1. Structural formula of trichloroisocyanuric acid.

The efficiency of the chemical treatment of the support was evaluated from the change in the contact angle. The contact angle values were measured by via the conventional sessile drop technique using the LK-1 goniometer (Open Science, Russia). Measurement error was ± 2°.

The second method was made in two ways. The first method consisted in sequentially treating of the precipitated membranes using ethanol and then hexane; and the second is the treatment of the precipitated composite membranes with 20% aqueous solution of glycerol.

The filtration experiments were carried out in the set-up with dead-end filtration cells described elsewhere [24]. Helium was used to pressurize the liquid above the membrane due to insignificant difference in solubility of the gas in ethanol. The chamber above the membrane was filled with liquid (solvent or its dye solution). The permeate collector was arranged so as to minimize solvent evaporation; the volume of each collected liquid sample was 2.5±0.5 ml. The permeation of membranes was analyzed with respect to the aprotic solvent – DMF. To determine the separation properties of the resulting membranes, anionic dye – Remazol Brilliant Blue (Sigma-Aldrich) – were used.

3. Results and discussion

The main problem of the obtained membranes was the detachment of the deposited layer due to weak adhesion between the material of the support and the cellulose layer, and also due to high cellulose shrinkage under drying, which in summation led to the adhesive separation of the cellulose – polyester support system after dimethylformamide filtration (Fig. 2).

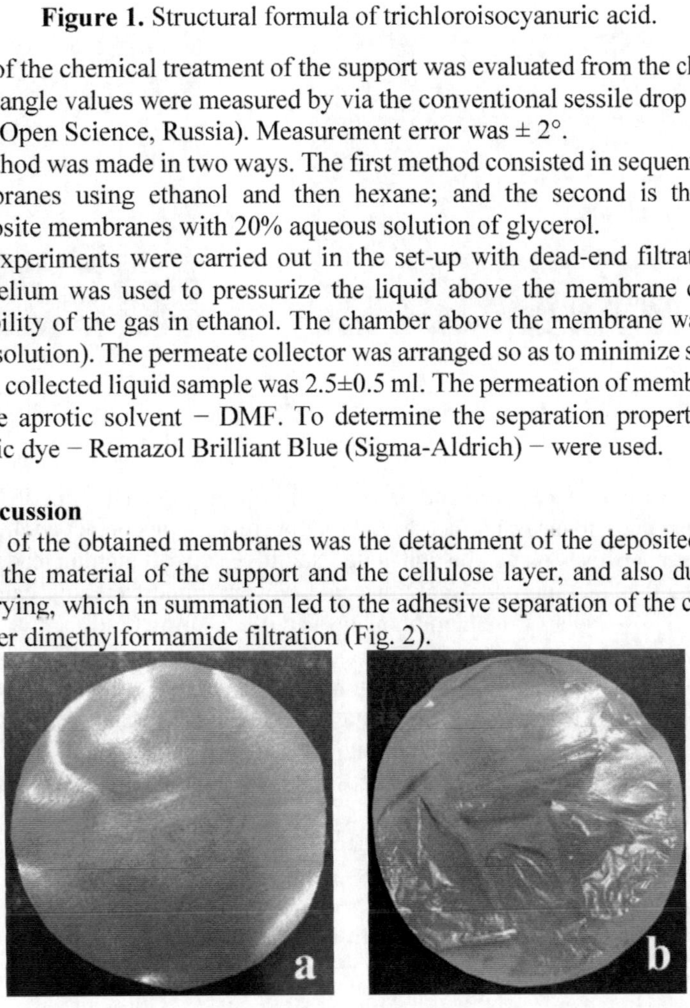

Figure 2. The composite cellulose membrane sample: a) freshly prepared; b) after DMF filtration.

To prevent cellulose layer detachment from the support, two approaches were analyzed:
- the preliminary etching of the polyester support before cellulose deposition;
- the decrease of cellulose shrinkage by post-treatment of membrane.

The images of the water and [Emim]OAc/DMSO mixture drops on initial and etched PET films (Fig. 3) as well as contact angles are given below.

IOP Conf. Series: Journal of Physics: Conf. Series **1099** (2018) 012039 doi:10.1088/1742-6596/1099/1/012039

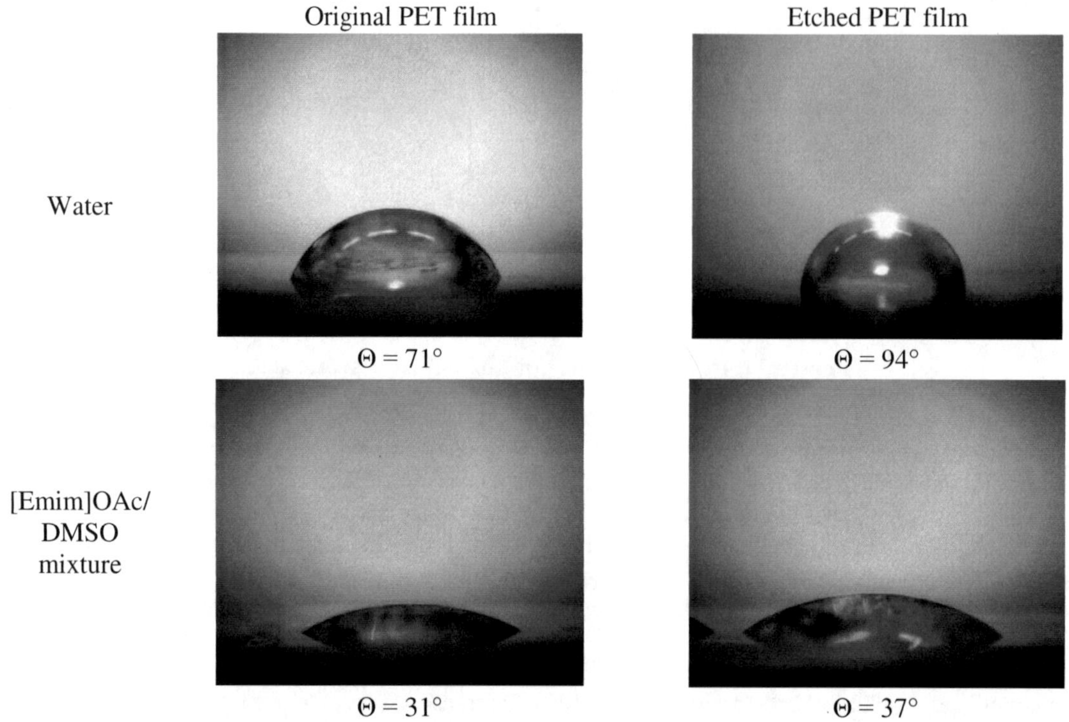

Figure 3. The images of the water and [Emim]OAc/DMSO mixture drops on initial and etched PET films.

It can be noticed that both water drop and drop of [Emim]OAc solution in DMSO spreads worse on the etched PET film, which indicates surface wettability decrease. The contact angle between the liquid and the support also enhances, so it can be concluded that PET support etching leads to hydrophobicity increase. This means that the approach including non-woven support etching can degrade adhesion between PET and cellulose layer of the membrane instead of elevating it.

The composite membrane post-treatment was carried out by two methods. The first method involved successive shrinkage of membrane, which was immersed successively into the liquids with different affinity to cellulose; whereas the second method included cellulose plasticization in order to decrease the level of internal stresses of the polymer. The results of the testing of composite cellulose membranes treated for preventing of the detachment of the deposited layer from the support are shown in Table 1.

Table 1. The comparison of the effectiveness of approaches to the suppress of the delamination

Etched PET	Coagulation bath	Post processing		P*, kg/m^2 h bar	R**, %	Adhesion failure
		bath 1	bath 2			
No	Water	Water	Water	0.40	76	Yes
Yes	Water	Water	Water	0.26	75	Yes
No	Water	Ethanol	Hexane	0.27	51	No
No	Water	Water	20% aqueous solution of glycerol	2.40	8	No

*Permeation of DMF
**Rejection of Remazol Brilliant Blue R

These data indicate that DMF permeation of composite cellulose membranes obtained by using etched support is the same as in case of ethanol and hexane pretreatment. Although the dye rejection coefficient for the etched membrane reached the highest values in comparison with other samples (75%), however, the cellulose layer exfoliated from non-woven polyester support. The post-treatment with 20% glycerol solution led to cellulose shrinkage decrease in the same way as for successive treatment with ethanol and hexane that also prevented the detachment of the deposited layer. On the other hand, the treatment with aqueous glycerol solution suppressed cellulose layer pores contraction during the drying, as evidenced by high value of DMF permeation through the obtained membranes. Permeation was approximately in 10-times higher than it was in case of membranes treated with ethanol and hexane. However, the rejection coefficient in case of membrane soaked in glycerol solution fails to meet the requirements of nanofiltration process.

4. Conclusion

Techniques were developed for the preparation of delamination-resistant cellulosic membranes. Two approaches have been analyzed to eliminate the adhesive peeling of the upper cellulose layer from the polyester nonwoven support during the filtration of the aprotic DMF solvent. The first approach, consisting of etching the PET support with solutions of trichloroisocyanuric acid in acetone, had a negative effect on the formed membranes: the cellulose layer left the support after the filtration experiments. The negative effect was also confirmed by an increase in the contact angles of water or the mixture of [Emim]OAc and DMSO for the initial and etched PET, indicating an increase in the hydrophobicity of the surface of the support film. The contact angle of the water for the initial PET film is 71°, and for the etched is 94°. The use of the second approach, which consisted in the post-processing of molded membranes, eliminated peeling. Treatment of the membranes with aqueous solutions of glycerol led to an increase in the permeation of DMF in 10 times, but the dye rejection dropped sharply to 8%. Therefore, as an optimal technique, sequential post-processing with ethanol and hexane was chosen, which makes it possible to obtain composite cellulose membranes with a high value of the rejection coefficient of the anionic dye.

Acknowledgments
This work was supported by the Russian Foundation for Basic Research, project no. 17-08-00499 A. The authors thank D.S. Bakhtin for help in measuring contact angles.

References

[1] Sokolov V Z, Kharlampovich G D Production and use of aromatic hydrocarbons. - M: Chemistry, 1980. 336 p. (in Russia)
[2] Torres-Garcia E, Galano A, Rodriguez-Gattorno G 2011 *Journal of catalysis* **282** 201
[3] Swanson B N 1985 *Rev Clin Basic Pharm.* **5** 1
[4] http://biocentr.org/dimeksid-dmso.html, recourse: 29.03.2018
[5] Volkov A, Tereshchenko G, Korneeva G 2008 *Russian Chem. Rev.* 2008 **77** 983
[6] Fengel D, Wegener G Wood (Chemistry, Ultrastructure, Reactions). - Berlin, New York: Walter de Gruyter, 1984
[7] Wüstenberg T Cellulose and Cellulose Derivatives in the Food Industry, Wiley-VCH, Weinheim, 2015
[8] Klemm D Comprehensive Cellulose Chemistry. Wiley-VCH, Weinheim, New York 1998
[9] Yushkin A, Anokhina T, Volkov A 2015 *Petr. Chem.* **55** 746
[10] Anokhina T, Yushkin A, Volkov V, Antonov S, Volkov A 2015 *Physics Procedia* **72** 171
[11] Bang Y, Lee S, Park J, Cho H 1999 *J. Appl. Polym. Sci.* **73** 2681
[12] Golova L, Makarov I, Kulichikhin V, Matukhina E *2010 Polym. Sci. Series A* **52** 1209
[13] GB 1144759 19690312, Eastman Kodak Co., inv. D. L. Johnson; *Chem. Abstr.* 70, 107081 (1996)
[14] Yoshihiko A, Akira M 2002 *J. Appl. Polym. Sci* **84** 2302

[15] Lewandowski Z 2002 *J. Appl. Polym. Sci* **83** 2762
[16] Li H J , Cao T M, Qin J J, Jie X M, Wang T H, Liu J H, Yuan Q 2006 *J. Membr. Sci.* **279** 328
[17] Anokhina T, Yushkin A, Makarov I, Ignatenko V, Kostyuk A, Antonov S, Volkov A 2016 *Petr. Chem.* **56** 1085
[18] Ilyin S, Makarova V, Anokhina T, Ignatenko V, Brantseva T, Volkov A, Antonov S 2018 *Cellulose* **25** 2515
[19] Heinze T, Schwikal K, Barthel S 2005 *Macromol. Biosci.* 2005 **5** 520
[20] Zhang H, Wu J, Zhang J, He J S 2005 *Macromolecules* 2005 **38** 8272
[21] Sukma F M, Çulfaz-Emecen P Z 2018 *J. Membr. Sci.* **545** 329
[22] Anokhina T, Pleshivtseva T, Ignatenko V, Antonov S, Volkov A 2017 *Petr. Chem.* **57** 477
[23] Durmaz E N, Çulfaz-Emecen P Z 2018 *Chem. Engineer. Sci.* 2018 **178** 93
[24] Volkov A 2014 *Sep. Purif. Tech.* **124** 43

Development of hollow fiber reverse osmosis membranes and modules

G G Kagramanov, I V Shibanov and A M Blanko-Pedrekhon

D. Mendeleyev University of Chemical Technology of Russia, Moscow, Russian Federation

Abstract. Perspectives of the hollow fiber reverse osmosis membranes and modules development have been analysed. Comparative analysis with other types of membrane modules and tentative calculation of module's design considering the hydrodynamic parameters of separation has been made.

1. Introduction

Modern industrial enterprises must meet the principles of energy and resource conservation and environmental safety requirements. Membrane technology fully meets these criteria, it is an effective method of processing, separating and concentrating both liquid and gas mixtures. Reverse osmosis, as a separation method of inorganic compounds aqueous solutions, has found wide application in industries, in medicine and other fields [1].

To carry out the reverse osmosis process, membrane modules are used. They can have a different design. At the present stage, almost 90% of the market, both in kind and in money terms, is occupied by spiral-wound modules (SW) [2]. They have a sufficiently high packing density of membranes (up to 1000 m^2 / m^3), a well-established system of production, installation and maintenance. However, such a design has a number of drawbacks [3]. So, for the formation of high pressure and drainage channels special grids - separators are necessary. Their presence affects hydrodynamics, contributes to increased hydraulic losses and, as a result, increases both capital costs during production (complex stages of sealing and winding), and operational ones. These disadvantages can be solved by switching to a different type of module design using membranes in the form of hollow fibers (HF).

The first experience of creating HF membranes and modules based on them was made by the American company DuPont in 1974 [4]. Module B-9 measuring 245 x 1016 (diameter x length) mm contained approximatively 4.4 million fibers, the membranes operational surface being 1200 m^2. Fibers with an outer diameter of 85 μm and an inner one of 42 μm were made of an aromatic polyamide [4, 5]. However, the development have not took into account the hydrodynamic parameters of liquid flow. The high hydraulic resistance of the drainage channel reduced the process' driving force - the pressure difference on both sides of membrane. Despite the high packing density of the hollow fibers (~ 20,000 m^2/m^3), the module's performance remained fairly low (~ 0,03 $\frac{l}{m^2 \cdot bar \cdot h}$) [5, 6].

Comparative analysis of advantages and disadvantages of SW and HF reverse osmosis modules allows drawing a conclusion about the prospects of modules based on HF. The estimated calculation was carried out taking into account the following assumptions:

- filtration from membrane outside space to inside one;
- specific productivity of both modules (SW and HF) remains equal;
- membrane specific productivity along the length of the high pressure channel (HPC) is constant;
- the density and viscosity of the liquid to be separated are constant along the length of the HPC;
- cocurrent flow of concentrate and permeate;

- standard overall dimensions of the module (module 8040: lengh 1016 mm, diameter 200 mm).

2. Calculation method

Optimization of membrane module design consisted in choosing the diameter of the fibers and, consequently, membranes packing density.

The flow velocity w_{in} (m / s) at the module input is calculated as:

$$w_{in} = \frac{L_{in}}{S} = \frac{L_{in}}{\frac{\pi \cdot D^2}{4} - \frac{\pi \cdot n_{HF} \cdot d_{od}^2}{4}} \tag{1}$$

where:

L_{in} - volumetric feed flux, m³/s;

S - the cross section area, m²;

D - the internal diameter of the membrane module, m;

n_{HF} – total number of HF in module;

d_{od}- outer diameter of the fiber, m.

The fluid velocity in an arbitrary cross section is equal to:

$$w = \frac{L}{S} = w_{in} - \frac{4 \cdot G \cdot d_{od} \cdot l \cdot n_{HF}}{D^2 - d_{od}^2 \cdot n_{HF}} \tag{2}$$

where:

L - change in the liquid flow rate along the length of HPC, m³ / s. L is calculated from the material balance equation;

G - specific productivity of the membranes, $\mathrm{m}^3/_{\mathrm{m}^2 \cdot s}$.

The general expression for calculating the hydraulic losses along the length of the module HPC are determined by the Darcy equation, which, after substituting expressions for all the quantities entering and integrating it, has the form of:

$$\Delta P_a = \frac{67 \cdot \mu \cdot w_{in} \cdot l}{d_{eq}^2} - 133 \cdot \frac{\mu \cdot G \cdot d_{od} \cdot l^2 \cdot n_{HF}}{d_{eq}^2 \cdot (D^2 - n_{HF} \cdot d_{od}^2)} \tag{3}$$

where:

ΔP_a - hydraulic resistance of the HPC, Pa;

d_{eq} - equivalent diameter of the HPC, m;

μ - viscosity of the liquid, Pa · s.

Similarly, when integrating the Darcy equation, an expression is derived for calculating the hydraulic resistance of the drainage channel ΔP_d (Pa):

$$\Delta P_d = 64 \cdot \mu \cdot G \cdot \frac{l^2}{d_{id}^3} \tag{4}$$

where:

d_{id} - internal diameter of the fiber, m;

The velocity can be calculated as the ratio of the permeate flux to the cross section of the fiber:

$$w = \frac{P}{S} = \frac{G \cdot F}{S} = \frac{4 \cdot G \cdot l}{d_{id}} \tag{5}$$

where:

P - flux of permeate, m³ / s;

S - cross-sectional area of the drainage channel, m²;

F - surface area of the drainage channel, m².

3. Results and discussion

According to the relationships obtained, the dependence of the HPC hydraulic resistance on the membranes operational area and the optimum range of geometric dimensions (diameters) of HF membranes were established (fig.1). Relationship between of the drainage channel's hydraulic resistance ΔP_d and the packing density of membrane in module for different HF diameters (from 0,8 to 1,6 mm) is shown on fig.2.

The obtained dependences allow to determine the driving force of the separation process (pressure drop on both sides of the membrane), which could be related to exploitation costs. For comparison, the the pressure channel's hydraulic resistance of SW modules is comparable or slightly lower than this

IOP Conf. Series: Journal of Physics: Conf. Series **1099** (2018) 012040 doi:10.1088/1742-6596/1099/1/012040

value for HF. However, the drainage grid significantly increases the hydraulic resistance in the drainage channel, which leads to a decrease in the driving force of the process (increase in operating costs). To sum up, these relationships allow developers to choose the optimal construction and parameters of membrane module.

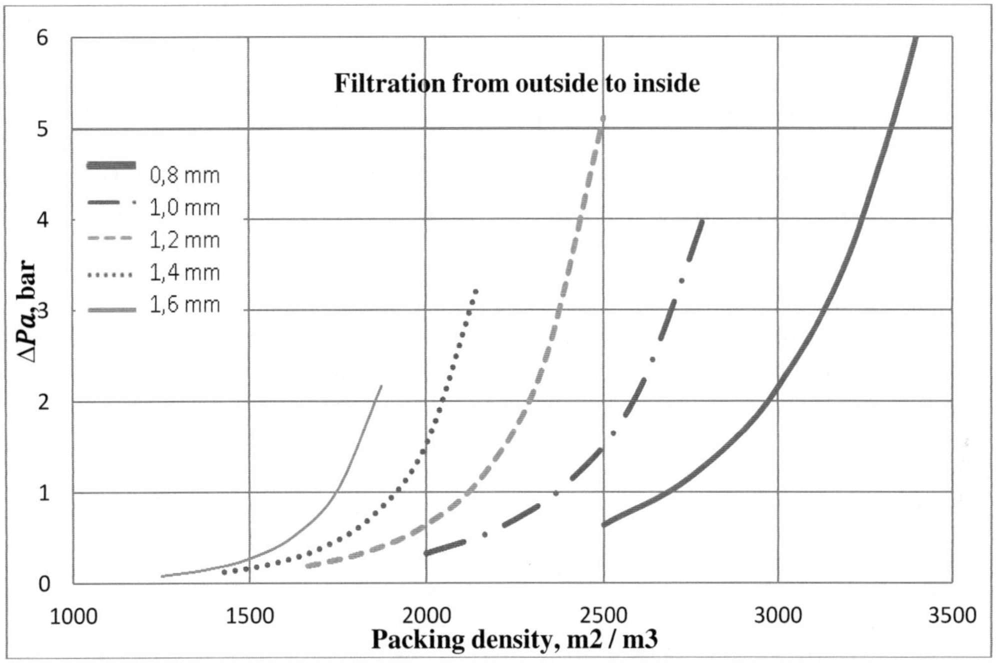

Figure 1. Relationship between of the HPC hydraulic resistance ΔP_a and the packing density of membrane module (m^2 / m^3).

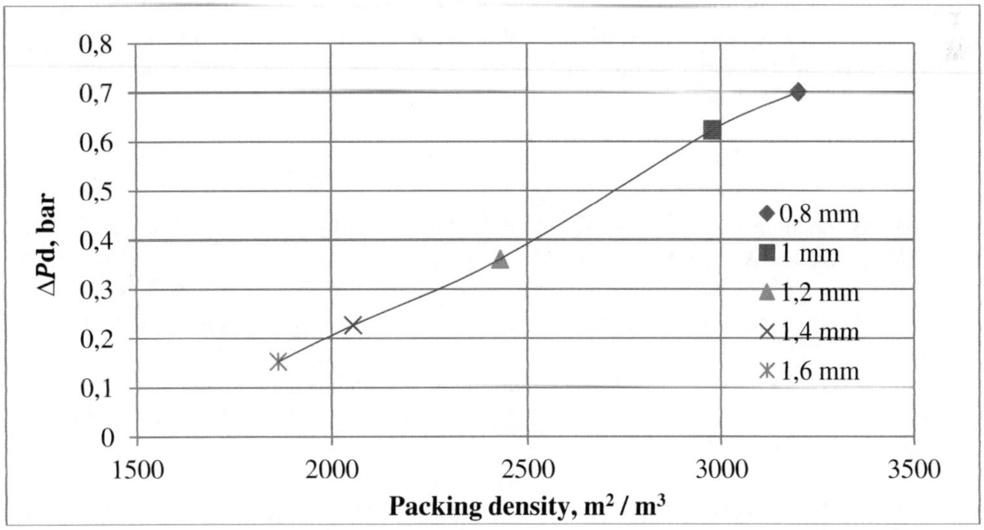

Figure 2. Relationship between of the drainage channel's hydraulic resistance ΔP_d and the packing density of membrane module for different HF diameters.

Based on the results of the calculation and analysis of industrial experience, the advantages of hollow fiber reverse osmosis membranes and modules are:

- *High packing density of membranes (m^2 / m^3).* With the same hydraulic resistance and operating pressure for SW and HF membranes, the packing density of HF is higher, which makes it possible to increase productivity in the same overall module dimensions by 1.5-2 times compared to the SW or to reduce the module dimensions while maintaining productivity.

- *Hydrodynamics in the HPC.* The hydrodynamic regime in the HPC of the HF module is turbulent, the transition region. Thus, Reynolds number value for HF remains approximately constant for different fiber diameters and is equal to 500. According to the paper [7], the critical value of the Reynolds number for a cramped channel at which the hydraulic regime becomes turbulent, is 100. HF module is the self-supporting construction and is not complicated by separating and drainage grids, liquid mixing occurs due to the flow around the fibers and their own oscillations. Thus, in the HPC of the HF module, the turbulence degree of the flow increases compared with SW one, hence an increase in the mass-transfer coefficient and a decrease in the concentration-polarization (CP) value could be achieved.

- *High resistance to contamination and economical washing stages.* Due to better hydrodynamics in HPC, the requirements for the feed water are less rigid for HF elements, compared with SW one - the layer of impurities forming on the fiber surface breaks down easily. As a consequence, the consumption of reagents and water for washing decreases. Comparative characteristics of SW and HF modules is shown in tab.1.

Table 1. Comparative characteristics of SW and HF modules (size 8040)

Parameter	SW	HF
Permeate productivity, m^3 / h	Up to 2	3 - 4
Operational surface, m2 (packing density, m^2 / m^3)	35 – 40 (~ 1200)	60 – 90 (~ 2500)
Fluid hydrodynamic mode in HPC	laminar	turbulent, transition area

4. Conclusions

The optimal geometric dimensions of the hollow fibers and the packing densities of them are discussed. Analysis of the manufacturing of HF and flat composite membranes and modules [2] and operational characteristics shows the perspectives of HF modules. The ability to reduce the size of hollow fiber modules while maintaining high performance makes them promising to use especially in limited spaces: compact desalination plants on sea vessels and submarines; mobile water treatment plants; water treatment plants for the aerospace industry.

References

[1] Zheng X, Zhang Z, Yu D, Chen X, Cheng R, Min S, Wang J, Xiao Q, Wang J 2015 *Resources, Conservation and Recycling* **105** 1–10

[2] Lee K P, Arnot T C, Mattia D 2011 *Journal of Membrane Science* **370** 1–22

[3] Kucera J 2010 Reverse Osmosis, Design, Process, and Applications for Engineers *Wiley-Scrivener* p 416

[4] Gorenflo A, Redondo J A, Reverberi F 2005 *Desalination* **178** 247-260

[5] Chatterjee A, Ahluwalia A, Chatterjee A, Senthilmurugan S, Gupta Sh K 2004 *Journal of Membrane Science* **236** 1–16

[6] Ukai T, Nimura Y, Ukai T, Hamada T, Matsui H 1980 *Desalination* **32** 169 – 178

[7] Happel J, Brenner H 1965 Reynolds number hydrodynamics *Published by Prentice Hall, Englewood Cliffs, N.J.* p 630

IOP Conf. Series: Journal of Physics: Conf. Series **1099** (2018) 012041 doi:10.1088/1742-6596/1099/1/012041

Research of the component of the hybrid alumosilicic reagent - aluminum sulfate in the process of water treatment

S N Tikhonov, R A Alexandrov, I M Kurchatov, E A Ananyeva and D Yu Feklistov

National Research Nuclear University MEPhI, Kashirskoe sh. 31, Moscow, 115409, Russia

Email: tisenik@mail.ru

Abstract. The work is devoted to theoretical and experimental studies of the process of hydrolysis of an aluminum salt in an aqueous solution. Quantitative dependences of secondary contamination by aluminum ions in the process of hydrolysis of aluminum sulfate on the initial salt concentration and *pH* of the solution are derived, and the range of *pH* values at which the largest number of uncharged aluminum hydroxide complexes are formed is determined. Experimental studies have confirmed the theoretical dependence of the *pH* value of the solution on the amount of aluminum sulfate dissolved in water.

1. Introduction

At present, one of the urgent tasks is to obtain clean drinking water. One of the most popular methods of water purification is a reagent method based on the processes of coagulation and flocculation, in particular, using aluminum-containing reagents. The use of such reagents can lead to an increase in the concentration of residual aluminum, which leads to secondary contamination of the water being purified by aluminum ions. However, when using a hybrid alumosilicic coagulant-flocculant FCS - 9, the residual aluminum concentration is significantly reduced due to the implementation of the sorption mechanism in self-assembled mesoporous alumosilic structures [1].

The method of obtaining the FCS - 9 reagent is based on the treatment with sulfuric acid of the aluminosilicate raw material and is described in. The mechanism of water purification by the FCS - 9 reagent includes the process of formation of aluminum hydroxide. In this paper, a numerical study of the mathematical model of the process of hydrolysis of a given component of a reagent is conducted, whose behavior is similar to the process of hydrolysis of aluminum sulfate [2, 3]. Consequently, the subject of this study is $Al_2(SO_4)_3$.

A comparison is also made between the theoretical change in the *pH* of the solution, depending on the amount of aluminum sulfate dissolved in the aqueous solution and experimental. Confirmation of the correctness of the mathematical model allows us to use numerical studies to determine the optimal dose of the reagent in the process of water purification.

2. Experiment

In the aluminum sulfate solution, stepwise hydrolysis takes place along a multiply charged cation:

$$2Al_2(SO_4)_3 + 4H_2O \rightleftharpoons 4AlOHSO_4 + 2H_2SO_4 \tag{1}$$

$$2AlOHSO_4 + 2H_2O \rightleftharpoons (Al(OH)_2)_2SO_4 + 2H_2SO_4 \tag{2}$$

$$(Al(OH)_2)_2SO_4 + 2H_2O \rightleftharpoons 2Al(OH)_3 + H_2SO_4 \tag{3}$$

In the process of hydrolysis, concentration of uncharged compounds $Al(OH)_3$, which have a strong coagulating ability, arise [2]. In this case, the total aluminum concentration Al_Σ as monomers will be determined by the sum of all the concentrations of various aluminum hydro complexes [4]:

$$Al_\Sigma = [Al^{3+}] + [AlOH^{2+}] + [Al(OH)_2^+] + [Al(OH)_3^0] + [Al(OH)_4^-] \tag{4}$$

When using the general equilibrium equations of chemical reactions to describe real solutions, it is necessary to use the ion activity method, according to which certain numbers are introduced into these equations, instead of the concentration to take into account the interaction of solution ions. In accordance with this, we rewrite equation (4) taking into account the activities of the components of the hydrolysis process, where in parentheses we denote the activities of the corresponding substances:

$$Al_\Sigma = (Al^{3+}) \times \left\{\frac{1}{\gamma_{Al^{3+}}} + \frac{K_1}{(H^+)\gamma_{(AlOH)^{2+}}} + \frac{K_2}{(H^+)^2\gamma_{Al(OH)_2^+}} + \frac{K_3}{(H^+)^3\gamma_{Al(OH)}} + \frac{K_4}{(H^+)^4\gamma_{Al(OH)_4^-}}\right\}, \tag{5}$$

where γ_i – coefficient of activity i – th solution component ($log(\gamma_i) = -Az_i^2\sqrt{I}$, $I_c = \frac{1}{2}\sum_{i=1}^{n} c_i z_i^2$), A = – 0,511 – a constant that does not depend on the charge of the ion and the ionic strength of the solution, but depends on the dielectric constant of the solvent and the temperature, z_i – ion charge i – th solution component, I – ionic strength of solution, c_i – molar concentration [5], and the corresponding equilibrium constants: $log_{10}(K_1) = -5.02$, $log_{10}(K_2) = -9.3$, $log_{10}(K_3) = -15$, $log_{10}(K_4) = -23.57$.

Using the equilibrium constants and equation (5), one can determine the concentrations of the hydrolysis products by the general formula:

$$\left(Al_m(OH)_n^{(3m-n)+}\right) = \frac{K_{mn} \times (Al^{3+})^m}{(H^+)^n \times \gamma_{Al_m(OH)_n^{(3m-n)+}}}, \tag{6}$$

where the constant K_{mn} corresponds to K_1 for m = 1, n = 1. In a similar way, correspondence with the remaining equilibrium constants is carried out.

On the basis of equation (6), the ratios of various products of hydrolysis of aluminum in solution relative to the total amount of dissolved aluminum are plotted as a function of the equilibrium *pH*.

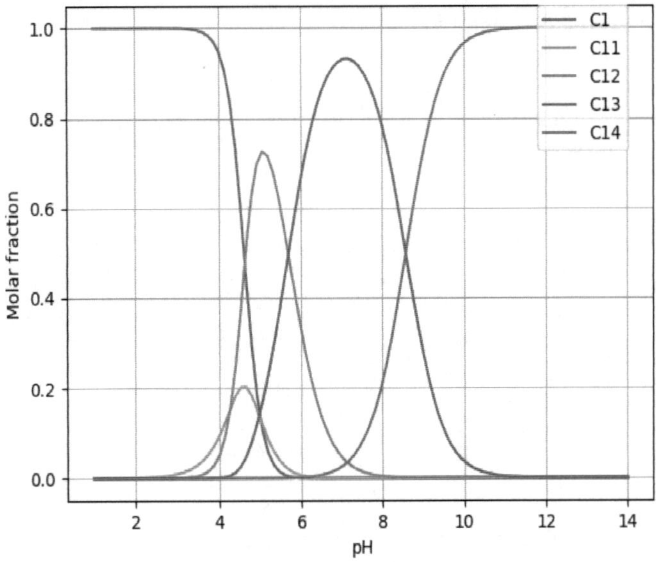

Figure 1. Molar fractions of dissolved hydrolysis products of mononuclear aluminum hydroxides in equilibrium with amorphous hydroxides; *C1* corresponds to the concentration of aluminum ions Al^{3+}, *C11* – concentration of $AlOH^{2+}$, *C12* – concentration of $Al(OH)_2^+$, *C13* – $Al(OH)_3^0$, *C14* – concentration of $Al(OH)_4^-$

As can be seen from figure 1, at $pH < 4$, aluminum in solution is predominantly in the form of a highly hydrated aluminum ion, the $Al(H_2O)_6^{3+}$ aqua complex, and in the range of *pH* values from 9 to 10, practically all of the aluminum in the solution is aluminate ion $Al(OH)_4^-$. Under neutral conditions, the predominant aluminum compounds in solution are uncharged molecules of $Al(OH)_3^0$.

Using the equation of equilibrium of the hydrolysis process and equation (6), graphs of the distribution of aluminum hydrolysis products are plotted as a function of the equilibrium *pH* and the amount of dissolved aluminum salt (figures 3 a, b, c) from which the quantitative effect of the $Al_2(SO_4)_3$ salt on the solution can be determined.

a) b)

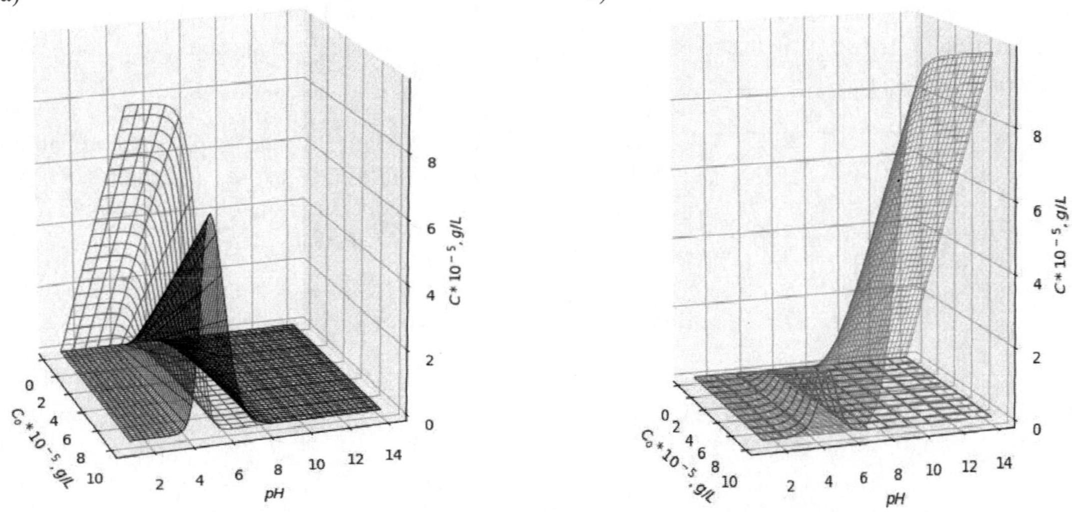

Figure 2 a, b The graph of the distribution of aluminum hydrolysis products as a function of the equilibrium *pH* and the amount of dissolved aluminum salt; *C0* is the concentration of aluminum salt, blue – *C1* corresponds to the concentration of aluminum ions Al^{3+}, red – *C11* is the concentration $AlOH^{2+}$, black – *C12* is the concentration $Al(OH)_2^+$, green – *C14* - concentration $Al(OH)_4^-$

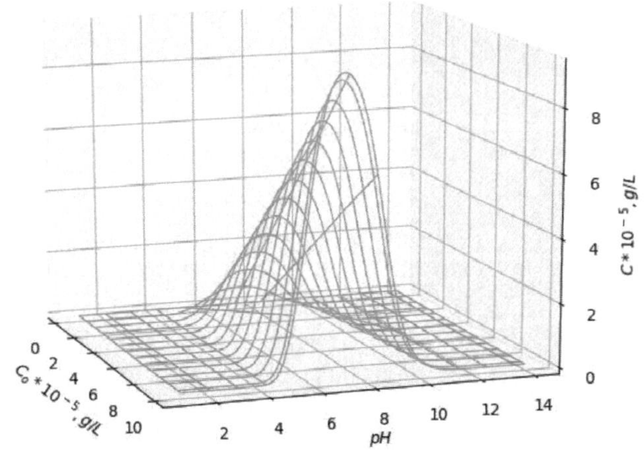

Figure 3. The graph of the distribution of aluminum hydrolysis products as a function of the equilibrium *pH* and the amount of dissolved aluminum salt; *C0* is the concentration of aluminum salt, *C1* corresponds to the concentration of aluminum ions Al^{3+}, *C11* is the concentration $AlOH^{2+}$, *C12* is the concentration $Al(OH)_2^+$, *C13* - $Al(OH)_3$, *C14* - concentration $Al(OH)_4^-$

It can be seen from figures 2 a, 2 b and 3 that the character of the molar concentration dependence at low concentrations of C_0 corresponds to the dependence shown in figure 1.

The most interesting is the dependence of the amount of formed uncharged $Al(OH)_3$ molecules on the initial pH_0 of the medium and on the C_0 concentration of the aluminum salt $Al_2(SO_4)_3$ in solution, shown in figure 3. This product of hydrolysis - colloidal particles, is an active component of the coagulation process and affects the aggregative stability of the solution.

As a result of the stepwise reactions described by equations (1 – 3), monomeric aluminum compounds are formed, a proton is released, and the *pH* of the solution decreases. Each subsequent stepwise reaction shifts the *pH* of the solution to the acid zone. In view of this, it is possible to describe the change in the *pH* of the medium from the amount of dissolved aluminum salt. Figure 3 shows the theoretical and experimental plot of the *pH* of the solution as a function of the aluminum salt concentration $C(Al_2(SO_4)_3$ in the solution. The error of the experimental results does not exceed 5% of the theoretical results.

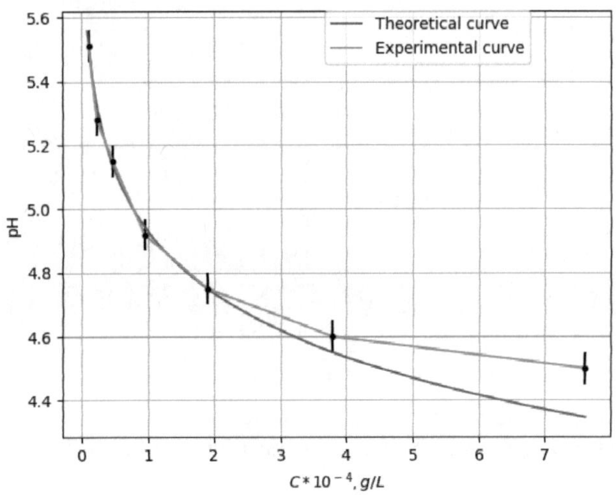

Figure 4. The graph of the *pH* of the solution as a function of the concentration of the aluminum salt $C(Al_2(SO_4)_3)$ in solution

The first experimental point was obtained in a solution with $7{,}5\times10^{-4}$ kg of aluminum sulphate per m3, after which it was diluted 2 times to obtain a solution with an aluminum sulfate content of $3{,}75\times10^{-4}$ kg / m^3. The process was repeated until the $Al_2(SO_4)_3$ content reached $0{,}12\times10^{-4}$ kg / m^3.

3. Conclusions

According to the numerical studies of the mathematical model of the process of hydrolysis of an aluminum salt in an aqueous solution, a theoretical dependence of the amount of hydrolysis products on the equilibrium *pH* of the solution and the amount of dissolved aluminum sulfate, which was experimentally confirmed, was constructed. The obtained data allow to determine the range of *pH* at which the largest amount of uncharged aluminum hydroxide molecules - $Al(OH)_3(H_2O)_3$ complexes - is formed and the water purification process is intensified.

Experimental studies on the *pH* of the solution on the amount of aluminum sulfate dissolved in aqueous solution are consistent with numerical studies of the mathematical model of the hydrolysis process, which allows determining the optimal dose of the reagent for various initial concentrations of contaminants and increasing the efficiency of the reagent under study.

Acknowledgements

The work was financially supported by the Ministry of Education and Science, agreement №14.575.21.0159. Unique project identifier RFMEFI57517X0159.

References

[1] Laguntsov N I and Neschimenko Y P 2008 Moscow: *Proceedings of Nanotechnology International Forum «Rusnanotech 08»* 609 -611
[2] Feng X and Baojie Z 2008 *Journal of Environmental Sciences* **20** 907 - 914
[3] Saukkoriipi J 2010 *Theoretical study of the hydrolysis of aluminum complexes* (Oulu: University of Oulu) p 90
[4] Benschoten J E V 1990 *Water Res.* **24** 1519 - 1526
[5] Qian Z 2009 *Dalton Trans.* **3** 521 - 529
[6] Kuan W 2005 *Water Res.* **39** 3457 - 66
[7] Hu C 2006 *Environ. Sci. Technol.* **40** 325 - 331

IOP Conf. Series: Journal of Physics: Conf. Series **1099** (2018) 012042 doi:10.1088/1742-6596/1099/1/012042

Solid-phase transformations in beryllium of technical purity with the participation of the Mössbauer iron isotope

S S Martynenko and V I Petrov

National Research Nuclear University MEPHI (Moscow Engineering Physics Institute), 115409, 31 Kashirskoe shosse, Moscow, Russian Federation

E-mail: vipetrov@mephi.ru

Abstract. New data on the redistribution of impurities in beryllium of technical purity were obtained by Mössbauer spectroscopy. The metal with 0.15% aluminum, 0.11% iron and other impurities was investigated. The introduction of iron enriched with isotope ^{57}Fe in addition to natural iron as an impurity provided the magnitude of the Mössbauer effect, sufficient for resolving the experimental spectra and determining the parameters of the supersaturated solution that decomposed and the secondary phases that appear. Isothermal annealing at 600 °C leads first to the precipitation of the $AlFeBe_4$ phase, which predominates for 1,700 hours. During the solid-phase sequential reaction, the release of its final product of the $FeBe_{11}$ type gradually develops. Modifications of the process, the necessary level of enrichment, as well as the possibility of regulating the distribution of impurities that affect the properties of beryllium are discussed. The results obtained can be useful in analyzing the behavior of impurities during heat treatment and operation of reactor beryllium and other materials.

1. Introduction

Beryllium is used in a number of nuclear reactors and other types of equipment. In beryllium of technical purity, it is customary to take into account the separation of secondary phases and the interrelation of impurities [1-6]. Over time, the material gradually degrades, in particular, thermal conductivity deteriorates [7]. Among a number of factors that determine the stability of the properties of the material, the behavior of the impurities is quite important.

In the present paper, Mössbauer studies of technical beryllium with a limiting content of aluminum impurities were carried out to evaluate possible structural-phase changes during long-term operation of the metal at an elevated temperature.

2. Experimental

The Mössbauer spectra (i. e., the dependence of the Mössbauer effect on the velocity of the sample under study) are obtained for an alloy simulating the composition of technical beryllium. Measurements of the effect on transmission were carried out on a spectrometer MC1101E with a resonant detector of the RSDU type (Rostov), a ^{57}Co radiation source in chromium (Ritvertz) was used. Samples were in the form of plates with a thickness of 1.5-2 mm. The gamma radiation source, detector and sample were placed in a thermostat in which a constant temperature of 25.00 ± 0.05 °C was maintained. The known Mössbauer parameters of the solid solution of iron in beryllium and the secondary phases of $FeBe_{11}$ and $AlFeBe_4$, and other possible iron-containing phases were used for identification [8-11]. The spectra were processed according to the methods [12, 13], developed with the participation of the authors.
Beryllium of technical purity usually has about 10 significant impurities [1-5], in particular in reactor beryllium [5], among important impurities can be noted oxygen - up to 2 wt. %, carbon - up to 0.1,

silicon - up to 0.03, iron - up to 0.17, aluminum - up to 0.14, magnesium and other metals - in total up to 0.05 wt. %. In the initial state (after preparatory technological annealing), such alloys have, as a rule, a Mossbauer spectrum consisting of doublet lines of a solid solution and one or two secondary phases, with characteristic parameters of the fine structure. In natural iron, 2% of the Mössbauer isotope ^{57}Fe is contained. This small content does not always allow us to determine the phase composition, and to enhance the effect, iron needs to be enriched.

Earlier, processes of precipitation of various phases in beryllium of technical grade during heat treatment were observed [1, 3, 11, 13, 14]. In addition to the known data, let us consider the case when the aluminum content is 0.15 wt. %, that is, it is at the upper level of the interval typical for the technical metal. The material also contained 0.11 wt. % of iron enriched with the ^{57}Fe isotope. This enrichment was obtained by fusing beryllium containing a small amount, about 0.05%, of natural iron, with the addition of iron, consisting of 90% of the Mössbauer isotope. The content of the remaining metallic impurities did not exceed 0.05 wt. %.

The alloy was homogenized at 1000 °C. Further, a long (more than 2,100 h) isothermal annealing at 600 °C was carried out. The temperature was chosen in order to accelerate diffusion processes. The spectra were taken at intermediate stages and upon completion of annealing with rapid cooling, fixing the high-temperature phase composition of the alloy.

3. Results and discussion

After homogenization, the Mössbauer spectrum of the investigated material showed only the presence of a solid solution of beryllium-iron, the other iron-containing phases were absent; the weak solubility of aluminum did not affect the spectrum.

At the beginning of isothermal annealing, as a result of the decomposition of the supersaturated solid solution, only the secondary phase of $AlFeBe_4$ was detected. Then, with some delay (not less than 50 hours), a phase of the $FeBe_{11}$ type began to appear. The spectrum of the sample after one of the annealing is shown in Fig. 1.

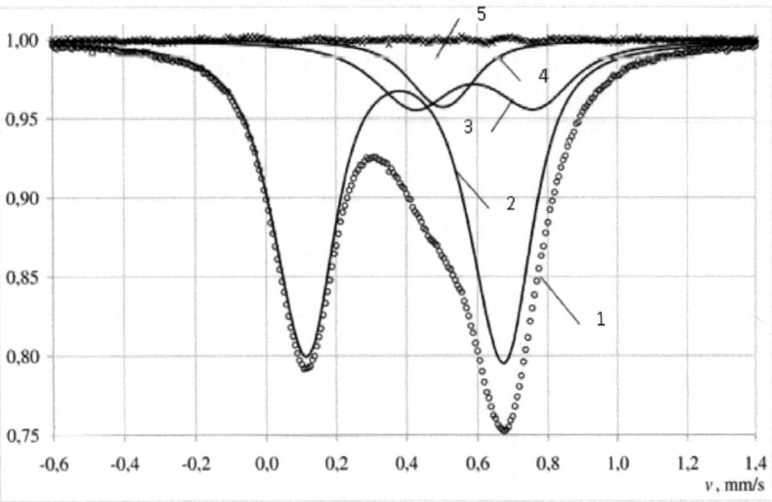

Figure 1. Mössbauer transmission spectrum for beryllium with 0.15% aluminum after 138 hours of annealing at 600 °C: 1 - experimental spectrum; 2 - doublet of a solid solution; 3 - doublet of the secondary phase of $AlFeBe_4$; 4 - doublet of the phase of the $FeBe_{11}$ type; 5 is the difference spectrum

Long-time isothermal annealing has made major changes. Fig. 2 represents the iron content in different phases during the process.

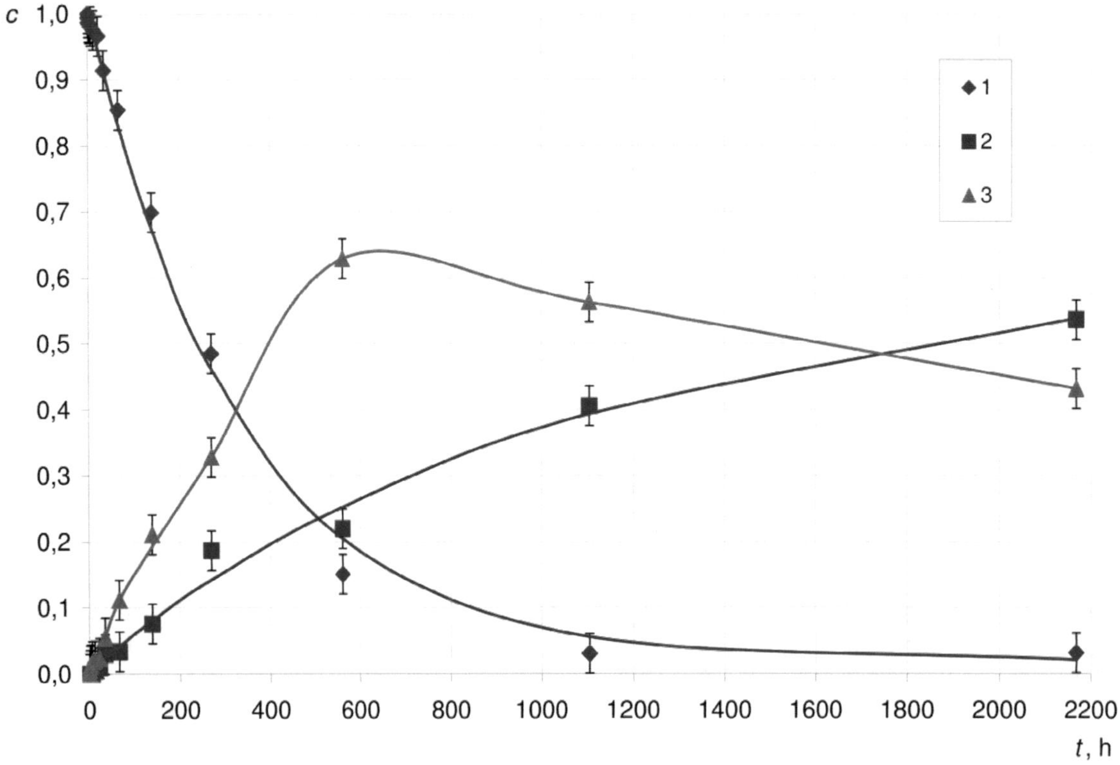

Figure 2. The relative iron content in various phases of beryllium with isothermal annealing at 600 °C: 1 - in a solid solution; 2 - in the $FeBe_{11}$ phase; 3 - in the $AlFeBe_4$ phase.

The processing of the spectra made it possible to reveal the two secondary phases. The $AlFeBe_4$ phase with an isomer shift of 0.57 ± 0.01 mm/s and a quadrupole splitting of 0.28 ± 0.01 mm/s can be satisfactorily identified from literature data, the second iron-containing phase with an isomer shift of 0.54 ± 0.01 mm/s and a quadrupole splitting of 0.14 ± 0.01 mm/s, taking into account the literature data, we are approximately classed as$FeBe_{11}$ type, but it has a somewhat higher isomeric shift, which may indicate partial substitution of iron for other impurities and the complication of the lattice.

In beryllium, the precipitation of the $AlFeBe_4$ phase is the most rapid during annealing, and it has proved to be the most preferable for a long time. Only after 1,700 hours it yielded to another secondary phase, but even with an exposure of 2,167 hours it contained about 40% of the total amount of iron impurity. As a result, the relatively high content of the ternary phase was observed during a time interval exceeding the usual one corresponding to a reduced content of aluminum.

After passing through the maximum, the fraction of the triple phase began to decrease, which indicates that it plays the role of an intermediate in a multistage solid reaction. Apparently, the marked excess of the triple phase and its gradual decay are a characteristic feature for a given impurity composition, taking into account somewhat higher aluminum content.

During annealing, the phase of $FeBe_{11}$ type initially lagged behind, and then gradually became dominant. We applied a diffusion model developed earlier [13] to this process, and estimated the time of possible output to a constant phase ratio. It turned out that this could occur no earlier than 4,000 hours of annealing. This significantly distinguishes the material in question from those studied previously. Its properties, possibly, will be stable for a longer time of exploitation. In contrast, the behavior of ordinary beryllium of technical purity for various isothermal anneals, even with exposure to 500-1000 h [13, 14], was characterized by the establishment of a certain ratio of secondary phases. With longer exposures, this ratio was almost unchanged.

IOP Conf. Series: Journal of Physics: Conf. Series **1099** (2018) 012042 doi:10.1088/1742-6596/1099/1/012042

In this work, by the nature of the change in the relative iron content in different phases during the annealing (Figure 2), the entire process can be attributed to a successive solid-state reaction of the type

$$A \rightarrow B \rightarrow C$$

The following notations are used here: A-system, which includes the initial supersaturated solid solution $(Be + Fe)_0$ and an admixture of poorly soluble aluminum at the grain boundaries; B - system including $AlFeBe_4$ phase and depleted solid solution $(Be + Fe)_1$; C is a system including a $FeBe_{11}$ type phase, free aluminum and a depleted solid solution $(Be + Fe)_2$. In the course of this multistage reaction, the following transformations occur gradually:

- first, with the inclusion of an admixture of iron diffusing from the solid solution to the boundary, that is, to aluminum, a triple phase is formed:

$$(Be + Fe)_0 + Al \rightarrow AlFeBe_4 + (Be + Fe)_1;$$

- then, with the precipitation of a new phase containing iron, redistribution of iron and aluminum begins

$$AlFeBe_4 + (Be + Fe)_1 \rightarrow FeBe_{11} + Al + (Be + Fe)_2$$

Since other metal impurities from the solution (denoted by the letter M) can be reactive [3], the reactions can also be represented as:

$$(Be + Fe + M)_0 + Al \rightarrow AlFeBe_4 + (Be + Fe + M)_1,$$
$$AlFeBe_4 + (Be + Fe + M)_1 \rightarrow (Fe, M)Be_x + Al + (Be + Fe + M)_2,$$

where the index x is close to 11.

The possibility of redistribution during long-time annealing helps to uncover the mechanism of complex processes during operation and to approach their management. With a decrease in the proportion of the triple phase, its effect on the physical and mechanical properties of the material is gradually weakened. Apparently, there is some critical portion of this phase that should be supported. One can find this fraction with allowance for the thermodynamic conditions for the existence of phases in beryllium alloys [1, 3, 15] and the features of successive reactions, and to maintain it, it should probably be ensured that the aluminum content exceeds the iron content.

It can be suggested in practice to test several alloys with reduced iron content to select the optimal process. We do not touch the fraction of the natural impurity and believe that the adopted technology does not allow it to be significantly reduced to a negligible value. At the same time, it is possible to preserve a sufficiently high value of the Mössbauer effect by providing the necessary minimum of enrichment with the Mössbauer isotope. To estimate this minimum, we use our experimental data on the quality of resolution of the spectra [12]. It has been established that, without enrichment, the difference spectrum is up to 10%, with an enrichment of 50%, it averages 3%. The 10% difference spectrum does not allow to properly resolve the complex spectrum, since it is necessary to single out 3 doublets with close lines. For an acceptable resolution, the difference spectrum should not exceed 5%.

Thus, with a decrease in enrichment, that is in the magnitude of the effect, the fraction of the difference spectrum that prevents resolution is growing. With a rough approximation, we can assume that the difference spectrum increases linearly with decreasing enrichment. Using the linear approximation, we find that the 5% difference spectrum is reached when enriching about 30%. Hence, to control the maintenance of the required amount of ternary phase, 30% enrichment of the iron with a Mössbauer isotope is sufficient. This value can be achieved with the preparation of alloys.

To simplify the preparation of samples with enriched iron, it is possible to use the ion implantation method based on the experience of introducing iron and aluminum from films into beryllium [16] and taking into account the general mechanochemical regularities [17]. Iron has several chemical compounds with beryllium and a relatively high solubility [1, 3]. To interact with beryllium, therefore, it requires a smaller dose of mechanical energy. Iron is excited by an ion beam with low losses on atomization and is energetically introduced. During the implementation, it is usually expected that phases will be created according to the type of stable and metastable formations existing under normal conditions. For the Be-Fe system, this is a solid solution, as well as compounds of the $FeBe_{11}$ and $AlFeBe_4$ type. Aluminum is slightly soluble in beryllium, does not have binary compounds with it, but diffuses better at elevated temperatures than iron. When implantation, this or that penetration depth of aluminum is realized. It depends on the ratio of the number of atoms of iron and aluminum in the matrix and the film. Subsequent heat treatment can equalize the concentration of impurities by volume and create the necessary amount of a triple phase.

Similarly, reactor irradiation can act on the material prepared with the optimal composition, which gradually, with the accumulation of a dose [18], creates conditions for the movement and redistribution of impurities and, accordingly, changes in the properties of the material. Thus, in principle it is possible to obtain a material with long stable properties

4. Conclusion

According to Mössbauer spectroscopy data, a long-term annealing of technical beryllium with an increased aluminum content leads to a sequential solid-phase reaction. This leads to a controlled redistribution of iron and aluminum impurities in favor of the intermediate secondary phase of $AlFeBe_4$, which has a significant amount for more than 2,100 hours.

The $AlFeBe_4$ phase is well identified with Mössbauer parameters, which coincide with the literature data. To control the maintenance of the required amount of triple phase, 30% enrichment of the iron with a Mössbauer isotope $_{57}Fe$ is sufficient.

The results obtained can be useful in analyzing the behavior of impurities during heat treatment and long-time operation of reactor beryllium and other materials.

References

[1] The Beryllium science and technology 1979, eds D Webster et al. (New York: Plenum Press)
[2] Azhazha V M, Babun A V, Kovtun K V, Sanin F P and Sanin A F 2005 *Berilii – konstruktsiinii material aerokosmichnoi tekhniki* (Beryllium is the structural material of air-space technique) (Dnepropetrovsk: ART-PRES, in Ukrainian)
[3] Papirov I I 1981 *Struktura i svoistva splavov berilliya. Spravochnik* (Structure and properties of beryllium alloys. Handbook) (Moscow: Energoizdat, in Russ.)
[4] Chaouadi R, Leenaerts A, Puzzolante J L and Scibetta M 2000 *Proceedings of the 4th International Workshop on Beryllium Technology for Fusion*. Karlsruhe, Sept. 1999. FZKA-6462 (Karlsruhe: Fusion Technology) 233-246
[5] Posevin A O and Chakin V P 2007 *Berillii pod oblucheniem: obzor* (Beryllium under irradiation: review) (Dimitrovgrad: NIIAR, in Russ.)
[6] Chakin V, Moeslang A, Svyatkin M, Posevin A, Vladimirov and Latypov R 2008 *Proceedings of the International Symposium on Materials Testing Reactors*, JAEA Oarai R@D Center, Japan, July 16-17, 2008 (Oarai: R@D Center) 107-116
[7] Latypov R N 2012 *Teploprovodnost berilliya posle oblucheniya do vysokoi povrezhdayuschei dozy* (Heat conductivity of the beryllium after a high doze) Thesis (Moscow: VNIINM, 29 p, in Russ.)
[8] Janot C and Gibert H 1972 *Mater. Sci. Eng.* **10** 23-31
[9] Janot C and Delcroix P 1974 *Phil. Mag.* **30** 651-655
[10] Kistner O C and Mozer B 1962 *Bull. Amer. Phys. Soc.* **7** 505-508
[11] Belozersky G N, Grigoriev V A, Ivanov V A, Semenov V G, Sokolov A Yu and Aleksandrov A A 1987 *Pis'ma Zh. Tekh. Fiz.* (Technical Physics Letters) **13** 531-534 (in Russ.)
[12] Petrov V I, Martynenko S S and Khmelevskaya N D 2018 *Journal of Applied Spectroscopy* **84** 1014-1018
[13] Filippov V P, Gladkov V P, Martynenko S S and Petrov V I 2014 *Hyperfine Interactions* **226** 365-373
[14] Filippov V P, Petrov V I, Martynenko S S, Salomasov V A 2016 *Hyperfine Interactions* **237** pp 1-7
[15] Myers S M and Smuguresky J E 1978 *Metal. Trans.* **9**A pp 1789-1794
[16] Volkov N V and Kalin B A 2003 *Poverkhnost. Rentgenovskie, sinkhrotronnye i neitronnye issledovaniya* (Journal of Surface Investigation: X-ray, Synchrotron and Neutron Techniques) **5** 43-47 (in Russ.)
[17] Yelsukov E P and Dorofeev G A 2004 *J. Mater. Sci.* **39** pp 5071-5079
[18] Was G S 2007 *Fundamental of radiation materials science* (Berlin, Heidelberg: Springer)